Pharmacology and Toxicology: Basic and Clinical Aspects

Mannfred A. Hollinger, Series Editor
University of California, Davis

Published Titles

Inflammatory Cells and Mediators in Bronchial Asthma, 1990,
  Devendra K. Agrawal and Robert G. Townley
Pharmacology of the Skin, 1991, Hasan Mukhtar
In Vitro *Methods of Toxicology*, 1992, Ronald R. Watson
Basis of Toxicity Testing, 1992, Donald J. Ecobichon
Human Drug Metabolism from Molecular Biology to Man, 1992, Elizabeth Jeffreys
Platelet Activating Factor Receptor: Signal Mechanisms and Molecular Biology,
  1992, Shivendra D. Shukla
Biopharmaceutics of Ocular Drug Delivery, 1992, Peter Edman
Beneficial and Toxic Effects of Aspirin, 1993, Susan E. Feinman
Preclinical and Clinical Modulation of Anticancer Drugs, 1993, Kenneth D. Tew,
  Peter Houghton, and Janet Houghton
Peroxisome Proliferators: Unique Inducers of Drug-Metabolizing Enzymes, 1994,
  David E. Moody
*Angiotensin II Receptors, Volume I: Molecular Biology, Biochemistry,
  Pharmacology, and Clinical Perspectives*, 1994, Robert R. Ruffolo, Jr.
Angiotensin II Receptors, Volume II: Medicinal Chemistry, 1994,
  Robert R. Ruffolo, Jr.
Chemical and Structural Approaches to Rational Drug Design, 1994,
  David B. Weiner and William V. Williams
Biological Approaches to Rational Drug Design, 1994, David B. Weiner and
  William V. Williams
Direct Allosteric Control of Glutamate Receptors, 1994, M. Palfreyman,
  I. Reynolds, and P. Skolnick
Genomic and Non-Genomic Effects of Aldosterone, 1994, Martin Wehling
Stealth Liposomes, 1995, Danilo Lasic and Frank Martin

T0136192

Pharmacology and Toxicology: Basic and Clinical Aspects

Mannfred A. Hollinger, Series Editor
University of California, Davis

STEALTH®
LIPOSOMES

Edited by
Danilo Lasic and Frank Martin

Stealth® is a registered trademark of Liposome Technology, Inc. (LTI).

CRC Press
Taylor & Francis Group
Boca Raton London New York

CRC Press is an imprint of the
Taylor & Francis Group, an **informa** business

Cover design: An artist's impression of sterically stabilized liposomes with encapsulated anticancer drug doxorubi-cin. Liposome is covered by a surface attached polymer which brushes away incoming liposomicidal macromolecules and agents. Polymer conformation and properties of polymer-lipid systems were studied by many theoretical and experimental methods. Additionally, an extremely efficient drug loading method which forces drug molecules into preformed liposomes due to a special gradient to such a high concentration that they precipitate in the liposome interior, as can be inferred from the painting, is also described in this book. It seems that both factors are a necessity for effective liposomal systemic anticancer chemotherapy and several results of the application of this formulation in laboratory animals as well as in human patients are described in this volume. Artist: Alenka Dvorzak Lasic. Technique: Mixed media. Spring 1993.

CRC Press
Taylor & Francis Group
6000 Broken Sound Parkway NW, Suite 300
Boca Raton, FL 33487-2742

First issued in paperback 2019

© 1995 by Taylor & Francis Group, LLC
CRC Press is an imprint of Taylor & Francis Group, an Informa business

No claim to original U.S. Government works

ISBN-13: 978-0-8493-8383-0 (hbk)
ISBN-13: 978-0-367-40180-1 (pbk)

Library of Congress Cataloging-in-Publication Data

Stealth liposomes / edited by Danilo Lasic and Frank Martin.
    p.   cm. — (Pharmacology and toxicology)
  Includes bibliographical references and index.
  ISBN 0-8493-8383-8
  1. Liposomes.   2. Drug carriers.   3. Drug targeting.   I. Lasic,
D.D.   II. Martin, F.J.   III. Series: Pharmacology & toxicology
(Boca Raton, Fla.)
    [DNLM: 1. Liposomes — pharmacology. 2. Drug Carriers.   QU 93 S799 1995]
RS201.L55S74 1995
615'.7—dc20
DNLM/DLC
                                                                              94-32683
Library of Congress Card Number 94-32683

Visit the Taylor & Francis Web site at
http://www.taylorandfrancis.com

and the CRC Press Web site at
http://www.crcpress.com

CONTENTS

PREFACE

The commercial prospects of liposomes in drug delivery have waxed and waned in the quarter of a century since these fatty vesicles were first proposed for this purpose. The enthusiasm of those who first proclaimed liposomes as the most likely player to fill the role of Ehrlich's "magic bullet", i.e., a vehicle able to deliver encapsulated drugs directly to sites of disease, soured soon after animal experimentation began in the early 1970s and showed that intravenously injected liposomes failed to reach their intended target sites in significant numbers.

Although these early liposomes fell short of Ehrlich's lofty target, important therapeutic value was achievable when both the weaknesses and strengths of the system were considered. In fact, liposomes lived up to expectations in several important respects. They are indeed safe. The intuitive belief that small cell-like structures composed of natural phospholipids would be well-tolerated when injected intravenously at high doses has proven to be correct. Liposomes also proved to be quite versatile with respect to what they can carry. Literally hundreds of agents have been either encapsulated in the aqueous compartment of liposomes or intercalated in the bilayer membrane itself. But, to paraphrase a passage from Alexander Pope, with these goals attained, "we tremble to survey the growing labours of the lengthen'd way." Soon after the biochemists and biophysicists handed their liposomes over to pharmacologists for testing *in vivo,* severe obstacles to the magic bullet concept were identified. The most discouraging of these was the empirical observation that the biological milieu in which liposomes were intended to act was a hostile place indeed. In the first place, components of blood, the medium through which liposomes were expected to swim to their target, attached to and destabilized liposomes within minutes of injection. Although a whole cadre of early liposome scientists attacked this problem with vengeance, the best that could be done was represented by small (<50 nm) vesicles composed of high phase transition lipids plus cholesterol — proverbial microscopic rocks — that were found to resist extraction of lipid molecules from the liposome matrix by plasma proteins and survive in circulation for a few hours. The next obstacle to overcome was even more daunting. Liposomes injected intravenously were rapidly recognized and removed by elements of the host defense system known as the reticuloendothelial system, or RES for short. This collection of macrophages, which line blood vessels in organs such as liver and spleen, are exquisitely designed to remove foreign bodies from the blood stream. As much as we would like to think that liposomes are "natural", these macrophages regard them as garbage and devour liposomes with abandon. Early attempts to circumvent RES uptake included flooding the circulation with kamikaze liposomes which were designed to occupy the macrophages while the drug-loaded liposomes slipped by. But a hefty pre-dose of sacrificial liposomes is simply not practical in the real world of clinical practice.

Refusing to give up, the true believers turned RES uptake into a good thing for some encapsulated agents. Relying on the RES as an intermediate "depot", designed to sequester the majority of an injected dose of highly noxious drugs within the macrophages and thereby avoid exposure of potential toxic sites to high levels of unencapsulated drug, provided meaningful improvements in safety for amphotericin B and doxorubicin, two important drugs. Encapsulation of anti-infective and antiparasitic agents designed to treat intracellular infections of the RES remains to this day a rational approach.

But the goal of sending encapsulated drugs directly to non-RES sites of disease such as tumors to achieve greater therapeutic efficacy (relative to an equal dose of the unencapsulated agent) or to widen the spectrum of activity of existing agents remained elusive indeed.

In the late 1980s, the liposome story took a dramatic turn for the better. Taking a hint from mother nature, a group of die-hard liposomologists began to modify liposomes with specific carbohydrate groups in order to make them resemble erythrocytes, relatively large blood cells which manage to evade RES uptake for many weeks. The first breakthroughs were reported by Allen, Papahadjopoulos, and Gabizon in 1987. These seminal investigations proved that RES-avoiding liposomes were possible. Later developments with polymer-stabilized "Stealth" liposomes which are chronicled in the forthcoming chapters improved the longevity of liposomes in the blood stream even further. Today's stealth liposomes have been shown to circulate in patients for over a week and to deliver encapsulated cancer agents selectively to tumors.

This volume is dedicated to those tenacious liposome engineers who, despite mounting skepticism on the part of the pharmaceutical industry and their academic colleagues, kept the dream alive. Stealth liposome technology has reopened the doors to a variety of therapeutic applications that RES uptake had slammed shut. These include ligand targeting, cell fusion, and gene therapy. "Hills peep o'er hills and Alps on Alps arise!"

D.D. Lasic
F.J. Martin

FOREWORD

Anonymity is difficult to sustain if one enters the bloodstream, or for that matter, the tissues of an animal. Almost all surfaces are tagged within milliseconds of being exposed to plasma which, after all, is an abundant source of animal glue! Both the tagging and the glueing are spontaneous events and a result of the release of electrostatically bound water as the ionic groups of surface and protein mutually satisfy their respective charge. The driving force responsible for the spontaneity of the adsorption/glueing process is, in addition to various attractive forces, the disordering of bound water.

But suppose the surface of the material introduced into the plasma already looks like the surface of a slice of (bulk) water, then the polyionic proteins will not find anything to tag, there is no entropy gain, no attractive interactions, and the interloper is invisible.

This book is about the development of the surface of a surrogate cell membrane (liposome) which achieves a remarkable degree of anonymity, a consequence of understanding the fundamental thermodynamics of adhesion. As to how non-stealthy surfaces, covered with glue, are recognized by the living defensive cells is another matter, but to avoid them one must simply look like water.

A.D. Bangham

INTRODUCTION

Pharmaceutical industry spokesmen are not shy to point out the time and expense involved in developing new pharmaceutical agents, be they made by medicinal chemistry, extracted from natural products, or modeled by computer. The current figures are 10 years and $100 to $200 million per agent. The enormity of the task is reflected in the paucity of new agents that have been introduced over the last 20 years. In this setting high hopes were set for drug delivery systems ("DDS"), which were believed to represent a fast-track, significantly less expensive route through development and regulatory review. Marketing groups pushed hard because DDS offered the possibility of capturing premium pricing for off-patent agents. DDS were promoted based on the somewhat naive belief that the therapeutic value of existing drugs could be improved substantially by changing their spatial and temporal distribution in the body, thus improving efficacy and reducing toxic side effects.

Among many drug delivery and carrier systems proposed, liposomes were regarded as an early front runner. Liposomes were considered safe for parenteral use by virtue of their biocompatible lipid matrix and evidence suggesting that they could be engineered to have specified sizes and permeability properties. However, these wildly optimistic expectations which emerged during the 1970s and 1980s have not materialized. Only one liposome product has been approved in several European countries and this one, Ambisome, relies upon active sequestration of drug-loaded liposomes in the reticuloendothelial system (RES) for its improved therapeutic performance. This dismal pace is due in part to the usual hurdles that must be overcome in any drug development program, including securing sources of raw materials, product and component quality control, manufacturing scale-up, shelf-life issues, clinical trial design, and execution and regulatory approval. But under-appreciated (unrecognized) biological limitations that plagued the first generation of liposomes are primarily responsible for slow progress. Perhaps the most serious of these are the inherent instability of liposomes in biological fluids and lack of control of their fate after injection into the bloodstream. Simply stated, liposomes as well as other colloidal particles are recognized as foreign bodies by the host's immune system and quickly cleared by phagocytic cells residing in liver and spleen. This simple fact, apart from any beneficial effect that such immune system targeting may provide, has led to a protracted period of disillusionment regarding the commercial prospects of liposomes in general and their broad medical utility in particular.

The recent development of liposomes that evade immune system uptake, fortunately for those of us in the field, has reversed this trend. Now it is possible to deliver intact drug-laden liposomes directly to sites of pathology. Moreover, the once popular concepts of long circulating liposome reservoirs releasing drugs into the bloodstream and ligand targeting have been revived. New areas of investigation will soon take the spotlight, such as engineering a greater level of control on the kinetics of drug release after reaching a pathological target tissue. Thus, research and development will concentrate on special lipid compositions, possibly with specially designed lipids, which will be capable of programmable change of properties or influenced by external triggering devices, giving rise to fusogenic, endocytotic, or suicidal liposomes. In other words, the scientific base will switch from thermodynamics to kinetics.

Leaving these speculations to the future, in this book we set out to present current knowledge and achievements. Because the appearance and understanding of long circulating liposomes and their applications involves several different areas of scientific endeavor, from theoretical physics, colloid and interface science, organic synthesis, biochemistry, biology, pharmaceutics, and medicine to oncology and anatomy, we have grouped articles into several sections, beginning with the theoretical, moving into organic chemistry, pharmaceutical characterization, preclinical studies, and finally human clinical experience.

This book chronicles efforts that have culminated in successful stabilization of liposomes in biological environments, such as the bloodstream. Lessons from this research, combined with other properties of membranes, as well as experiences from improving physical and chemical stability of liposomes, can be used to increase liposome stability also for oral intake and other applications. In this volume, however, we shall be concerned mostly with the increased liposome stability in blood circulation.

The human body protects itself with an elaborate host defense system which removes any non-self colloidal particles very quickly and efficiently. This includes conventional liposomes and so to explore their potential as intravenous and systemic drug carriers researchers have actively tried to engineer around this uptake. It was found that liposomes coated with appropriate polymer at appropriate surface density greatly reduce this process. Because they are virtually invisible to defense mechanisms of the body, one of the editors coined the name Stealth® liposomes.* The origin of the effect was attributed to steric stabilization by the other editor and so "sterically stabilized" liposomes are often used as a synonym. For these long-circulating liposomes other researchers use a variety of other, albeit much less frequently used, names, including surface-modified liposomes, Ninjasomes, cryptosomes, RES-avoiding liposomes, etc.

In organizing this volume we made every effort to invite contribution from all the major laboratories involved in this area. With a few exceptions we believe the contributed chapters capture the state of the art. In the following introduction we shall try to cover some important references which may not have been otherwise covered in this volume. Otherwise, some parts of the field were reviewed in References 1 to 6.

Without going into details of who did what when, we shall briefly introduce the field. Practically all of the long circulating liposomes now use polyoxyethylene polymer covalently attached to the lipid anchor, phosphatidylethanolamine. The approach of various groups vary predominantly in the polymer chain length, nature of chemical linkage, acyl chain composition of hydrophobic anchor, and bilayer composition.

We at LTI have arrived at the optimal long circulating formulation as others have, empirically. We found the best circulation times were seen with compositions containing about 5 mol% of ^{2000}mPEG conjugated via carbamate bond and anchored on two distearoyl chains in bilayers containing long saturated chains and high fractions of cholesterol. This type of formulation is used in most of our preclinical and exclusively in clinical product. Liposomes can be produced by most of the commonly used preparation methods. Volumes range from milliliters to hundreds of liters. Other labs use very similar formulations, from 5 to 10 mol% of PEG-lipid in the membrane and possibly with PEG 1000 and 5000 Da. We are sure that other polymers that can prolong liposome circulation times will be discovered.

Most researchers believe that the origin of extended blood circulation times is in reduced opsonin adsorption due to the steric stabilization of the particles provided by the polymer layer.

Preclinical and clinical studies have largely concentrated on the passive targeting of encapsulated drugs to perivascular sites around damaged or leaky blood vessels. Several reports of successful targeting with ligand-bearing liposomes have also appeared. This work is being done mostly in the laboratories of L. Huang, T. Allen, A. Gabizon, Y. Barenholz, D. Papahadjopoulos, D. Crommelin, and at LTI. Future applications will likely apply sterically stabilized liposomes also to *in vitro* and *in vivo* diagnostics and dermatics, as well as in fundamental studies of cell function. Interactions in living systems are based on specific attraction and nonspecific repulsion and sterically stabilized liposomes undoubtedly will be an important tool in such research. Also, their potential in the delivery of cytokines, recombinant proteins, and other biopharmaceuticals, as well as in gene therapy, has not been exploited yet.

* Stealth® is a registered trademark of Liposome Technology, Inc. (LTI).

Independently of coating liposomes with synthetic polymers, almost exclusively with polyethylene oxide with various termini, several groups have focused their research on coating liposomes with various polysaccharides, with the hope of achieving site-specific delivery.[7] A similar approach using various glycolipids is described in Chapter 3. Obviously, if nonrecognizable sugars are selected such liposomes may have enhanced stability, probably also in biological fluids. In addition to simple polysaccharides, such as cellulose, also other biopolymers, such as dextrans, xanthanes, pullulan, amylopectin, mannans, levans, chitins, and peptideglycans were studied.

Another approach to increase liposome stability is polymerization.[8] However, none of the many scenarios that have greatly improved their physical, colloidal, or chemical stability was found to be effective against macrophage attack.

In this book, however, we shall concentrate mostly on liposomes with grafted polyoxyethylene chains. We have grouped the chapters into what we hope is a logical sequence. In Chapter 1, Papahadjopoulos presents introductory remarks. Next, Gregoriadis provides a historical perspective with a description of liposome behavior in biological systems, and Ghosh and Bacchawat review the role of glycolipid-containing liposome systems. Then, theoretical considerations are presented followed by contributions that provide experimental verification of the various theoretical models. This physics is followed by chemistry, biology, and pharmacology. After preclinical studies in various tumors, sterically stabilized liposomes with attached antibodies and some other applications are presented. We conclude with the use of sterically stabilized liposomes loaded with anticancer agent doxorubicin in humans.

Theoretical aspects are covered in the contributions of Borisov et al. and Hristova and Needham. They show that the scaling concepts, as developed by de Gennes, can be applied to explain the stabilizing property of surface grafted polymers. In addition to these two theoretical papers a rather simple model, an extension of the original hypothesis,[9] is presented by Torchilin et al. in Chapter 6.

Experimental measurements have shown the validity of the hypothesis that prolonged presence of Stealth liposomes in the bloodstream is due to reduced opsonization.[9] Repulsive pressure measurements by osmotic stress technique and surface force apparatus, as we shall see in the contributions by McIntosh et al. and Kuhl et al., have shown strong and long-range steric repulsion. The latter chapter also presents monolayer studies which give good estimates of the size and behavior of polymer. Along these lines we can add that measurements of protein adsorption were shown to correlate with the inverse of blood circulation times.[10] Still unpublished results also show reduced interaction of proteins with black lipid membranes (BLM)-containing PEG-lipids and reduced adsorption on the monolayers containing PEG-lipids.[11]

Chemical synthesis, pros and cons of various chemical linkages and synthetic routes, as well as conjugation of antibodies on the far end of polyethylene polymer are reviewed by Zalipsky. We should add that in the case of coating liposome surface with [2000]PEG, one can use conventional conjugation techniques[12] due to the large size of the antibody.

Pharmacokinetics and biodistribution are reviewed by Woodle et al. Here we can see the real changes in temporal and spatial distribution of various liposome-associated markers brought by Stealth liposomes. Concomitantly, the microscopic localization of labeled Stealth liposomes is described in the paper by Huang et al. The improved extravasation of sterically stabilized liposomes containing various markers, including the drug doxorubicin, can be followed *in vivo*. Direct observation of the fate of liposomes upon intravenous administration in tumors and healthy tissue, an indispensable method for modeling of the treatment and understanding of the drug action, is presented in Chapter 12 by Dewhirst et al.

Preclinical studies are described in the chapters by Lasic, Vaage and Barbera, Allen, and Uster in the case of anticancer therapy, and Bakker-Woudenberg et al. in the case of infections.

The first two chapters concentrate on anthracyclines, the last two on cytosinarabinose and vincristine, respectively. While in the case of epirubicin and doxorubicin liposomes passive targeting of the drug to the tumor is the operative mechanism, the mode of action in the case of ara-C liposomes is that of a long circulating microreservoir. We should add that in a recent meeting[13] (Liposomes in Drug Delivery, Nineties and Beyond, London, December 1993) a few novel preclinical applications were described, including a 20-fold enhanced drug (doxorubicin) accumulation in brain tumors demonstrating that sterically stabilized vesicles can bypass the blood-brain barrier at sites of trauma.[13] Another therapeutic application of Stealth liposomes is in the delivery of biopharmaceuticals, such as various polypeptides and proteins.[14] The preclinical group of articles is concluded with the chapter by Zheng et al. on the use of hemoglobin-laden long circulating liposomes as a blood substitute and by the application of Stealth liposomes containing signal carriers in diagnostics by Torchilin et al.

The next stage in the development of Stealth liposomes certainly includes attachment of targeting ligands on the liposomes, as described by Allen et al.

The last section describes applications of Stealth liposomes in humans. Gabizon et al. present a detailed description of the pharmacokinetics of DOX-SL (doxorubicin-loaded Stealth liposomes) in cancer patients, pointing out that not only do the liposomes circulate for protracted periods (up to a week) but that the liposomes retain their encapsulated drug and selectively enter tumors intact. Northfelt et al. follow with an examination of the pharmacokinetics and tumor localization of DOX-SL in AIDS patients with Kaposi's sarcoma (KS). Bogner and Goebel round out this section with efficacy results for DOX-SL in AIDS KS.

In conclusion, we have attempted to present the current state of this rapidly evolving field. We have tried to show the importance of multi- and intradisciplinary approaches in solving complex problems of liposomes in medical uses. The success of first formulations is therefore due to the input of many sciences, from theoretical physics, colloid and surface science, chemical synthesis, biology, biochemistry, and pharmacology to medicine, anatomy, and oncology. We would like to conclude by observing that the work in this field is not rewarded only by the quality of the science but also by the knowledge that severely ill people are deriving clinical benefit from this new liposome technology. Our fervent hope is that many more will follow.

REFERENCES

1. **Lasic, D.D.,** *Liposomes: from Physics to Applications,* Elsevier, Amsterdam, 1993.
2. **Woodle, M.C. and Lasic, D.D.,** Sterically stabilized liposomes, *Biochim. Biophys. Acta,* 1113, 171, 1992.
3. Several Authors, in Forum on "Covalently Attached Polymers and Glycans to Alter the Biodistribution of Liposomes", Huang, L., Ed., *J. Liposome Res.,* 3, 289, 1992.
4. **Allen, T.M. and Papahadjopoulos, D.,** Sterically stabilized (stealth) liposomes: pharmacokinetics and biodistribution, in *Liposome Technology* Vol. 3, Gregoriadis, G., Ed., CRC Press, Boca Raton, FL, 1993, 59.
5. **Papahadjopoulos, D., Allen, T.M., Gabizon, A., Mayhew, E., Matthay, K., Huang, S.K., Woodle, M.C., Lasic, D.D., Redemann, C., and Martin F.,** Sterically stabilized liposomes: improvements in pharmacokinetics and antitumor therapeutic efficacy, *Proc. Natl. Acad. Sci. U.S.A.,* 88, 11460, 1991.
6. **Lasic, D.D.,** Sterically stabilized vesicles, *Angew. Chem. Int. Ed. Eng.,* 33, 1685, 1994.
7. **Sato, T. and Sunamoto, J.,** Recent aspects in the use of liposomes in biotechnology and medicine, *Prog. Lip. Rep.,* 31, 345, 1992.
8. **Ringsdorf, H., Schlarb, B., and Venzmer, J.,** Molecular architecture of polymeric oriented systems: models to study organization, surface recognition and dynamics of biomembranes, *Angew. Chem. Int. Ed. Eng.,* 27, 153, 1988.
9. **Lasic, D.D., Martin, F.J., Gabizon, A., Huang, S.K., and Papahadjopoulos, D.,** Sterically stabilized liposomes: a hypothesis on the molecular origin of extended circulation times, *Biochim. Biophys. Acta,* 1070, 187, 1991.
10. **Chonn, A., Semple, S.C., and Cullis P.R.,** Association of blood proteins with large unilamellar liposomes in vivo, *J. Biol. Chem.,* 267, 1992.

11. **Winterhalter, M., Benz, R., Klotz, K.H., and Lasic, D.D.,** Electric field induced breakdown in lipid membranes: influence of polymer and protein adsorption, *Trans. Coll. Interface Sci.,* in press, 1995..

12. **Heath, T.D. and Martin, F.J.,** The development and application of protein liposome conjugation techniques, *Chem. Phys. Lip.,* 40, 347, 1986.

13. **Siegal, T.,** Doxorubicin encapsulated in long-circulating liposomes in an experimental brain tumor model: tissue distribution and efficacy studies, in Liposomes in Drug Delivery, Nineties and Beyond, London, Dec. 1993, book of abstracts.

14. **Woodle, M.C., Storm, G., Newman, M.S., Jekot, J., Collins, L., Martin, F.J., and Szoka, F.C.,** Prolonged systemic delivery of peptide drugs by long circulating liposomes: illustration with vasopressin in the Brattleboro rat, *Pharm. Res.,* 9, 260, 1992.

THE EDITORS

Danilo D. Lasic, Ph.D. is Senior Scientist at MegaBios, San Carlos, California, where he is currently working with cationic liposomes in gene delivery. Previously, he was a senior scientist at Liposome Technology, Inc., where he led studies for theoretical understanding of long circulating liposomes as well as developed the first formulations for preclinical studies. In addition, he also actively participated in the scale-up of the preparation of stealth liposomes laden with the anticancer agent doxorubicin.

Dr. Lasic graduated from University of Ljubljana, Slovenia, in 1975 with a degree in physical chemistry. He received his M.Sci in 1977 from University of Ljubljana. He obtained his Ph.D. at Institute J.Stefan, Solid Physics Department in Ljubljana in 1979. After post-doctoral work with Dr. Charles Tanford at Duke and Dr. Helmut Hauser at ETH Zurich, he was a research fellow at Institute J.Stefan in Ljubljana, and a visiting lecturer in the Department of Chemistry and a visiting scientist in the Department of Physics at University of Waterloo in Canada . Then he joined Liposome Technology, Inc., in Menlo Park, California. He is also a professor of Colloid Science at University of Ljubljana.

He has published more than 110 research papers as well as a book on liposomes (*Liposomes: From Physics to Applications*, Elsevier, 1993). His best known papers are the ones dealing with thermodynamics and the mechanisms of liposome formation and the origin of "stealth effect" as well as the applications of drug laden stealth liposomes.

Frank Martin, Ph.D., is Vice President and Chief Scientific Officer at Liposome Technology, Inc. (LTI), a Menlo Park, California based firm dedicated to the development of liposome therapeutic products. Dr. Martin has over twenty years experience with liposome systems. As a graduate student at Northwestern University, he investigated the attachment and uptake of liposomes by cells. He expanded his interest to antibody-targeting of liposomes during his postdoctoral training with Wayne Hubbell at the University of California at Berkeley and Demetrios Papahadjopoulos at the University of California at San Francisco. Dr. Martin is co-inventor of methods to produce liposomes on large scale and played a key role in the development of "Stealth"® liposomes. He is currently responsible for identifying new technologies that complement his company's product development strategy.

THE CONTRIBUTORS

T.M. Allen
Department of Pharmacology
University of Alberta
Edmonton, Alberta, Canada

Phillipe Auroy
Laboratoire Léon Brillouin (CEA-CNRS)
Saclay, France

Loïc Auvray
Laboratoire Léon Brillouin (CEA-CNRS)
Saclay, France

Bimal K. Bachhawat
Liposome Research Centre
Department of Biochemistry
University of Delhi South Campus
New Delhi, India

Irma A.J.M. Bakker-Woudenberg
Department of Clinical Microbiology
Erasmus University
Rotterdam, The Netherlands

Emilio Barbera
Hipple Cancer Research Center
Dayton, Ohio

Yechezkel Barenholz
Department of Membrane Biochemistry
Hebrew University-Hadassah Medical
    School
Jerusalem, Israel

Richard Beissinger
Department of Chemical Engineering
Illinois Institute of Technology
Chicago, Illinois

A.A. Bogdanov
MGH-NMR Center
Department of Radiology
Massachusetts General Hospital-East
Charlestown, Massachusetts

Oleg Borisov
Institute for Macromolecular Compounds
St. Petersburg, Russia
Laboratoire Léon Brillouin (CEA-CNRS)
Saclay, France

Mark Dewhirst
Department of Radiation Oncology
Duke University
Durham, North Carolina

Daniel S. Friend
Department of Pathology
University of California
San Francisco, California

Alberto Gabizon
Sharett Institute of Oncology
Hadassah University Hospital
Jerusalem, Israel

Prahlad C. Ghosh
Liposome Research Centre
Department of Biochemistry
University of Delhi South Campus
New Delhi, India

Gregory Gregoriadis
Centre for Drug Delivery Research
The School of Pharmacy
University of London
London, England

C.B. Hansen
Department of Pharmacology
University of Alberta
Edmonton, Alberta, Canada

Kalina Hristova
Department of Mechanical Engineering
    and Materials Science
Duke University
Durham, North Carolina

Anthony Huang
Liposome Technology, Inc.
Menlo Park, California

Shi Kun Huang
Liposome Technology, Inc.
Menlo Park, California

J.N. Israelachvili
Department of Chemical and Nuclear
    Engineering
University of California
Santa Barbara, California

Lawrence Kaplan
San Francisco General Hospital
AIDS Program
Oncology Division
San Francisco, California

A.K. Kenworthy
Department of Cell Biology
Duke University Medical Center
Durham, North Carolina

B.A. Khaw
Center for Drug Targeting and Analysis
Northeastern University
Boston, Massachusetts

T.L. Kuhl
Department of Chemical and Nuclear
    Engineering
University of California
Santa Barbara, California

D.D. Lasic
MegaBios Corp.
Burlingame, California

D.E. Leckband
Department of Chemical Engineering
State University of New York
Buffalo, New York

Frank J. Martin
Liposome Technology, Inc.
Menlo Park, California

T.J. McIntosh
Department of Cell Biology
Duke University Medical Center
Durham, North Carolina

J. Narula
Center for Drug Targeting and Analysis
Northeastern University
Boston, Massachusetts

David Needham
Department of Mechanical Engineering
    and Materials Science
Duke University
Durham, North Carolina

Mary S. Newman
Liposome Technology, Inc.
Menlo Park, California

Donald W. Northfelt
San Francisco General Hospital
AIDS Program
Oncology Division
San Francisco, California

V.G. Omelyanenko
Department of Pharmaceutics
The University of Utah
Salt Lake City, Utah

Demetrios Papahadjopoulos
Department of Pharmacology and Cancer
    Research Institute
University of California
San Francisco, California

M.I. Papisov
MGH-NMR Center
Department of Radiology
Massachusetts General Hospital-East
Charlestown, Massachusetts

Julie Russell
San Francisco General Hospital
AIDS Program
Oncology Division
San Francisco, California

G. Storm
Department of Pharmaceutics
University of Utrecht
The Netherlands

V.P. Torchilin
Center for Imaging and Pharmaceutical
    Research
Department of Radiology
Massachusetts General Hospital-East
Charlestown, Massachusetts

V.S. Trubetskoy
Center for Imaging and Pharmaceutical
    Research
Department of Radiology
Massachusetts General Hospital-East
Charlestown, Massachusetts

Paul S. Uster
Liposome Technology, Inc.
Menlo Park, California

Jan Vaage
Department of Molecular Immunology
Roswell Park Cancer Institute
Buffalo, New York

Paul A. Volberding
San Francisco General Hospital
AIDS Program
Oncology Division
San Francisco, California

Martin C. Woodle
Genta, Inc.
San Diego, California

Peter K. Working
Liposome Technology, Inc.
Menlo Park, California

Samuel Zalipsky
Liposome Technology, Inc.
Menlo Park, California

Shuming Zheng
Department of Chemical Engineering
Illinois Institute of Technology
Chicago, Illinois

Yaoming Zheng
Department of Chemical Engineering
Illinois Institute of Technology
Chicago, Illinois

Chapter 1

STEALTH LIPOSOMES: FROM STERIC STABILIZATION TO TARGETING

Demetrios Papahadjopoulos

Our understanding of the fate of liposomes *in vivo* took a sharp turn in the late 1980s, following the work from two independent laboratories, T. M. Allen's at the University of Alberta[1-5] and my own, at the University of California, San Francisco (UCSF).[6-10] These studies demonstrated that the inclusion of a small percentage of some specific glycolipids and phospholipids had a dramatic effect in increasing the circulation time of liposomes in blood $T_{1/2}$. Moreover, it was also shown that the long $T_{1/2}$ was correlated with higher uptake by implanted tumors in mice.[6-10] This has greatly expanded the potential of liposomes as drug carriers, and some formulations have already shown promise for increasing the efficacy of chemotherapeutic agents in recent clinical trials.[11-13] Such liposomes were initially given the name "Stealth®"[2,14] on the basis of their avoidance of rapid detection and uptake by the reticuloendothelial system (RES). More recently they have been characterized as "sterically stabilized",[15,16] on the basis of their enhanced stability and reduced reactivity to plasma proteins and cell surface receptors.

Prior to the advent of sterically stabilized liposomes (SL) most of the published studies on *in vivo* disposition had used liposomes composed of neutral phospholipids, such as phosphatidylcholine or sphingomyelin, mixed with varying amounts of cholesterol and sometimes including a small percentage of an acidic phospholipid.[17,18] Cholesterol was added to increase the stability of liposomes in the presence of plasma proteins[18,19] and the negative charge in order to avoid aggregation and to increase encapsulation efficiency.[20,21] Such compositions are now referred to as "conventional" liposomes (CL). As it happens, they are recognized by the phagocytic cells of the RES and are removed from the circulation fairly rapidly, accumulating mostly in the liver and spleen within a few minutes to a few hours.[17,18] The rate of accumulation depends on a variety of parameters, such as particle size, surface charge, and bilayer fluidity.[22-25] Generally, the $T_{1/2}$ decreases with increasing diameter, negative charge density, and fluidity. Thus, for conventional liposomes the longest $T_{1/2}$ can be achieved using small unilamellar vesicles composed of distearoyl phosphatidylcholine (or sphingomyelin) and cholesterol, which provide a rigid neutral bilayer.[18,23,25,26]

The exact mechanism for the effect of each of these parameters is not well understood. It is known, however, that various plasma proteins adsorb on the liposome surface, and the degree of adsorption has been found to correlate with the rate of removal from blood.[27] Negative surface charge can usually be recognized by receptors on a variety of cells, and most importantly by macrophages.[5,28,29] The mechanism for the effect of size is more difficult to define. It could be related to the strength of multiligand interactions[16] or to the specificity of complement proteins for large liposomes.[30] A correlation between plasma destabilization and rate of removal had been observed.[18] However, we have recently observed that rigid neutral liposomes are specifically destabilized by complement factor C3,[31] which has been implicated in liposome clearance. Generally speaking, it would be expected that a rigid neutral bilayer would be less susceptible to hydrophobic penetration or charge-charge interactions with either plasma proteins or cell surface ligands.[16] The opposite would be true for a fluid and either negatively or positively charged lipid bilayer. The mode of interaction of complement and other plasma proteins with various liposomes is still not understood. The first observations on the ability of specific glycolipids such as GM1 ganglioside (GM1), cerebroside sulfate, or phosphatidylinositol (PI) to promote increased $T_{1/2}$ were interpreted in terms of a "shielded charge",[7,32] since all these molecules had a headgroup with negative charge which was shielded by a large carbohydrate residue, possibly providing steric inhibition for specific interactions. Later, we generalized this to the concept of steric stabilization, which includes steric inhibition of a variety of interactions at the bilayer surface, including hydrophobic penetration by bulky proteins.[16]

Although the most widely used molecule for steric stabilization is presently a synthetic phospholipid derivative of polyethylene glycol (see below), most of the early observations with SL were obtained with either GM1[1,7] or PI[6,7,9] and these studies have defined a variety of their basic properties. Long $T_{1/2}$ was found to be correlated with lower uptake by liver and spleen and the kinetics were dose-independent.[15,33] This meant that the percentage of liposomes in blood or liver at any particular time was the same, irrespective of the amount injected. This was in contrast to observations with conventional liposomes (CL), even those composed of sphingomyelin and cholesterol, which had relatively long $T_{1/2}$, but still showed saturation in liver uptake at high dosages.[26] This differentiates clearly the behavior of SL from CL, indicating the absence of a saturable component of uptake by the liver, presumably the Kupffer cells. The difference between SL and CL has been substantiated also *in vitro* with cultured cell lines, even in the absence of serum components.[5,28]

Most of the early observations with SL were obtained with isotopically labeled aqueous or lipid markers for liposome contents.[1,7,34,35] It was, therefore, significant to establish that an encapsulated chemotherapeutic drug, such as doxorubicin, also showed a remarkably long $T_{1/2}$ in blood[9] as well as an increased accumulation in mouse tumors.[15,36] These results were the prelude to a variety of preclinical and clinical studies to be discussed in detail later.

An important development in steric stabilization was the synthesis and use of PEG-PE as the stabilizing component[37-40] which eventually superceded early compositions containing GM1 or PI. This synthetic phospholipid was a much more acceptable preparation for clinical applications compared to GM1 derived from bovine brain, or hydrogenated plant PI. In addition, it was reported recently[41] that liposomes containing PEG-PE show a useful flexibility in composition for the bulk bilayer lipid, which can be either solid or fluid and can include negatively charged phospholipids without a decrease in $T_{1/2}$. The length of the PEG chain and the density on the surface are both important, as indicated by a variety of studies which have established that the optimal effect for a long $T_{1/2}$ is produced by a chain approximately 2000 Da, at a density of about 5 to 8% of total phospholipid.[41-43] A number of physicochemical studies have attempted to determine the conformation of the chains above the liposome surface,[44] the effect on the repulsive forces between liposomes,[45] the effect on surface charge and zeta potential,[46] the adsorption of various proteins on the surface,[27] and the reactivity of protein and other ligands co-incorporated on the liposome surface.[47,48]

Although it was established early that SL of relatively small size can accumulate in implanted mouse tumors,[6-8,10,54,62] it was not known whether this involved extravasation beyond the endothelial barrier within the tumor mass. More recent work using either fluorescence microscopy[15] or a newly established method of liposome-encapsulated colloidal gold[53] have shown that SL can indeed extravasate beyond the endothelial barrier mainly in postcapillary venules[49,50] where liposome material can be observed extracellularly among tumor cells. Video-enhanced intra-vital fluorescence from tumors growing in skin-flap observation windows in living rats[51] and mice[52] have substantiated the earlier observations and have provided a quantitative approach to the permeability of tumors to SL. Most importantly, therapeutic experiments with a variety of mouse tumors have demonstrated enhanced antitumor efficacy and lower toxicity for the encapsulated drugs[15,54-59,68] even with tumors that are not responsive to the free drug or the same drugs encapsulated in conventional liposomes. Although the mechanism for the enhanced antitumor effect is not known at present, it seems likely that increased tumor accumulation of drug-loaded liposomes provides a local depot source for the drug, which can then diffuse further into the tumor mass and affect many tumor cells in the vicinity.

An important question in relation to further medical applications was whether SL could be directed to recognize and bind to specific cells. This could be accomplished by incorporating surface ligands such as antibodies if there was no interference with the long $T_{1/2}$. An early observation with SL was that their accumulation in implanted mouse tumors could be further increased when tumor-recognizing antibodies were conjugated to the liposome surface.[6] The ability to target SL to other specific tissues by specific ligands was clearly demonstrated by using antibody-recognizing lung endothelial cells.[34] Recent studies from several laboratories using SL stabilized with PEG-PE have indicated that such liposomes can indeed be targeted to specific cells without losing their property of long $T_{1/2}$ in blood and can show antitumor effects in animal model systems if injected at an early stage in tumor development.[35] This is quite encouraging for future possibilities of *in vivo* targeting, and of obvious significance for situations in which there is accessibility to the target cells. These would include most cells within the vascular compartment, as well as various cells in tissues where SL can penetrate to a certain extent, such as liver, spleen, areas of inflammation, and possibly tumors.

Although SL have been observed to localize within implanted tumors beyond the endothelial layer,[15,49,52] it is not possible to predict whether ligand-directed targeting will greatly enhance their accumulation and/or their efficacy in solid tumors. This is because solid tumors do not have an efficient drainage system and allow only limited diffusion for large macromolecular assemblies such as liposomes.[52,53] Any extravasated particles, therefore, are not expected to have access to the majority of tumor cells for direct delivery of their contents. In this context, the release of small diffusible drug molecules from the SL interior would be important for the delivery of cytotoxic material to other proximal tumor cells in the surrounding area. Results on increased antitumor efficacy in therapeutic experiments against mouse tumors not sensitive to the free drug confirm this hypothesis and have provided the basis for current clinical trials.[15,54-59]

A variety of specific subjects covering the properties and uses of SL will be dealt with in the chapters following in this volume and have been reviewed elsewhere extensively.[60-65] At this point the most significant results have been obtained in clinical trials with cancer patients in studies of pharmacokinetics, as well as toxicity and efficacy of SL loaded with doxorubicin. The remarkably long $T_{1/2}$(~44 h) of doxorubicin in patients' blood,[11] the enhanced accumulation of the drug in biopsy specimens and tumor exudates,[11,12] and the enhanced efficacy in patients with Kaposi's sarcoma[13] are an excellent indication of SL's potential in chemotherapy and other medical applications. Extending their usefulness would involve new methods for efficient encapsulation of drugs other than doxorubicin[66,67] or vincristin,[68] further development of SL that are able to release their contents at low pH following endocytosis[69] or during temperature increase produced by local hyperthermia[31,59] or external irradiation. An ultimate goal for present research is the incorporation of a fusogenic molecule that would induce fusion of SL following their binding to the target cells.[71,72]

REFERENCES

1. **Allen, T.M. and Chonn, A.,** Large unilamellar liposomes with low uptake by the reticuloendothelial system, *FEBS Lett.*, 223, 42, 1987.
2. **Allen, T.M.,** Stealth liposomes: avoiding reticuloendothelial uptake, in *Liposomes in the Therapy of Infectious Diseases and Cancer*, Vol. 89, UCLA Symp. on Molecular and Cellular Biology, Lopez-Berestein, G. and Fidler, I., Eds., Alan R. Liss, New York, 1988, 405.
3. **Allen, T.M. and Mehra, T.,** Recent advances in sustained release of antineoplastic drugs using liposomes which avoid uptake into the reticuloendothelial system, *Proc. West. Pharmacol. Soc.*, 32, 111, 1989.
4. **Allen, T.M., Hansen, C., and Rutledge, J.,** Liposomes with prolonged circulation times: factors affecting uptake by reticuloendothelial and other tissues, *Biochim. Biophys. Acta*, 981, 27, 1989.
5. **Allen, T.M., Austin, G.A., Chonn, A., Lin, L., and Lee, K.C.,** Uptake of liposomes by cultured mouse bone marrow macrophages: influence of liposome composition and size, *Biochim. Biophys. Acta*, 1061, 56, 1990.
6. **Papahadjopoulos, D. and Gabizon, A.,** Targeting of liposomes to tumor cells *in vivo*, *Ann. NY Acad. Sci.*, 507, 64, 1987.

7. **Gabizon, A. and Papahadjopoulos, D.,** Liposome formulations with prolonged circulation time in blood and enhanced uptake by tumors, *Proc. Natl. Acad. Sci. U.S.A.,* 85, 6949, 1988.

8. **Gabizon, A., Huberty, J., Straubinger, R.M., Price, D.C., and Papahadjopoulos, D.,** An improved method for *in vivo* tracing and imaging of liposomes using a gallium 67-deferoxamine complex, *J. Liposome Res.,* 1, 123, 1988.

9. **Gabizon, A., Shiota, R., and Papahadjopoulos, D.,** Pharmacokinetics and tissue distribution of doxorubicin encapsulated in stable liposomes with long circulation times, *J. Natl. Cancer Inst.,* 81, 1485, 1989.

10. **Gabizon, A., Price, D.C., Huberty, J., Bresalier, R.S., and Papahadjopoulos, D.,** Effect of liposome composition and other factors on the targeting of liposomes to experimental tumors: biodistribution and imaging studies, *Cancer Res.,* 50, 6371, 1990.

11. **Gabizon, A., Catane, R., Uziely, B., Kaufman, B., Safra, T., Cohen, R., Martin, F., Huang, A., and Barenholz, Y.,** Prolonged circulation time and enhanced accumulation in malignant exudates of doxorubicin encapsulated in polyethylene-glycol coated liposomes, *Cancer Res.,* 54, 987, 1994.

12. **Northfelt, D.W., Martin, F.J., Kaplan, L.D., Russell, J., Anderson, M., Lang, J., and Volberding, P.A.,** Pharmacokinetics, tumor localization, and safety of Doxyl in AIDS patients with Kaposi's sarcoma, *Proc. Am. Soc. Clin. Oncol.,* 12, 51, 1993.

13. **Bogner, J.R. and Goebel, F.D.,** Efficacy of DOX-SL (Stealth Liposomal Doxorubicin) in the treatment of Advanced AIDS-Related Kaposi's Sarcoma, Chapter 23 in this volume.

14. **Martin, F.J.,** Liposome Technology, Inc., Internal Report, 1987.

15. **Papahadjopoulos, D., Allen, T.M., Gabizon, A., Mayhew, E., Matthay, K., Huang, S.K., Lee, K.-D., Woodle, M.C., Lasic, D.D., Redemann, C., and Martin, F.J.,** Sterically stabilized liposomes: improvements in pharmacokinetics, and anti-tumor therapeutic efficacy, *Proc. Natl. Acad. Sci. U.S.A.,* 88, 11460, 1991.

16. **Lasic, D. D., Martin, F. J., Gabizon, A., Huang, S.K., and Papahadjopoulos, D.,** Sterically stabilized liposomes: a hypothesis on the molecular origin of the extended circulation times, *Biochim. Biophys. Acta,* 1070, 187, 1991.

17. **Gregoriadis, G., Ed.,** *Liposomes as Drug Carriers: Recent Trends and Progress,* John Wiley & Sons, New York, 1988.

18. **Senior, J.H.,** Fate and behaviour of liposomes *in vivo*: a review of controlling factors, *Crit. Rev. Therap. Drug Carrier System,* 3, 123, 1987.

19. **Mayhew, E., Rustum, Y., Szoka, F., and Papahadjopoulos, D.,** Role of cholesterol in enhancing the antitumor activity of cytosine arabinoside entrapped in liposomes, *Cancer Treat. Rep.,* 63(11–12), 1923, 1979.

20. **Szoka, F. and Papahadjopoulos, D.,** Liposomes: preparation and characterization, in *Liposomes: From Physical Structure to Therapeutic Applications,* Knight, G., Ed., Elsevier/North Holland Biomedical Press, 1981, 51.

21. **Woodle, M.C. and Papahadjopoulos, D.,** Liposome preparation and size characterization, in *Methods in Enzymology,* Vol. 171, Biomembranes, Part R, Fleischer, S. and Packer, L., Eds., 1989, 193.

22. **Juliano, R.L. and Stamp, D.,** Pharmacokinetics of liposome-encapsulated anti-tumor drugs, *Biochem. Pharmacol.,* 27, 21, 1978.

23. **Gregoriadis, G., Kirby, C., Large, P., Meehan, A., and Senior, J.,** Targeting of liposomes: study of influencing factors, in *Targeting of Drugs,* Gregoriadis, G., Senior, J., and Trouel, A., Eds., Plenum Press, New York, 1982, 155.

24. **Spanjer, H.H., Van Galen, M., Roerdink, F.H., Regts, J., and Scherphof, G.L.,** Intrahepatic distribution of small unilamellar liposomes as a function of liposomal lipid composition, *Biochim. Biophys. Acta,* 863, 224, 1986.

25. **Hwang, K.J., Luk, K.S., and Beaumier, P.L.,** Hepatic uptake and degradation of unilamellar sphingomyelin/cholesterol liposomes: a kinetic study, *Proc. Natl. Acad. Sci. U.S.A.,* 77, 4030, 1980.

26. **Hwang, K.J. and Beaumier, P.L.,** Disposition of liposomes *in vivo*, in *Liposomes as Drug Carriers: Recent Trends and Progress,* Gregoriadis, G., Ed., John Wiley & Sons, New York, 1988, 19.

27. **Chonn, A., Semple, S.C., and Cullis, P.R.,** Association of blood proteins with large unilamellar liposomes *in vivo*, *J. Biol. Chem.,* 267, 1992.

28. **Lee, K.-D., Hong, K., and Papahadjopoulos, D.,** Recognition of liposomes by cells: *in vitro* binding and endocytosis mediated by specific lipid headgroups and surface charge density, *Biochim. Biophys. Acta,* 1103, 185, 1992.

29. **Allen, T.M., Wiliamson, P., and Schlegel, R.A.,** Phosphatidylserine as a determinant for reticuloendothelial recognition of model erythrocyte membranes, *Proc. Natl. Acad. Sci. U.S.A.,* 85, 8067, 1988.

30. **Funato, K., Harashima, H., Sakata, K., and Kiwada, H.,** Enhanced hepatic uptake of liposomes through complement activation depending on the size, *Pharmaceut. Res.,* 11, 402, 1994.

31. **Gaber, M.H., Hong, K., Huang, S.K., and Papahadjopoulos, D.,** Thermosensitive liposomes: release of Doxorubicin by bovine serum and human plasma in relation to hyperthermia, (submitted), 1994.

32. **Gabizon, A. and Papahadjopoulos, D.,** The role of surface charge and hydrophilic groups on liposome clearance *in vivo*, *Biochim. Biophys. Acta,* 1103, 94, 1992.

33. **Allen, T. M. and Hansen, C.,** Pharmacokinetics of stealth vs conventional liposomes: effect of dose, *Biochim. Biophys. Acta,* 1068, 131, 1991.
34. **Maruyama, K., Kennel, S., and Huang, L.,** Lipid composition is important for highly efficient target binding and retention of immunoliposomes, *Proc. Natl. Acad. Sci. U.S.A.,* 87, 5744, 1990.
35. **Ahmad, I., Longenecker, M., Samuel, J., and Allen, T.M.,** Antibody-targeted delivery of doxorubicin entrapped in sterically stabilized liposomes can eradicate lung cancer in mice, *Cancer Res.,* 55, 1484, 1993.
36. **Gabizon, A.,** Selective tumor localization and improved therapeutic index of anthracyclines encapsulated in long-circulating liposomes, Cancer Res., 52, 891, 1992.
37. **Woodle, M.C., Neuman, M., Collins, L.R., Redemann, C., and Martin, F.J.,** Improved long circulating liposomes "Stealth" using synthetic lipids, *Proc. Int. Symp. Control Release Bioact. Mater.,* 17, 77, 1990.
38. **Klibanov, A.L., Maruyama, K., Torchilin, V.P., and Huang, L.,** Amphipathic polyethyleneglycols effectively prolong the circulation time of liposomes, *FEBS Lett.,* 268, 235, 1990.
39. **Blume, G. and Cevc, G.,** Liposomes for sustained drug release *in vivo, Biochim. Biophys. Acta,* 1029, 91, 1990.
40. **Allen, T.M., Martin, F.J., Redemann, C., Hansen, C., and Yau-Young, A.,** Liposomes containing synthetic lipid derivatives of polyethylene glycol show prolonged circulation $T_{1/2}$ *in vivo, Biochim. Biophys. Acta,* 1066, 29, 1991.
41. **Woodle, M.C., Matthay, K.K., Newman, M.S., Hadiyat, J.E., Collins, L.R., Redemann, C., Martin, F.J., and Papahadjopoulos, D.,** Sterically stabilized liposomes: versatility of lipid compositions with prolonged circulation, *Biochim. Biophys. Acta,* 1105, 193, 1992.
42. **Klibanov, A.L., Maruyama, K., Beckerleg, A.M., Torchilin, V.P., and Huang, L.,** Activity of amphipathic poly(ethylene glycol) 5000 to prolong the circulation time of liposomes depends on the liposome size and is unfavorable for immunoliposome binding to target, *Biochim. Biophys. Acta,* 1162, 142, 1991.
43. **Litzinger, D.C. and Huang, L.,** Amphipathic poly(ethylene glycol) 5000-stabilized dioleoylphosphatidyl-ethanolamine liposomes accumulate in spleen, *Biochim. Biophys. Acta,* 1127, 249, 1992.
44. **Kuhl, T.L., Leckband, D.E., Lasic, D.D., and Israelachvili, J.N.,** Modulation of interacting forces between bilayers exposing short-chained ethylene oxide groups, *Biophys. J.* 66, 1479, 1994.
45. **Needham, D., McIntosh, T.J., and Lasic, D.D.,** Repulsive interactions and mechanical stability of polymer-grafted lipid membranes, *Biochim. Biophys. Acta,* 1108, 40, 1992.
46. **Woodle, M.C., Collins, L.R., Sponsler, E., Kossovsky, N., Papahadjopoulos, D., and Martin, F.J.,** Sterically stabilized liposomes: reduction in electrophoretic mobility but not electrostatic surface potential, *Biophys. J.,* 61, 902, 1992.
47. **Mori, A., Klibanov, A.L., Torchilin, V.P., and Huang, L.,** Influence of the steric barrier activity of amphipathic poly(ethyleneglycol) and ganglioside GM1 on the circulation time of liposomes and on the target binding of immunoliposomes *in vivo, FEBS Lett.,* 284, 2263, 1991.
48. **Maruyama, K., Mori, A., Bhadra, S., Ravi Subbiah, M.T., and Huang, L.,** Proteins and peptides bound to long-circulating liposomes, *Biochim. Biophys. Acta,* 1170, 246, 1991.
49. **Huang, S.K., Lee, K.-D., Hong, K., Friend, D.S., and Papahadjopoulos, D.,** Microscopic localization of sterically stabilized liposomes in colon carcinoma-bearing mice, *Cancer Res.,* 52, 5135, 1992.
50. **Huang, S.K., Martin, F.J., Jay, G., Vogel, J., Papahadjopoulos, D., and Friend, D.S.,** Extravasation and transcytosis of liposomes in Kaposi's sarcoma-like dermal lesions of transgenic mice bearing the HIV tat gene, *Am. J. Pathol.,* 143, 10, 1993.
51. **Wu, N.W., Da, D., Rudoll, T.L., Needham, D., Whorton, A.R., and Dewhirst, M.W.,** Increased microvascular permeability contributes to preferential accumulation of stealth liposomes in tumor tissue, *Cancer Res.,* 53, 3765, 1993.
52. **Yuan, F., Leunig, M., Huang, S.K., Berk, D.A., Papahadjopoulos, D., and Jain, R.K.,** Microvascular permeability and interstitial penetration of sterically stabilized (Stealth) liposomes in a human tumor xenograft: an intravital microscopy study, *Cancer Res.,* 54, 3352, 1994.
53. **Huang, S.K., Lee, K.-D., Hong, K., Friend, D.S., and Papahadjopoulos, D.,** Microscopic localization of sterically stabilized liposomes in colon carcinoma-bearing mice, *Cancer Res.,* 52, 5135, 1992.
54. **Huang, S.K., Mayhew, E., Gilani, S., Lasic, D.D., Martin, F.J., and Papahadjopoulos, D.,** Pharmacokinetics and therapeutics of sterically stabilized liposomes in mice bearing C-26 colon carcinoma, *Cancer Res.,* 52, 6774, 1992.
55. **Mayhew, E., Lasic, D., Babbar, S., and Martin, F.J.,** Pharmacokinetics and antitumor activity of epirubicin encapsulated in long-circulating liposomes incorporating a polyethylene glycol-derivatized phospholipid, *Int. J. Cancer,* 51, 302, 1992.
56. **Vaage, J., Mayhew, E., Lasic, D., and Martin, F.J.,** Therapy of primary and metastatic mouse mammary carcinomas and doxorubicin encapsulated in long circulating liposomes, *Int. J. Cancer,* 51, 942, 1992.
57. **Williams, S.S., Alosco, T.R., Mayhew, E., Lasic, D.D., Martin, F.J., and Bankert, R.B.,** Arrest of human lung tumor xenograft growth in severe combined immunodeficient mice using doxorubicin encapsulated in sterically stabilized liposomes, *Cancer Res.,* 53, 3964, 1993.

58. **Allen, T.M., Mehra, T., Hansen, C., and Chin, Y.C.,** Stealth liposomes: an improved sustained release system for arabinofuranosylcytosine, *Cancer Res.,* 52, 2431, 1992.

59. **Huang, S.K., Stauffer, P.R., Hong, K., Guo, J.W.H., Phillips, T.L., Huang, A., and Papahadjopoulos, D.,** Liposomes and hyperthermia in mice: increased tumor uptake and therapeutic efficacy of doxorubicin in sterically stabilized liposomes, *Cancer Res.,* 54, 1994.

60. **Allen, T.M. and Papahadjopoulos, D.,** Sterically stabilized (stealth) liposomes: pharmacokinetic and therapeutic advantages, in *Liposome Technology,* 2nd ed., Vol. 3, Gregoriadis, G., Ed., CRC Press, Boca Raton, FL, 1992, 59.

61. **Woodle, M.C. and Lasic, D.D.,** Sterically stabilized liposomes, *Biochim. Biophys. Acta,* 1113, 171, 1992.

62. **Huang, L., Ed.,** Covalently attached polymers and glycans to alter the biodistribution of liposomes, *J. Liposome Res.,* 2, 289, 1992.

63. **Papahadjopoulos, D. and Gabizon, A.,** Sterically stabilized liposomes: pharmacological properties and drug carrying potential, in, *Liposomes as Tools in Basic Research and Industry,* Philippot, J.R. and Schuber, F., Eds., CRC Press, Boca Raton, FL, 1994.

64. **Papahadjopoulos, D.,** A personal perspective of liposomes as a drug carrier system, *J. Liposome Res.,* 2 (3), ii–xviii, 1992.

65. **Lasic, D.D.,** *Liposomes: From Physics to Applications,* Elsevier, Amsterdam, 1993, chap. 11 and chap. 14.

66. **Lasic, D.D., Frederik, P.M., Stuart, M.C.A., Barenholz, Y., and MacIntosh, T.J.,** Gelation of liposome interior: a novel method for drug encapsulation, *FEBS Lett.,* 312, 255, 1992.

67. **Haran, G., Cohen, R., Bar, L.K., and Barenholz, Y.,** Transmembrane ammonium sulfate gradients in liposomes produce efficient and stable entrapment of amphipathic weak bases, *Biochim. Biophys. Acta,* 1151, 201, 1993.

68. **Vaage, J., Donovan, D., Mayhew, E., Uster, P., and Woodle, M.,** Therapy of mouse mammary carcinomas with vincristine and doxorubicin encapsulated in sterically stabilized liposomes, *Int. J. Cancer,* 54, 959, 1993.

69. **Liu, D. and Huang, L.,** pH-sensitive, plasma-stable liposomes with relatively prolonged residence in circulation, *Biochim. Biophys. Acta,* 1022, 348, 1990.

70. **Liu, D., Mori, A., and Huang, L.,** Role of liposome size and RES blockade in controlling biodistribution and tumor uptake of GM1-containing liposomes, *Biochim. Biophys. Acta,* 1104, 95, 1992.

71. **Wilschut, J. and Hoekstra, D., Eds.,** *Membrane Fusion,* Marcel Dekker, New York, 1991.

72. **Duzgunes, N.,** Membrane fusion techniques, in *Methods in Enzymology,* Vol. 220, Academic Press, San Diego, CA, 1993.

Chapter 2

FATE OF LIPOSOMES *IN VIVO* AND ITS CONTROL: A HISTORICAL PERSPECTIVE

Gregory Gregoriadis

TABLE OF CONTENTS

I. INTRODUCTION

The liposome drug-carrier concept, proposed and tested a quarter of a century ago,[1-3] has undergone a variety of stages which have recently culminated in the marketing of life-saving pharmaceutical products. A bird's eye view of the liposome story and its impact on drug delivery has been outlined elsewhere.[4] Here, I shall confine myself to key developments in my laboratory which, in the course of time, contributed to the understanding of vesicle fate *in vivo* and, eventually, its control. The latter refers to the tailoring of the structural characteristics of liposomes in ways that ensure (1) quantitative retention of entrapped drugs during exposure of the carrier to the biological milieu (e.g., blood) en route to the target, and (2) vesicle clearance rates from the circulation (or other injected sites) that are conducive to optimal carrier distribution, for instance uptake by target tissues.

II. EARLY YEARS (1970–1974)

Following work during 1970–1973 with agent-containing multilamellar vesicles (MLV),[1-3,5-11] several basic aspects of their behavior became apparent. For instance, vesicle rate of clearance from the blood of intravenously injected rats was rapid, dose-dependent, and biphasic.[2,3] It was subsequently observed[5] that neutral and positively charged (with stearylamine)* MLV exhibited a slower rate of clearance than negatively charged MLV. A year later[12] the effect of surface charge on vesicle clearance was found to apply to small unilamellar vesicles (SUV) as well, with the same work[12] also showing that SUV had a longer residence time than MLV. Another observation[6,7] made at that time was that small, water-soluble drugs (e.g., 5-fluoro-uracil and penicilin G) leaked considerably from circulating MLV. In view of what we know today, this is hardly surprising, as MLV used were cholesterol-poor.[6,7] On the other hand,

---

* Recent experiments[20] with an alternative positively charged lipid and liposomes of varying bilayer fluidity have shown that a slow clearance for such positively charged liposomes is probably an artefact, masking a normally rapid clearance for vesicles which retain their charge in the circulation.

entrapped larger solutes such as albumin and amyloglycosidase did not appear to leak.[2] Finally and perhaps more importantly, liposomes with entrapped materials were shown to end up in the fixed macrophages of the reticuloendothelial system (RES), mainly in the liver and spleen, through the lysosomotropic pathway.[3,8-11] Within lysosomes, vesicles were seen[11] to lose their organized structure (presumably through the action of phospholipases) and release their contents. Depending on molecular size and ability to withstand the hostile milieu of the organelles, drugs could then act either locally (e.g., hydrolysis of stored sucrose by liposomal β-fructofuranosidase)[9] or, after diffusion through the lysosomal membrane, in other cell compartments (e.g., inhibition of DNA-directed RNA synthesis by liposomal actinomycin D in partially hepatectomized rats).[10] Such knowledge of liposomal fate and behavior (remarkably detailed for this early stage of liposome research) enabled us[3,6,13-15] to test, propose, or anticipate a number of applications, including the treatment of certain inherited metabolic disorders, intracellular infections, cancer, gene therapy, and, because of antigen-presenting cell involvement in vesicle uptake *in vivo*, immunopotentiation and vaccine delivery.[16]

III. CRITERIA FOR VESICLE TARGETING

The demonstration[17,18] early on in the 1970s that liposomes coated with cell-specific ligands (e.g., antibodies and asialoglycoproteins) can interact with and be internalized by alternative (other than macrophages) cells expressing appropriate receptors, both *in vitro* and *in vivo*, led to the concept[17,18] of vesicle targeting and to novel avenues of liposome research and potential uses. Perhaps more encouraging also were indications[19] that even plain liposomes could enhance the delivery of drugs to solid tumors *in vivo*. It was soon predicted,[18] however, that success in this area would not only require quantitative retention of entrapped drugs by vesicles (also needed for "passive" targeting to fixed macrophages) but also sufficiently long time for circulating (targeted) vesicles to encounter, and associate with, the target. Since vesicle behavior *in vivo* resulting from a given structural feature of the system must be closely related to the particular environment in which the system exists, knowledge of the influence of the biological milieu on both drug leakage and vesicle clearance rates appeared essential.

A. DRUG RETENTION BY LIPOSOMES

The first indication[21,22] as to why liposomes might become permeable or destabilized in the presence of blood, thus allowing entrapped solutes to leak out, came from the finding that plasma high density lipoproteins (HDL) remove phospholipid molecules from the vesicle bilayer. It appeared logical to assume[23] that prevention of such HDL action by, for instance, rendering bilayers packed would allow vesicles to remain intact. An earlier observation[24] that addition of cholesterol into liposomes reduces protein-induced vesicle permeability in buffer was significant in this respect. Experiments to test this assumption were thus carried out, initially with MLV[23] and later on with SUV.[25] In both cases excess cholesterol (e.g., a phospholipid-to-cholesterol molar ratio of 1:1) stabilized bilayers, with very little of the entrapped water-soluble marker or phospholipid being lost in the presence of blood[23,25-27] or *in vivo* after intravenous,[25,26] intraperitoneal,[25] or subcutaneous[28] injection. A role for HDL in the destabilization of liposomes was further substantiated by (1) the demonstration[29] that cholesterol-free SUV exposed to blood plasma from lipoprotein-deficient mice or a patient with congenital lecithin-cholesterol acyltransferase deficiency (characterized by low HDL levels), were more stable than when incubated with plasma from normal mice or humans; (2) the identification[29] of HDL as the only lipoprotein, among a number of species added to lipoprotein-deficient plasma, that could restore vesicle-destabilizing activity.

In view of these findings it was of interest to see whether reduction of vesicle leakage to solutes by cholesterol-induced bilayer packing could also be achieved with bilayers made rigid

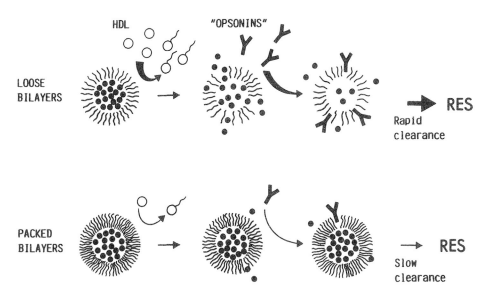

FIGURE 1. Correlation between stability of liposomes and clearance from the circulation. In the scheme proposed, extent of bilayer porosity (and leakage of solutes in blood) is dependent on the facility with which high density lipoproteins (HDL) remove phospholipid molecules from the bilayer. "Loose" bilayers *(top)* are attacked by HDL more effectively than packed bilayers *(bottom)*. The greater the "gaps" on the bilayer after HDL attack, the more extensive the opsonin adsorption onto vesicles and uptake by the reticuloendothelial system (RES). (From Gregoriadis, G., *News Physiol. Sci.*, 4, 146, 1989. With permission.)

by substituting phosphatidylcholine (PC) with high-melting phospholipids. As anticipated,[27] SUV or MLV made of hydrogenated lecithin, distearoyl phosphatidylcholine (DSPC), or sphingomyelin (SM) were less permeable to entrapped solutes in the presence of blood plasma, with bilayer stability becoming even greater when excess cholesterol was also present.[28-33] Thus, liposomes composed of equimolar DSPC and cholesterol were able,[31] in our hands, to completely retain their solute (carboxyfluorescein) content even after 48 h of exposure to plasma at 37°C.

B. VESICLE CLEARANCE RATES: THE ROLE OF LIPID COMPOSITION

In the course of work on liposomal stability it was observed[30] that solute retention by cholesterol-rich SUV, although always pronounced during relatively short (e.g., up to 2 h) periods of exposure to blood plasma, was reduced significantly with some formulations (e.g., those made of PC) on incubation for longer periods, whereas with others (e.g., SM) it remained unchanged. It soon turned out[30] that this phenomenon correlated with the residence time in the blood circulation of the corresponding (doubly labeled) vesicles, e.g., the greater the vesicle stability was the longer the half-life.[31] It was disappointing to note,[32] however, that this relationship between vesicle stability and clearance observed for neutral SUV (of a size up to 80 nm diameter in our work) did not apply to very stable negatively charged SUV or, more importantly, to larger neutral vesicles. In the latter case clearance rates were found[32] to increase progressively with increasing size, even though bilayer stability remained high for prolonged periods of incubation with plasma. It thus appears that vesicle size and surface charge both override the state of bilayer stability in determining vesicle clearance from the circulation.

On the basis of such findings, a tentative hypothesis[34] was put forward to unify observations of the role of liposomal phospholipids and cholesterol in promoting bilayer stability *in vitro* and (neutral) vesicle half-lives *in vivo*. Figure 1 proposes that, for liposomes that survive HDL attack and complete disintegration, bilayer destabilization facilitates the adsorption onto or insertion into it of plasma proteins (e.g., opsonins) thought to be responsible[35] for the presentation of vesicles and other particles in a form recognizable by macrophages. The hypothesis predicts that the lesser the destabilization of the bilayer the fewer or smaller the "gaps" on it (and the lesser the leakage of solutes) and, therefore, the smaller the amount of opsonins capable of associating with the vesicles. Thus, vesicles with decreasing amounts of opsonin on their surface would be expected to exhibit a proportionally reduced uptake by the RES. It is recognized, however, that the scheme in Figure 1, although probably valid in general terms, is too simple to reflect real events. Indeed, since 1976 a number of workers have shown that a variety of plasma components (including α_2-macroglobulin,[36] blood coagulation factors,[37] components of the complement cascade,[38] C reactive protein,[39] and fibronectins[40]) can bind to liposomes depending on their lipid composition and surface properties. More recent work[41] suggests that liposomes interact (or fail to interact) with tissue-specific opsonins that are capable of facilitating or inhibiting (dysopsonins) uptake of vesicles with differing lipid compositions by the macrophages of the liver, spleen, or bone marrow. The ways by which the bilayer surface in its varying states of fluidity or destabilization encourages or discourages interaction with specific opsonins is unknown. Nonetheless, the failure[41] of a liver-specific opsonin to enhance uptake of liposomes rich in cholesterol or composed of "high melting" phospholipids by Kupffer cells, is compatible with the sequence of events in Figure 1, especially if the operative term "gaps" (see above) is substituted with the more prudent "opportunities" (for protein insertion or absorption).

C. TOWARDS "INVISIBLE" LIPOSOMES

The importance of designing liposomes in ways that will help them avoid early interception by the RES and thus exhibit prolonged residence times in the blood circulation cannot be overestimated. Such liposomes could, as already alluded to, interact effectively with a variety of cell targets both intravascularly and, when possible, extravascularly (e.g., to treat tumors) or duplicate the function of erythrocytes in the form of hemosomes. To that end, the ability of long-lived cholesterol-rich SUV made of high melting phospholipids to deliver significant portions of the injected drug dose to solid tumors[42] or to cure solid-tumor-bearing mice[43] seems promising. Indeed, it may be that the simplicity of the approach, coupled with novel technology that achieves[44] a high (entrapped) drug-to-lipid mass ratio in small vesicles, will render especially designed SUV a preferred carrier for uses where a minimum vesicle size is a prerequisite for success (e.g., imaging of tumors, delivery of drugs to the bone marrow, etc.). Nonetheless, the case of long-lived larger vesicles that can accommodate realistic quantities of drugs (without increasing the lipid dose) or molecules such as proteins (e.g., hemoglobin and cytokines) and enzymes still remains as strong as ever.

Having realized[32] that manipulation of the lipid composition of large liposomes (over 80 to 100 nm in diameter) does not improve vesicle half-life in the circulation significantly, we[45] and others (see this book) have recently, and independently, attempted to mimick Nature. It is known, for instance, that certain pathogenic bacteria with a highly hydrophilic surface evade the immune system in animals. It is also probable that the high concentration of sugars on the surface of circulating cells (e.g., erythrocytes) contributes to their life span. It was thought likely that liposomes with a sufficiently hydrophilic surface would repulse opsonin adsorption and thus curtail interception by the RES. The molecule of choice that eventually proved successful in this respect was polyethyleneglycol (PEG) and its various derivatives. There is now overwhelming evidence that pegylated liposomes of about 100 nm in diameter circulate in the blood for longer periods than similar non-pegylated vesicles. As the present book attests, interest in this area is mushrooming with numerous applications in various stages of progress. As always, there is room, however, for improvement, in terms, for instance, of further augmenting vesicle half-lives, applying the concept to even larger vesicles or perhaps replacing PEG with a natural, biodegradable molecule. There is evidence[46] to suggest that polysialic acids (easily harvested from the cultures of certain nonpathogenic bacteria) may be an attractive alternative.

REFERENCES

1. **Gregoriadis, G., Leathwood, P.D., and Ryman B.E.,** Enzyme entrapment in liposomes, *FEBS Lett.,* 14, 95, 1971.
2. **Gregoriadis, G. and Ryman, B.E.,** Fate of protein-containing liposomes injected into rats. An approach to the treatment of storage diseases, *Eur. J. Biochem.,* 24, 485, 1972.
3. **Gregoriadis, G. and Ryman, B.E.,** Lysosomal localization of β-fructofuranosidase-containing liposomes injected into rats. Some implications in the treatment of genetic disorders, *Biochem. J.,* 129, 123, 1972.
4. **Gregoriadis, G.,** Liposomes: a tale in drug targeting, *J. Drug Targeting,* 1, 3, 1993.
5. **Gregoriadis, G. and Neerunjun, D.E.,** Control of the rate of hepatic uptake and catabolism of liposome-entrapped proteins injected into rats. Possible therapeutic applications, *Eur. J. Biochem.,* 47, 179, 1974.
6. **Gregoriadis, G.,** Drug entrapment in liposomes, *FEBS Lett.,* 36, 292, 1973.
7. **Gregoriadis, G.,** Drug entrapment in liposomes. Possibility for chemotherapy, *Biochem. Soc. Trans.,* 2, 117, 1974.
8. **Gregoriadis, G., Putman, D., Louis, L., and Neerunjun, D.,** Comparative fate and efect of non-entrapped and liposome-entrapped neuraminidase injected into rats, *Biochem. J.,* 140, 323, 1974.
9. **Gregoriadis, G. and Buckland, R.A.,** Enzyme-containing liposomes alleviate a model for storage disease, *Nature (London),* 244, 170, 1973.
10. **Black, C.V.D. and Gregoriadis, G.,** Intracellular fate and effect of liposome-entrapped actinomycin D injected into rats, *Biochem. Soc. Trans.,* 2, 869, 1974.
11. **Segal, A.W., Wills, E.J., Richmond, J.E., Slavin, G., Black, C.D.V., and Gregoriadis, G.,** Morphological observations on the cellular and subcellular destination of intravenously administered liposomes, *Br. J. Exp. Pathol.,* 55, 3207, 1974.
12. **Juliano, R.L. and Stamp, D.,** Effects of particle size and charge on the clearance of liposomes and liposome-encapsulated drugs, *Biochem. Biophys. Res. Commun.,* 63, 651, 1975.
13. **Gregoriadis, G.,** The carrier potential of liposomes in biology and medicine, *N. Engl. J. Med.,* 295, 704, 1976.
14. **Neerunjun, D.E. and Gregoriadis, G.,** Prolonged survival of tumour bearing mice treated with liposome-entrapped actinomycin D, *Biochem. Soc. Trans.,* 2, 868, 1974.
15. **Belchetz, P.E., Braidman, I.P., Crawley, J.C.W., and Gregoriadis, G.,** Treatment of Gaucher's disease with liposome-entrapped glucocerebroside β-glucosidase, *Lancet,* ii, 116, 1977.
16. **Allison, A.C. and Gregoriadis, G.,** Liposomes as immunological adjuvants, *Nature (London),* 252, 252, 1974.
17. **Gregoriadis, G. and Neerunjun, D.E.,** Homing of liposomes to target cells, *Biochem. Biophys. Res. Commun.,* 65, 537, 1975.
18. **Gregoriadis, G.,** Structural requirements for the specific uptake of macromolecules and liposomes by target tissues, in *Enzyme Therapy in Lysosomal Storage Disease,* Tager, J.M., Hooghwinkel, G.J.M., and Daems, W. Th., Eds., North-Holland, Amsterdam, 1974, 131.
19. **Gregoriadis, G., Neerunjun, D.E., and Hunt, R.,** Fate of a liposome-associated agent injected into normal and tumour-bearing rodents. Attempts to improve localization in tumour tissues, *Life Sci.,* 21, 357, 1977.
20. **Tan, L. and Gregoriadis, G.,** The effect of positive surface charge of liposomes on their clearance from blood and its relation to vesicle lipid composition, *Biochem. Soc. Trans.,* 17, 690, 1989.
21. **Krupp, L., Chobanian, A.V., and Brecher, J.P.,** The in-vivo transformation of phospholipid vesicles to a particle resembling HDL in the rat, *Biochem. Biophys. Res. Commun.,* 72, 1251, 1976.
22. **Scherphof, G., Roerdink, G., Waite, M., and Parks, J.,** Disintegration of phosphatidylcholine liposomes in plasma as a result of interaction with high-density lipoproteins, *Biochim. Biophys. Acta,* 542, 296, 1978.
23. **Gregoriadis, G. and Davis, C.,** Stability of liposomes *in vivo* and *in vitro* is promoted by their cholesterol content and the presence of blood cells, *Biochem. Biophys. Res. Commun.,* 89, 1287, 1979.
24. **Papahadjopoulos, D., Cowden, M., and Kimelberg, H.,** Role of cholesterol in membranes. Effects of phospholipid-protein interactions, membrane permeability and enzyme activity, *Biochim. Biophys. Acta,* 310, 8, 1973.
25. **Kirby, C., Clarke, J., and Gregoriadis, G.,** Effect of the cholesterol content of small unilamellar liposomes on their stability *in vivo* and *in vitro, Biochem. J.,* 186, 591, 1980.
26. **Kirby, C. and Gregoriadis, G.,** The effect of the cholesterol content of small unilamellar liposomes on the fate of their lipid components *in vivo, Life Sciences,* 27, 2223, 1980.
27. **Kirby, C., Clarke, J., and Gregoriadis, G.,** Cholesterol content of small unilamellar liposomes controls phospholipid loss to high density lipoproteins in the presence of serum, *FEBS Lett.,* 111, 324, 1980.
28. **Tumer, A., Kirby, C., Senior, J., and Gregoriadis, G.,** Fate of cholesterol-rich unilamellar liposomes containing ^{111}In-labelled bleomycin after subcutaneous injection into rats, *Biochim. Biophys. Acta,* 760, 119, 1983.
29. **Senior, J., Gregoriadis, G., and Mitropoulos, K.,** Stability and clearance of small unilamellar liposomes: studies with normal and lipoprotein-deficient mice, *Biochim. Biophys. Acta,* 760, 111, 1983.

30. **Gregoriadis, G. and Senior, J.,** The phospholipid component of small unilamellar liposomes controls the rate of clearance of entrapped solutes from the circulation, *FEBS Lett.,* 119, 43, 1980.

31. **Senior, J. and Gregoriadis, G.,** Is half-life of circulating small unilamellar liposomes determined by changes in their permeability?, *FEBS Lett.,* 145, 109, 1982.

32. **Senior, J., Crawley, J.C.W., and Gregoriadis, G.,** Tissue distribution of liposomes exhibiting long half-lives in the circulation after intravenous injection, *Biochim. Biophys. Acta,* 839, 1, 1985.

33. **Wolff, B. and Gregoriadis, G.,** The use of monoclonal anti-Thy_1 IgG_1 for the targeting of liposomes to AKR-A cells *in vitro* and *in vivo, Biochim. Biophys. Acta,* 802, 259, 1984.

34. **Gregoriadis, G.,** The physiology of the liposome, *News Physiol. Sci.,* 4, 146, 1989.

35. **Absolom, D.,** Opsonins and dysopsonins — an overview, *Methods Enzymol.,* 132, 28, 1986.

36. **Black, C.D.V. and Gregoriadis, G.,** Interaction of liposomes with blood plasma proteins, *Biochem. Soc. Trans.,* 4, 253, 1976.

37. **Bonte, F. and Juliano, R.L.,** Interaction of liposomes with serum proteins, *Chem. Phys. Lipids,* 40, 359, 1986.

38. **Loughrey, H.C., Bally, M.B., Reinish, L.W., and Cullis, P.R.,** The binding of phosphatidylglycerol liposomes to rat platelets is mediated by complement, *Thromb. Haemost.,* 64, 172, 1990.

39. **Richards, R.L., Gewurz, H., Siegel, J., and Alving, C.R.,** Interaction of C-reactive protein and complement with liposomes, *J. Immunol.,* 112, 1185, 1979.

40. **Rossi, J.D. and Wallace, B.A.,** Binding of fibronectin to phospholipid vesicles, *J. Biol. Chem.,* 258, 3327, 1983.

41. **Moghimi, S.M. and Patel, H.M.,** Techniques to study the opsonic effect of serum on uptake of liposomes by phagocytic cells from various organs of the RES, in *Liposome Technology,* Vol. 3, Gregoriadis, G., Ed., CRC Press, Boca Raton, 1993, 44.

42. **Presant, C.A., Proffitt, R.T., Turner, A.F., Williams, L.E., Winsor, D.W., Werner, J.L., Kennedy, P., Wiseman, C., Gala, K., McKenna, R.S. et al.,** Successful imaging of human cancer with indium-[111]-labeled phospholipid vesicles, *Cancer,* 62, 905, 1988.

43. **Large, P. and Gregoriadis, G.,** Phospholipid composition of small unilamellar liposomes containing melphalan influences drug action in mice bearing PC6 tumours, *Biochem. Pharmacol.,* 32, 1315, 1983.

44. **Gregoriadis, G., da Silva, H., and Florence, A.T.,** A procedure for the efficient entrapment of drugs in dehydration-rehydration liposomes (DRV), *Int. J. Pharm.,* 65, 235, 1990.

45. **Senior, J.H., Delgado, C., Fisher, D., Tilcock, C., and Gregoriadis, G.,** Influence of surface hydrophilicity of liposomes on their interaction with plasma proteins and clearance from the circulation: studies with polyethylene glycol-coated vesicles, *Biochim. Biophys. Acta,* 1062, 77, 1991.

46. **Gregoriadis, G., McCormack, B., Wang, Z., and Lifely, R.,** Polysialic acids: potential in drug delivery, *FEBS Lett.,* 315, 271, 1993.

Chapter 3

EFFECT OF SURFACE MODIFICATION WITH GLYCOLIPIDS AND POLYSACCHARIDES ON *IN VIVO* FATE OF LIPOSOMES

Prahlad C. Ghosh and Bimal K. Bachhawat

TABLE OF CONTENTS

I. INTRODUCTION

Since the discovery that liposomes can be used as effective *in vivo* delivery vehicles for biologically active molecules and drugs, a number of uses of liposomes as carriers of drug, enzymes, and DNA have been described.[1-3] However, the targeting of liposomes towards specific tissues is severely limited by the rapid sequestration of liposomes by the reticuloendothelial system (RES). Therefore, efforts have been made in various laboratories to formulate liposomes that bypass the RES with concomitant prolonged circulatory life. We approached this problem in two ways: (1) by cell-specific targeting of liposomes using various glycolipids; (2) by prolonging the circulatory life of liposomes by attaching polysaccharides. In this chapter we would like to focus on the work carried out by various investigators to alter the *in vivo* behavior of liposomes using glycolipids and polysaccharides.

II. ALTERATION OF *IN VIVO* DISTRIBUTION OF LIPOSOMES AFTER SURFACE MODIFICATION WITH GLYCOLIPIDS

In 1968, Ashwell and Morell demonstrated for the first time that survival of the glycoprotein in the circulation is dependent on the terminal sugar of the oligosaccharide chain of the protein. They reported the presence on the hepatocyte surface of a specific receptor for glycoproteins that has a β-linked terminal galactose or *N*-acetylgalactosamine oligosaccharide chains.[4] Another sugar-specific mechanism was identified by Stahl and Schlesinger,[5] who reported in 1978 the presence of mannose receptors on the cells of the RES, including the liver sinusoidal (Kupffer) cells and alveolar macrophages. These reports suggested to us that incorporation of various glycosides on the liposomal surface should permit the *in vivo* targeting of liposomes toward different cell types of liver.[6] This concept prompted us to devise initially an *in vitro* liposome model system whereby the receptor-ligand interaction could be thoroughly followed under defined experimental conditions.[7,8] In this study monosialo-ganglioside (GM1)-containing multilamellar as well as unilamellar liposomes were prepared and their interaction with the galactose-binding *Ricinus communis* lectin RCA1 were investigated. A number of interesting inferences were derived from this study: (1) about 60% of the galactose residues of GM1 is externally available on the surface of unilamellar liposomes, but in multilamellar liposomes, only 25% of such residues was exposed on the surface; (2) the rate of lectin-GM1 liposome interaction is markedly affected by the surface density of the sugar residues; (3) the binding of liposome-associated GM1 to RCA1 was 100 to 1000 times higher than those of free galactose and *p*-nitro-β-D-galactopyranoside; (4) phase transition tempera-ture (Tc) of the phospholipid component of liposomes, lengths of the surface-bound oligosac-charide chains, and cholesterol concentration affect the binding of terminal sugar residues with lectins.

With this information we have designed the *in vivo* experiment to deliver invertase to hepatocytes with the help of GM1. The GM1 was incorporated to achieve a high density on the surface of the liposomes.[9] A rapid uptake of this liposomal enzyme into the liver following intravenous (IV) administration was observed. This rapid uptake was found to be a galactose receptor-mediated endocytic process.[9] However, Gregoriadis and Neerunjun reported that GM1-liposomes exhibited a pattern of removal from plasma similar to that found for the negatively charged phosphatidic acid liposomes.[10] Similarly, Jonah et al. reported that GM1-liposomes have no specific affinity for liver cells.[11] These conflicting results prompted us to reinvestigate further the role of different glycoside residues on the surface of liposomes and on their uptake by various organs.[12-14] In these studies various glycolipids bearing no charge, such as asialoganglioside, galactoyl cerebroside, and a number of neutral glycosides e.g., *p*-aminophenyl derivatives of β-galactoside, α-mannoside, and β-*N*-acetylglucosaminide, were attached to the surface of liposomes.[12]

In another approach, we synthesized glycolipid having terminal α- and β-galactosyl residues by coupling melibiose and lactose with PE by reductive amination using sodium cyanoborohydride.[13] Using these various types of glycoside-bearing liposomes, we reported a number of interesting observations:

1. Both β-galactosyl- and α-mannosyl-bearing liposomes were removed rapidly from the circulation with concomitant enhanced uptake by the liver, and this uptake is mediated by hepatic receptors responsible for the removal of glycoproteins having terminal galactose and mannose residues, respectively

2. The anomeric form of the terminal sugar, as well as the chain length of the oligosaccha-rides, play an important role in the *in vivo* uptake of liposomes

3. The density of the glycoside residues on the surface of liposomes is the determining factor for the uptake of glycosylated liposomes by the liver

4. The uptake of liposome-entrapped material by hepatocytes was threefold higher than that by nonparenchymal cells from liposomes having β-galactoside on the surface, whereas the uptake by nonparenchymal cells was sevenfold higher from liposomes having α-mannoside on their surface

5. The recognition of liposomes having β-galactose residue on the surface by the hepatic asialoglycoprotein receptor is dependent on the lipid composition of the liposomes

We further confirmed these observations using small unilamellar liposomes.[15] From these studies we concluded that the composition of liposomes and critical receptor density and its accessibility on the surface of liposomes deserve active consideration — a fact that had not been taken into account by other investigators.[10,11]

Our observations on galactose- and mannose-specific uptake of liposomes by the hepatocyte and nonparenchymal cells, respectively, have since been confirmed in a number of laboratories both *in vivo*[16-20] and *in vitro*.[21-24] Glucose-mediated delivery of liposomal material to hepatic cells has also been reported.[25] However, this needs to be substantiated. Although several investigators have arrived at a similar conclusion regarding the *in vivo* localization of β-galactosylated and α-mannosylated liposomes, there seems to be no unanimity on the quantitative aspects. A number of investigators had reported a much slower rate of uptake of galactosylated liposomes,[16,17,19] compared with our results. This reported difference may be due to the use of synthetic phospholipids and lactosylceramide by these workers, since it has been reported by Ng and Heath that the extent of cell-associated liposomes from egg phosphatidylcholine (PC)-based liposomes is 16-fold higher than DSPG-based liposomes.[26] The mechanism of these results have been attributed to their differential processing by lysosomal enzymes. During our study with model membrane, we observed that with lectin-lactosylceramide liposomes the rate of interaction is only one tenth that of the GM1-liposomes. Therefore, this may be another factor contributing to this difference in the rate of uptake. It is tempting to speculate that the much higher uptake observed in our study may be due to the presence of a tetrasaccharide (GluNAc-Gal-GalNAC-Gal) in the asialoganglioside, which may somehow help in penetration of multilamellar asialoganglioside liposome through endothelial fenestration of liver.

A. EFFECT OF LIPID COMPOSITION ON HEPATIC UPTAKE OF GALACTOSYLATED LIPOSOMES

A galactose-specific receptor has been demonstrated on the surface of Kupffer cells.[27-30] This raises an interesting question with regard to various factors that regulate the vectorial transfer of galactosylated liposomes to hepatocytes. Spanjer et al. have reported that bulk phospholipid composition is the deciding factor for intrahepatic distribution of liposomes upon lactosylceramide incorporation.[31] Dimyristoyl-phosphatidylcholine-based liposomes were preferentially recognized by the galactose-receptor on hepatocytes, upon lactosylceramide or triantenary galactose-terminated cholesterol derivative incorporation, whereas the Kupffer cells were the preferred site for sphingomyelin.[31,32] The mechanism of preferential accumulation of sphingomyelin-based galactosylated liposomes by nonparenchymal cells mediated by a galactose-specific receptor still remains to be investigated. However, it has been reported that the spingomyelin has the ability to form intermolecular hydrogen bonding[33] and therefore vesicles prepared with sphingomyelin are presumed to possess a more rigid character than the phosphatidylcholine vesicles. The galactose residues exposed on this relatively rigid liposome bilayer may allow a firmer "grip" by the asialoglycoprotein receptor molecules in the plasma membrane, as has already been reported by Munn and Parce in a study of Fc receptor-mediated uptake of "solid" and "fluid" liposomes by macrophages.[34] Banno et al.[35] have studied the effects of the membrane fluidity of lactosylceramide-containing phosphatidylcholine liposomes on their uptake by hepatocytes *in vivo* and *in vitro*. They have reported that the less fluid DPPC-liposome is an appropriate matrix for lactosylceramide to act as an effective ligand for hepatic endocytosis. These results strongly suggest that the topographic orientation of the galactose residue on the surface of solid liposomes will evidently be different from that on PC-based fluid liposomes. Differences in the characteristics of the receptors on the hepatocyte and the Kupffer cell membrane may be another contributing factor.[30]

The hepatocyte receptors are randomly distributed throughout the plasma membrane and this distribution pattern of receptors on hepatocytes allows the efficient uptake of molecular ligands. On the other hand the Kupffer cell receptors, which are localized in preaggregated clusters all over the plasma membrane, do not mediate endocytosis of molecular ligands, but endocytose particulate ligands of all sizes.[29] The clustered arrangement of receptors on the Kupffer cell allows a multiple binding of receptors to particulate ligands that results in the formation of a stable complex, ensuring the transport to coated pits and into the cell interior. These latter properties of Kupffer cell receptors appear to play a very important role in enhanced uptake of lactosylceramide or *tris*-Gal-Chol-containing sphingomyelin-based liposomes. This, however, does not explain why asialoganglioside- or lactosylceramide-containing PC-based liposomes, which are a particulate ligand with a size similar to sphingomyelin liposomes, are preferentially taken up by hepatocytes.[14,16]

Haensler and Schuber,[36] on the other hand, reported that incorporation of triantennary galactosyl ligand on the surface of liposomes leads to a two- to threefold lesser interaction of liposomes with mouse resident peritoneal macrophages than a monogalactosyl ligand. Their method of synthesis of this ligand was different from that employed by Spanjer et al. Spanjer et al.[32] have synthesized a triantennary galactosyl ligand by coupling *tris*(galactosyloxy-methyl) aminomethane to cholesterol using glycyl and succinyl as intermediate spacer moieties.[32] On the other hand, Haensler et al. synthesized this ligand by coupling derivatives of 1-thio-β-D-galactose to the amino groups of lysyl-lysine dipeptide. The intergalactosyl distance between these two derivatives may evidently be different. As a result the triantennary galactosyl ligand present on lysyl-lysine dipeptide may not be able to adopt the adequate topology required for optimal binding to cellular lectins.[36] Lee et al.[37] also reported that intergalactosyl distances are also important for the binding of triantennary molecular ligands to the galactose receptors of hepatocytes.[37] The optimal binding takes place when the three galactosyl headgroups of a triantennary ligand are located at the apexes of a triangle of sides 1.5, 2.2, and 2.5 nm. It seems, therefore, that the topological characteristics of the different receptor binding domains at the surface of galactose receptor-bearing cells are different. From the available reports it can be proposed that liposomes can be preferentially diverted toward hepatocytes or Kupffer cells of liver by selecting appropriate lipid composition with an adequate oligomeric ligand possessing the right intrinsic structure on the surface for optimal binding.

B. ROLE OF MONOSIALOGANGLIOSIDE (GM1) IN MODULATING THE *IN VIVO* BEHAVIOR OF LIPOSOMES

Contrary to our observation on GM1 liposomes, Allen and Chonn first reported that incorporation of 5 to 7 mol% of GM1 into the liposomes composed of egg PC and cholesterol not only reduces the susceptibility of these liposomes to lysis by the plasma components, but also increases their circulatory levels by three- to tenfold.[38] However, uptake into the RES still remained high, with more liposomes present in liver and spleen than were present in blood 2 h postinjection. These results are very much similar to that reported earlier by Surolia and Bachhawat.[9] They have also reported that increasing the rigidity of the liposomal bilayer leads to a dramatic increase in the level of circulating liposomes and a concomitant decrease in uptake into liver and spleen. They have also examined the role of several other gangliosides and glycolipids, but none could substitute for GM1 in their ability to prolong circulatory life. Allen and Chonn have also demonstrated that the liposome composition that has been shown to dramatically reduce RES uptake *in vivo*[38] had similar pronounced effects in reducing macrophage uptake *in vitro*.[39] This provides experimental evidence that the mechanism for reduction of liposome uptake *in vivo* is through the ability of these liposome formulations to avoid uptake by macrophages.

Gabizan and Papahadjopoulas have shown that not only GM1 but certain other glycolipids, e.g., hydrogenated soybean phosphatidyl inositol (HPI) and sulfatides, also prolong the circulatory life of liposomes.[40] They have also reported that the phospholipid composition of the liposomes plays a significant role in enhancing the circulatory life of liposomes.

Phospholipids capable of conferring a rigid bilayer act synergistically with glycolipids in prolonging the circulatory life of liposomes. A strong correlation between the residence time of liposomes in blood and their uptake by tumors implanted in mice was observed. Liposomes exhibiting extended circulation residence time preferentially accumulated in the tumor as compared to the rest of the body.[40] This preferential accumulation of long circulating liposomes in tumor has been attributed to the efficient extravasation of these liposomes, due to their long blood residence time, from the leaky capillary vessel walls of tumor leading to their accumulation in the tumor. The tumor localization of various antitumor drugs delivered through these glycolipid-containing long circulating liposomes in different tumor-bearing mice has been confirmed by various investigators.[41-45] It has also been reported that the treatment of various kinds of tumor-bearing mice with anticancer drugs loaded in these liposomes resulted in more effective tumor growth retardation with concomitant increased survival time than the conventional liposomes.[40,41,43] Bakker-Woudenberg et al. have reported that administration of long circulating liposomes composed of HPI:HPC:Chol resulted in significant localization in *Klebsiella pneumoniae*-infected lung in mice than the conventional liposomes.[46] Litzinger and Huang[3] have shown that inclusion of 5 mol% of GM1 with PC:Chol liposomes containing 34A-antibody, which is specific to pulmonary endothelial cells, not only enhances the binding of these liposomes to lung endothelial cells but showed longer retention time than the conventional liposomes.

GM1 has been used for prolonging the residence time of relatively plasma-stable acid-sensitive liposomes in the circulation.[47] This makes these acid-sensitive liposomes potentially useful as a cytoplasmic delivery vehicle for drugs under *in vivo* conditions. Prolongation of the circulatory life of thermosensitive liposomes composed of DPPC:DSPC at 9:1 molar ratio has also been reported after GM1 incorporation.[44] The treatment of tumor-bearing mice with doxorubicin loaded in GM1:DPPC:DSPC liposomes in combination with hyperthermia resulted in more effective tumor growth retardation and increased survival time than DPPC:DSPC liposomes. The combination of drug-loaded, long circulating, thermosensitive liposomes with local hyperthermia at the tumor site could be clinically useful for delivering a wide range of chemotherapeutic agents in the treatment of solid tumors. Recently, Cullis et al. have reported that GM1 is incapable of inhibiting the ability of doxorubicin-loaded liposomes to impair or block RES function. It is concluded, therefore, that liposomes exhibiting extended circulation life can induce RES blockage and do not avoid uptake by liver phagocytes.[48]

1. Effect of Liposome Size

Liu et al.[49] have reported that biodistribution of GM1-liposomes is highly dependent on their size. They have reported that liposomes with a diameter less than 70 nm are rapidly removed from the circulation and mainly accumulated in the liver parenchymal cells. The blood clearance and the liver and spleen uptake of these liposomes were similar to those of the PS-containing liposomes.[49] These results are in agreement with an earlier observation by Surolia and Bachhawat,[9] who have shown that GM1-liposomes are rapidly removed from the circulation and mainly accumulated in the liver parenchymal cells via direct interaction with hepatic asialoganglioside receptors.[9] However, Liu et al. have not determined whether the uptake of small GM1-liposomes by the liver is mediated by hepatic asialoglycoprotein receptors. Although it has been reported that small rigid unilamellar liposomes could extravasate and accumulate in the parenchymal cells of the liver, it does not rule out the possibility that accumulation of GM1-liposomes in parenchymal cells is mediated by hepatic asialoglycoprotein receptors.

They have also reported that spleen uptake became predominant when the liposome diameter was 300 nm or greater.[51] It was found that the activity of GM1 in prolonging the circulatory half-life of liposomes is limited to a relatively narrow size range, i.e., 70 to 200 nm. These liposomes also accumulated in the tumor more readily. On the other hand Allen et al.

have reported that blood/RES ratios increased with decreasing liposome size up to 80 nm.[52] Maximum blood/RES ratios were obtained for liposomes extruded through 100 or 80 nm pore size filters. However, they have reported that blood/RES ratios for liposomes extruded through 50 nm pore size filters is much higher than liposomes extruded through 200 nm pore size filters. These results are inconsistent with the report that small rigid liposomes with or without GM1 are rapidly removed from circulation and presumably accumulated in liver parenchymal cells.[49,50] They have also reported significant elevated blood/RES ratios for liposomes extruded through 400 nm pore size filters. This is in contrast to the observation of Liu et al., who have shown that liposomes with diameters greater than 300 nm mainly accumulate in the spleen with concomitant decrease of the blood level.[51] The reasons for these discrepant results are not very clear. However, Liu et al. used egg PC-based fluid liposomes for their studies, whereas Allen et al. used mainly sphingomyelin (SM)-based solid liposomes. The orientation of GM1 in various sizes of liposomes prepared from fluid egg PC and their ability to be taken up into spleen may be different from liposomes prepared from solid phospholipids, which may account for their differential localization. Recently Ng and Heath have shown that the rate of uptake of egg PC liposomes is much faster than the DSPG liposomes in CV-1 P cells in culture.[26] They have also reported that the extent of cell-associated liposomes from egg PG-based liposomes is 16-fold higher than DSPG-based liposomes. The mechanism of these results has been attributed to their differential processing by lysosomal enzymes. This may be another factor contributing to the differential uptake of egg PC-based and SM-based GM1-containing long circulating liposomes in the liver and spleen.

2. Possible Mechanism of Prolongation of Circulatory Life of GM1-Liposomes

The mechanism by which carbohydrate-containing, negatively charged rigid liposomes remain in the circulation, avoiding RES uptake, is not yet clear. To explain this result, the hypothesis of "shielded negative" charge was proposed.[53] However, this hypothesis was proposed from the studies on headgroup conformation of GM1 carried out in dimethylsulfoxide (DMSO). In the presence of DMSO one cannot expect a true picture of the orientation of oligosaccharide chains in the liposomal membrane. We have observed that the orientation of oligosaccharides of GM1 in an aqueous environment is different from when it is incorporated in liposomal membrane.[7] Consequently GM1-liposomes are cleared from the circulation rapidly[9] by hepatic asialoglycoprotein receptors as suggested by Ashwell and Morell.[4] However, the biodistribution of GM1 liposomes, particularly with respect to different cell types of liver, is not carried out in detail by these investigators. Therefore, the mechanism by which the GM1 prolongs the circulatory life of liposomes remains unclear.

Recently Park and Huang[54] have chemically modified GM1 and synthesized various neoglycolipid analogs of GM1 and studied the effect of these lipids on the circulatory time of liposomes. They have reported that the negatively charged carboxyl group of GM1 is not critical to prolonging the liposome circulation time and showed that the polyhydroxyl side chain of the sialic acid is much more important than the carboxyl group. These results also argue against the shielded negative charge hypothesis for the activity of GM1. Further, they have reported that the GM1 analogs that lost the binding activity for cholera toxin are incapable of prolonging the circulatory life of liposome. It appears, therefore, that GM1-specific protein(s) may bind to GM1 in a manner similar to that of cholera toxin. Binding of these GM1-specific protein(s) to liposomes could lead to a reduced level of opsonization and/or inactivation of bound opsonin(s), resulting in a reduced uptake by the RES.[54]

Chonn et al. have observed a strong correlation between the residence time of liposomes in blood and their ability to bind serum protein both *in vivo* and *in vitro*.[55] Liposomes exhibiting very rapid clearance kinetics have the greatest ability to bind blood protein. In contrast, liposomes exhibiting extended circulation residence times have markedly reduced amounts of associated blood proteins. It was found that increasing the GM1 content of PC:CH large unilamellar vesicles (LUVs) progressively reduced the amount of total protein bound to PC:CH LUVs. Further, the decrease in protein binding was apparent for all blood proteins, suggesting a nonspecific effect. Presumably at least some of these proteins function as opsonins to promote liposome uptake by the RES. These results support a significant role of protein in determining the fate of liposomes in the circulation. It seems, therefore, that any molecule that is capable of reducing protein binding to liposomes, even in a relatively nonspecific manner, will prolong the circulation half-life of liposomes.

The activity of GM1 in reducing the affinity of liposomes to macrophages may also be related to the degree of lipid peroxidation in the liposome membrane. It has been reported that a very low amount of GM1 in biomembrane is able to protect the membrane from lipid peroxidation.[54] Products of lipid peroxidation are efficient chemoattractants for macrophages.[54] Inhibition of lipid peroxidation by GM1 would reduce the production of these chemoattractants and diminish the uptake of liposomes by macrophages.

III. *IN VIVO* FATE OF LIPOSOMES COATED WITH POLYSACCHARIDES

It has been reported that attachment of dextran to drugs and enzymes not only confers chemical and biological stability but also increases their circulatory lifetimes.[56-58] However, the stability, biological properties, and circulatory lifetimes of these conjugates are highly dependent on the size and charge of dextrans and the mode of conjugation.[59,60]

Keeping this in mind attempts have been made to prolong the circulatory lifetime of liposomes after conjugation with dextran.[61] Dextran was conjugated onto the surface of PE-liposomes by the cyanogen bromide activation method.

It was observed that conjugation of dextran on the surface of liposomes significantly prolonged their circulatory lifetime with concomitant decreased levels of liposomal accumulation in the liver. The rate of clearance of dextran-coated liposomes from the circulation was found to be dependent on the density of dextran molecules on the surface. At 18 mol% dextran on the surface, 50% of the injected dose remained in the circulation, which is 2.5-fold higher than the liposomal formulation without dextran. The enhanced circulatory lifetime of liposomes after conjugation with dextran may be due to increased hydrophilicity on the surface of liposomes which resulted in reduced binding of plasma protein(s) responsible for rapid removal of liposomes from the circulation. Another factor that may be contributing in the enhanced circulatory lifetime of these liposomes is the inability of dextran to cross the cell membrane. Recently, enhancement of the circulatory lifetime of liposomes by increasing the hydrophilicity on the surface by incorporating GM1 or DSPC-PEG has been reported by various investigators.[45,62]

On the other hand Takada et al.[63] have reported that polysaccharide (*O*-palmitoylamilopectin and *O*-palmitoyl mannan)-coated liposomes are cleared rapidly from the circulation in comparison with the conventional liposomes and accumulated preferentially in alveolar macrophages. This group has also reported that antibiotics entrapped in these liposomes were more effective in the treatment of infectious disease associated with macrophages.[64] However, when *o*-palmitoyl pullulan was employed to coat liposomes, no significant alteration of *in vivo* behavior of these liposomes was observed, although relatively stable liposomes were obtained. They have employed *o*-palmitoyl pullulan for binding of cell-specific monoclonal antibody fragments to the liposomes. These immunoliposomes were found to be very effective in transporting drug to specific cells *in vitro* as well *in vivo*.[65] All these studies suggested that *in vivo* disposition of liposomes could be modulated significantly by coating the liposome surface with various polysaccharides.

IV. APPLICATION OF GLYCOLIPID-TARGETED LIPOSOMES TO CHEMOTHERAPY AND IMMUNOTHERAPY

A. PREVENTION OF EXPERIMENTAL HEPATITIS

The selective uptake of galactosylated and mannosylated liposomes to hepatocytes and nonparenchymal cells of liver suggested that it could be possible to treat more effectively the diseased condition of the liver by delivering therapeutic agents through these liposomes. In order to study the efficacy of therapeutic agents entrapped in galactosylated liposomes, we chose D-galactosamine-induced hepatitis as a model because it specifically destroys hepatocytes and thus offers a suitable experimental model. This disease can be prevented experimentally by prior administration of uridine.[66,67] It was observed that the requirement of uridine was reduced to 10-fold when asialoganglioside liposomes were employed as a vehicle of uridine. However, uridine entrapped in α-mannoside liposomes had very marginal effect.[68] These results are quite conceivable, because the asialoglycoprotein receptors are present on the hepatocytes, which lack mannose receptors.

However, the regenerating effect of uridine in asialoganglioside liposomes was no longer observed after 6 h of D-galactosamine administration. This was attributed to the extensive damage of hepatic cells during the 6-h period. These results and the report that asialoglycoprotein receptor activity is virtually absent on hepatoma cells,[69] limits the potential application of asialoganglioside liposomes as carriers of drugs in the diseased liver condition.

However, Keegan-Rogers et al.[69] have taken advantage of the lack of asialoglycoprotein receptors on hepatoma cells and designed a novel chemotherapeutic model to rescue normal hepatocytes during chemotherapy of hepatoma. They have delivered a rescue agent covalently bound to an asialoglycoprotein specifically to normal hepatocytes prior to toxin treatment. Therefore, in the presence of a corresponding toxin, while the normal hepatocytes are specifically protected, the malignant cells remain vulnerable to the full effect of the toxin. Asialoganglioside liposome-entrapped rescue agents may find potential application in the rescue of normal hepatocytes during chemotherapy for hepatoma.

Another interesting development in the application of β-galactosylated liposomes emerges from the work of Nicolau and co-workers, who have successfully used lactosylceramide liposome to introduce the biologically active preproinsulin gene into hepatocytes.[70,71] This is a potentially significant finding considering the secretory nature of hepatocytes, which have the enzyme required for the processing of prepolypeptides to polypeptides.

B. TREATMENT OF INFECTIOUS DISEASES

Aspergillosis, a systemic fungal infection, is prevalent in immunosuppressed patients and is often severe and life-threatening for them. Reticuloendothelial organs such as liver, spleen, and lungs are frequently the targets of systemic fungal infections.[72] Amphotericin B is the drug of choice for the treatment of most systemic fungal infections. However, the severe adverse effects limit its potential utility in clinical practice. It has been demonstrated by various investigators, including our group, that intercalation of amphotericin B in liposomal membrane not only reduces the toxicity of the drug but also increases the therapeutic index.[72-74] However, no attempt has been made to deliver amphotericin-B to the infection site by using mannosylated liposomes, which are known to be taken up specifically by macrophages via the mannose receptors.[5] We have employed mannosylated-liposome intercalated amphotericin B and hamycin for the treatment of experimental aspergillosis in mice.[73,74] It was observed that incorporation of both amphotericin and hamycin into mannosylated liposomes resulted in reduced toxicity and enhanced therapeutic index of these drugs as compared to the usual liposomal formulations. The superior therapeutic efficacy of these drugs delivered through mannosylated liposomes was found to be due to the enhanced accumulation in the infection site.

Leishmaniasis is another parasitic disease associated with the RES. Mannosylated liposome-encapsulated antileishmanial drug, urea stibamine, has also been employed for the treatment of experimental leishmaniasis in hamster.[25] Medda et al.[25] compared the efficacy of urea stibamine in mannosylated liposomes as compared to the usual liposomal formulation for the treatment of experimental leishmaniasis. They reported a significantly higher efficacy of urea stibamine in mannosylated liposome.

C. AS IMMUNOADJUVANT AND IN IMMUNOTHERAPY

The immunopotentiating effect of liposomes to a number of antigens has been well established. The mechanism of cellular action of liposomes in stimulating antibody production remains to be established. However, it is believed that most adjuvants exert their major effect on macrophages, which are the site of *in vivo* accumulation of liposomes. The enhanced uptake of mannosylated liposomes specifically to macrophages led us to study the effect of surface sugars on the immune response of the entrapped antigen. It has been observed that liposomes carrying galactose on their surfaces induce an immune response comparable to sugar-free neutral liposomes. However, the immune response by mannose-coupled liposomes is almost equal to that of the free antigen without CFA.[75] On the other hand, Garcon et al. reported that the presence of a mannosylated ligand on the surface of tetanus toxoid-containing liposomes leads to enhanced adjuvancity.[76] This discrepant result may be due to a different antigen used by these workers.

Dumont et al. reported that lipopolysaccharide (LPS) encapsulated in mannosylated liposomes is less toxic and activates macrophages *in vitro* at nontoxic concentration.[77] They also observed that targeting of LPS to tissue macrophages through mannosylated liposomes induced regression of experimental solid tumor in mice and was very effective against lung metastases. However, it may be mentioned that the level of mannose receptors on activated macrophages is very low.[78] Thus, as mentioned in the case of galactose receptors, this fact should also be taken into consideration before designing a targeting experiment based on mannose receptors.

REFERENCES

1. **Ostro, M.J.,** *Liposomes from Biophysics to Therapeutics,* Marcel Dekker, New York, 1987, 277.
2. **Swenson, C., Popescu, M.C., and Ginsberg, R.S.,** Preparation and use of liposomes in the treatment of microbial infections, *Crit. Rev. Microbiol.,* 15, S1, 1988.
3. **Litzinger, D.C. and Huang, L.,** Phosphatidylethanolamine liposomes: drug delivery, gene transfer and immunodiagnostic applications, *Biochim. Biophys. Acta,* 1113, 201, 1992.
4. **Ashwell, G. and Morell, A.G.,** Role of surface carbohydrates in the hepatic recognition and transport of circulating glycoproteins, *Adv. Enzymol.,* 41, 99, 1974.
5. **Stahl, P.D. and Schlesinger, P.H.,** Receptor-mediated pinocytosis of mannose/N-acetylglucosamine-terminated glycoprotein and lysosomal enzymes by macrophages, *TIBS,* 5, 194, 1980.
6. **Bachhawat, B.K.,** Lysosomal acid hydrolases in health and diseases, *Neurol. India,* 22, 169, 1974.
7. **Surolia, A., Bachhawat, B.K., and Podder, S.K.,** Interaction between lectin from *Ricinus communis* and liposomes containing gangliosides, *Nature (London),* 257, 802, 1975.
8. **Surolia, A. and Bachhawat, B.K.,** The effect of lipid composition on liposome-lectin interaction, *Biochem. Biophys. Res. Commun.,* 83, 779, 1978.
9. **Surolia, A. and Bachhawat, B.K.,** Monosialoganglioside liposome-entrapped enzyme uptake by hepatic cells, *Biochim. Biophys. Acta,* 497, 760, 1977.
10. **Gregoriadis, G. and Neerunjun, E.D.,** Control of the rate of the hepatic uptake and catabolism of liposome-entrapped proteins injected into rats: a possible therapeutic application, *Eur. J. Biochem.,* 47, 178, 1974.
11. **Jonah, M.M., Cerny, E.A., and Rahman, Y.E.,** Tissue distribution of EDTA encapsulated within liposomes containing glycolipids or brain phospholipids, *Biochim. Biophys. Acta,* 541, 321, 1978.
12. **Ghosh, P. and Bachhawat, B.K.,** Grafting of different glycosides on the surface of liposomes and its effect on the tissue distribution of ^{125}I-labelled γ-globulin encapsulated in liposomes, *Biochim. Biophys. Acta,* 632, 562, 1980.

13. **Ghosh, P., Bachhawat, B.K., and Surolia, A.,** Synthetic glycolipids: interaction with galactose-binding lectin and hepatic cells, *Arch. Biochem. Biophys.,* 206, 454, 1981.
14. **Ghosh, P., Das, P.K., and Bachhawat, B.K.,** Targeting of liposomes towards different cell types of rat liver through the involvement of liposomal surface glycosides, *Arch. Biochem. Biophys.,* 213, 266, 1982.
15. **Dasgupta, P. and Bachhawat, B.K.,** Receptor-mediated uptake of asialoganglioside liposomes: subcellular distribution of the liposomal marker in isolated liver cell types, *Biochem. Int.,* 10, 327, 1985.
16. **Szoka, F.C., Jr. and Mayhew, E.,** Alteration of liposome disposition *in vivo* by bilayer situated carbohydrate, *Biochim. Biophys. Res. Commun.,* 110, 140, 1983.
17. **Spanjer, H.H. and Scherphof, G.L.,** Targeting of lactosylceramide containing liposomes to hepatocytes *in vivo*, *Biochim. Biophys. Acta,* 734, 40, 1983.
18. **Banno, Y., Ohki, K., and Nozawa, Y.,** Targeting of asialofetuin sugar chain bearing liposomes to hepatocytes *in vivo*, *Biochim. Biophys. Acta,* 7, 455, 1983.
19. **Gregoriadis, G. and Senior, J.,** Targeting of small unilamellar liposomes to the galactose receptor *in vivo*, *Biochem. Soc. Trans.,* 12, 337, 1984.
20. **Dragsten, P.R., Manzeil, D.B., Covert, G., and Baker, T.,** Drug targeting using vesicles targeted to the hepatic asialoglycoprotein receptor, *Biochim. Biophys. Acta,* 926, 270, 1987.
21. **Barratt, G., Tenu, J.P., Yapo, A., and Petit, J.F.,** Preparation and characterization of liposomes containing mannosylated phospholipids capable of targeting drugs to macrophage, *Biochim. Biophys. Acta,* 862, 153, 1986.
22. **Haensler, J. and Schuber, F.,** Preparation of neogalactosylated liposomes and their interaction with mouse peritoneal macrophages, *Biochim. Biophys. Acta,* 946, 95, 1988.
23. **Muller, C.D. and Schuber, F.,** Neo-mannosylated liposomes: synthesis and interaction with mouse Kupffer cells and resident peritoneal macrophages, *Biochim. Biophys. Acta,* 986, 97, 1989.
24. **Ponpipom, M.M., Shen, T.Y., Baldeschwieler, J.D., and Wu, P.S.,** Modification of liposome surface properties by synthetic glycolipids, in *Liposome Technology*, Vol. 3, Gregoriadis, G., Ed., CRC Press, Boca Raton, FL, 1983, 117.
25. **Medda, S., Mukherjee, S., Das, N., Naskar, K., Mahato, S.B., and Basu, M.K.,** Sugar-coated liposomes: a novel delivery system for increased drug efficacy and reduced toxicity, *Biotechnol. Appl. Biochem.,* 17, 37, 1993.
26. **Ng, K.Y. and Heath, T.D.,** Association of methotrexate encapsulated in negatively charged liposomes with cells of nanomolar lipid concentrations, *J. Liposome Res.,* 2, 217, 1992.
27. **Kolb-Bachofen, V., Schlepper-Schafer, J., and Jogell, W.,** Electron microscopic evidence for an asialoglycoprotein receptor on Kupffer cells, localization of lectin-mediated endocytosis, *Cell*, 29, 859, 1982.
28. **Kolb-Bachofen, V., Schlepper-Schafer, J., and Knob, H.,** Receptor-mediated particle uptake by liver macrophage: the galactose-particle receptor mediates uptake via coated and also non-coated structure, *Exp. Cell Res.,* 148, 173, 1983.
29. **Schlepper-Schafer, J., Hulsmann, D., Djovkar, A., Meyer, H.E., Herbertz, L., Kolb, H., and Kolb-Bachoten, V.,** Endocytosis via galactose receptor *in vivo*: ligand size directs uptake by hepatocytes and/or liver macrophage, *Exp. Cell Res.,* 165, 494, 1986.
30. **Kempka, G. and Kolb-Bachofen, V.,** Galactose-specific receptors on liver cells. I. Hepatocyte and liver macrophage receptors differ in their membrane anchorage, *Biochim. Biophys. Acta,* 847, 108, 1985.
31. **Spanjer, H.H., Morselt, H., and Scherphof, G.L.,** Lactosylceramide-induced stimulation of liposome uptake by Kupffer cells *in vivo*, *Biochim. Biophys. Acta,* 774, 49, 1984.
32. **Spanjer, H.H., van Berkel, T.J.C., Scherphof, G.L., and Kempen, H.J.M.,** The effect of water-soluble tris-galactoside terminated cholesterol derivative on the *in vivo* fate of small unilamellar vesicles in rats, *Biochim. Biophys. Acta*, 816, 396, 1985.
33. **Allen, T.M.,** A study of phospholipid interactions between high density lipoproteins and small unilamellar vesicle, *Biochim. Biophys. Acta*, 640, 385, 1981.
34. **Munn, M.W. and Parce, J.W.,** Antibody-dependent phagocytosis of haptenated liposomes by human neutrophils is dependent on the physical state of the liposomal membrane, *Biochim. Biophys. Acta,* 692, 101, 1982.
35. **Banno, Y., Ohki, K., Morita, T., Yoshioka, S., and Nozawa, Y.,** Involvement of membrane fluidity of lactosylceramide-targeted liposomes in their intrahepatic uptake, *Biochem. Int.,* 12, 865, 1986.
36. **Haensler, J. and Schuber, F.,** Influence of the galactosyl ligand structure on the interaction of galactosylated liposomes with mouse peritoneal macrophages, *Glycoconjugate J.,* 8, 116, 1991.
37. **Lee, Y.C., Townsend, R.R., Hardy, M.R., Lonngren, J., Arnap, J., Haraldsson, M., and Lonn, H.,** Binding of synthetic oligosaccharides to the hepatic Gal/GalNAc lectin, *J. Biol. Chem.,* 258, 199, 1983.
38. **Allen, T.M. and Chonn, A.,** Large unilamellar liposomes with low uptake into the reticuloendothelial system, *FEBS Lett.,* 223, 42, 1987.
39. **Allen, T.M., Austin, G.A., Chonn, A., Lin, L., and Lee, K.C.,** Uptake of liposomes by cultured mouse bone marrow macrophages: influence of liposome composition and size, *Biochim. Biophys. Acta,* 1061, 56, 1991.

40. **Gabizon, A. and Papahadjopoules, D.,** Liposome formulations with prolonged circulation time in blood and enhanced uptake by tumors, *Proc. Natl. Acad. Sci. U.S.A.,* 85, 6949, 1988.
41. **Huang, S.K., Lee, K.D., Hong, K., Friend, D.S., and Papahadjopoulos, D.,** Microscopic localization of sterically stabilized liposomes in colon carcinoma-bearing mice, *Cancer Res.,* 52, 5135, 1992.
42. **Allen, T.M., Mehra, T., Hansen, C., and Chin,** Stealth liposomes: an improved sustained release system for 1-β-D-arabinofuranosylcytosine, *Cancer Res.,* 52, 2431, 1992.
43. **Gabizon, A.A.,** Selective tumor localization and improved therapeutic index of anthracyclines encapsulated in long-circulating liposomes, *Cancer Res.,* 52, 891, 1992.
44. **Maruyama, K., Unezaki, S., Takahashi, N., and Iwatsuru, M.,** Enhanced delivery of doxorubicin to tumor by long-circulating thermosensitive liposome and local hyperthermia, *Biochim. Biophys. Acta,* 1149, 209, 1993.
45. **Allen, T.M.,** Stealth liposomes: five years on, *J. Liposome Res.,* 2, 289, 1992.
46. **Bakker-Woudenberg, I.A.J.M., Lokerse, A.F., Kate, M.T., and Storm, G.,** Enhanced localization of liposomes with prolonged blood circulation time in infected lung tissue, *Biochim. Biophys. Acta,* 1138, 318, 1992.
47. **Liu, D. and Huang, L.,** pH-sensitive, plasma stable liposomes with relatively prolonged residence time, *Biochim. Biophys. Acta,* 1022, 348, 1990.
48. **Parr, M.J., Bally, M.B., and Cullis, P.R.,** The presence of GM1 in liposomes with entrapped doxorubicin does not prevent RES blockage, *Biochim. Biophys. Acta,* 1168, 249, 1993.
49. **Liu, D., Mori, A., and Huang, L.,** Role of liposome size and RES blockade in controlling biodistribution and tumor uptake of GM1-containing liposomes, *Biochim. Biophys. Acta,* 1104, 95, 1992.
50. **Chow, D.D., Essien, H.E., Padki, M.M., and Hwang, K.J.,** Targeting small unilamellar liposomes to hepatic parenchymal cells by dose effect, *J. Pharmacol. Exp. Ther.,* 248, 506, 1989.
51. **Liu, D., Mori, A., and Huang, L.,** Large liposomes containing ganglioside GM1 accumulate effectively in spleen, *Biochim. Biophys. Acta,* 1066, 159, 1991.
52. **Allen, T.M., Hansen, C., and Rutledge, J.,** Liposomes with prolonged circulation times: factors affecting uptake by reticuloendothelial and other tissues, *Biochim. Biophys. Acta,* 981, 27, 1989.
53. **Scarsdale, J.N., Prestegard, J.H., and Yu, R.K.,** NMR and computational studies of interactions between remote residues in gangliosides, *Biochemistry,* 29, 9843, 1990.
54. **Park, Y.S. and Huang, L.,** Effect of chemically modified GM1 and neoglycolipid analogs of GM1 on liposome circulation time: evidence supporting the dysopsonin hypothesis, *Biochim. Biophys. Acta,* 1166, 105, 1993.
55. **Chonn, A., Semple, S.C., and Cullis, P.R.,** Association of blood proteins with large unilamellar liposomes *in vivo*: relation to circulation lifetimes, *J. Biol. Chem.,* 267, 18759, 1992.
56. **Marshall, J.J., Humphreys, J.D., and Abramson, S.L.,** Attachment of carbohydrate to enzymes increases their circulatory lifetimes, *FEBS Lett.,* 83, 240, 1977.
57. **Benbough, J.E., Wiblin, C.N., Rafter, T.N.A., and Lee, J.,** The effect of chemical modification of L-asparaginase on its persistence in circulating blood of animals, *Biochem. Pharmacol.,* 28, 833, 1979.
58. **Yasuda, Y., Fujita, T., Takakura, Y., Hashida, M., and Sezaki, H.,** Biochemical and biopharmaceutical properties of macromolecular conjugates of uricase with dextran and polyethylene glycol, *Chem. Pharm. Bull.,* 38, 2053, 1990.
59. **Seppala, I. and Makela, O.,** Antigenicity of dextran protein conjugates in mice: effect of molecular weight of the carbohydrate and comparison of two modes of coupling, *J. Immunol.,* 143, 1259, 1989.
60. **Fujita, T., Yasuda, Y., Takakura, Y., Hashida, M., and Sezaki, H.,** Tissue distribution of ^{111}In-labeled uricase conjugates with charged dextrans and polyethylene glycol, *J. Pharmacobiol-Dyn.,* 14, 623, 1991.
61. **Pain, D., Das, P.K., Ghosh, P., and Bachhawat, B.K.,** Increased circulatory half-life of liposomes after conjugation with dextran, *J. Biosci.,* 6, 811, 1984.
62. **Woodle, M.C. and Lasic, D.D.,** Sterically stabilized liposomes, *Biochim. Biophys. Acta,* 1113, 171, 1992.
63. **Takada, M., Yuzuriha, T., Katayama, K., Iwamoto, K., and Sunamoto, J.,** Increased lung uptake of liposomes coated with polysaccharides, *Biochim. Biophys. Acta,* 802, 237, 1980.
64. **Sunamoto, J. and Iwamoto, K.,** Protein-coated and polysaccharide-coated liposomes as drug carriers, *Crit. Rev. Ther. Drug Carrier Syst.,* 2, 117, 1986.
65. **Sunamoto, J., Sato, T., Hirota, M., Fukushima, K., Hiratani, K., and Hara, K.,** A newly developed immunoliposome — an egg phosphatidylcholine liposome coated with pullulan bearing both a cholesterol moiety and an IgMs fragment, *Biochim. Biophys. Acta,* 898, 323, 1987.
66. **Decker, D.,** Quantitative aspects of biochemical mechanism leading to cell death, in *Pathogenesis and Mechanism of Liver Cell Necrosis,* Keppler, D., Ed., University Park Press, Baltimore, 1975, 45.
67. **Keppler, D.,** Consequence of uridine triphosphate deficiency in liver and hepatoma cells, in *Pathogenesis and Mechanism of Liver Cell Necrosis,* Keppler, D., Ed., University Park Press, Baltimore, 1975, 87.
68. **Ghosh, P.C. and Bachhawat, B.K.,** Targeting of liposomes to hepatocytes, *in Liver Diseases: Targeted Diagnosis and Therapy Using Specific Receptor and 26 Ligands,* Wu, G.Y. and Wu, C.H., Eds., Marcel Dekker, New York, 1991, 87.

69. **Keegan-Rogers, U., Wu, C.H., and Wu, G.Y.,** Receptor-mediated protection of normal hepatocytes during chemotherapy for hepatocellular carcinoma, in *Liver Diseases: Targeted Diagnosis and Therapy Using Specific Receptor and Ligands*, Wu, G.Y. and Wu, C.H., Eds., Marcel Dekker, New York, 1991, 105.

70. **Soriano, P., Dijkstra, J., Legrand, A., Spanjer, H.H., Lardos-Gagliardi, D., Roerdink, F., Scherphof, G.L., and Nicolau, C.,** Targeted and nontargeted liposomes for *in vivo* transfer to rat liver cells of a plasmid containing the prepro-insulin I gene, *Proc. Natl. Acad. Sci. U.S.A.*, 80, 7128, 1983.

71. **Nandi, P.K., Legrand, A., and Nicolau, C.,** Biologically active, recombinant DNA in clathrin-coated vesicles isolated from rat livers after *in vivo* injection of liposome encapsulated DNA, *J. Biol. Chem.*, 261, 16722, 1986.

72. **Tremblay, C., Barza, M., Fiore, C., and Szoka, F.,** Efficacy of liposome-intercalated amphotericin-B in the treatment of systemic candidiasis in mice, *Antimicrob. Agents Chemother.*, 26, 170, 1984.

73. **Ahmad, I., Sarkar, A.K., and Bachhawat, B.K.,** Mannosylated liposomes-mediated delivery of amphotericin-B in the control of experimental *Aspergillosis* in Balb/c mice, *J. Clin. Biochem. Nutr.*, 10, 171, 1991.

74. **Monnis, M., Ahmad, I., and Bachhawat, B.K.,** Mannosylated liposomes as carriers for hamycin in the treatment of experimental *Aspergillosis* in balb/c mice, *J. Drug Targeting*, 1, 147, 1993.

75. **Latif, N. and Bachhawat, B.K.,** The effect of surface sugars on liposomes in immunopotentiation, *Immunol. Lett.*, 8, 75, 1984.

76. **Garcon, N., Gregoriadis, G., Taylor, M., and Summerfields, J.,** Mannose-mediated targeted immunoadjuvant action of liposomes, *Immunology*, 64, 743, 1988.

77. **Dumont, S., Muller, C.D., Schuber, F., and Bartholeyns, J.,** Antitumoral properties and reduced toxicity of LPS targeted to macrophages via normal or mannosylated liposomes, *Anticancer Res.*, 10, 155, 1990.

78. **Imber, M.J., Pizzo, S.V., Johnson, W.J., and Adams, D.O.,** Selective diminution of the binding of mannose by murine macrophage in the late stages of activation, *J. Biol. Chem.*, 257, 5129, 1982.

Chapter 4

STABILIZATION OF COLLOIDAL DISPERSIONS BY GRAFTED POLYMERS

Oleg Borisov, Philippe Auroy, and Loïc Auvray

TABLE OF CONTENTS

I. INTRODUCTION

Although the detailed mechanisms of the enhanced stability of Stealth® liposomes injected in the human body are not yet fully understood, there is no doubt that they have to be related to the layers of poly(ethyleneglycol) chains attached to the liposome membranes, which oppose a steric barrier against the attacks of the diverse elements of the immune system. These steric effects are well known in the field of colloid science and technology, where they are used to insure the stability of colloidal dispersions. The aim of this chapter is to review briefly the physical description of these effects.

Colloidal dispersions are fundamentally unstable. Colloidal particles in suspension attract each other by the ubiquitous van der Waals forces. If this attraction is not counterbalanced by any repulsive force, spontaneous aggregation of the particles occurs and the system flocculates. The strength of van der Waals attraction depends on the size, the geometry of the particles, and the chemical composition of the system. For two spheres of radius b separated by a distance h (smaller than b), the energy of interaction has the following form:[1-4]

$$V_A^s = -(A/12\pi)\,(b/h)$$

A is called the Hamaker constant, related to the polarizabilities of the particles and of the solvent. Typical values of A range from kT for organic dispersions to 100 kT for sols of metals.

One of the first ways to ensure the stability of colloidal suspensions in water (or in some polar medium) is provided by the Coulomb repulsion between electrostatic charges which may be present on the particle surfaces.[1] The conditions of stability then depend critically on several parameters: ionic strength, surface potential or charge, and particle size. As a result the method of electrostatic stabilization is not very flexible. In particular, it cannot be applied in many situations of interest, at high ionic strength in water and, of course, in apolar media, where there is no charge dissociation.

An alternative method is to use polymers.[2] This trick goes back to the ancient Egyptians, who made their ink in this way, and to Faraday, who showed that gelatin prevented the precipitation of his gold sols by addition of salt. It is now an essential process in paints and ink technology. The principle is simple: by attaching long flexible chains to the surface of colloidal particles, one creates a rather dilute corona of connected monomers around the particles. The overlap of the coronas of two different particles approaching each other confines the attached polymer chains, reduces their entropy of configuration, and increases locally the repulsion of excluded volume between the monomers. This provides a long-range repulsion between the particles, which may be sufficient to counterbalance the van der Waals forces and stabilize the suspension. The range of the repulsive force is directly related to the thickness of the corona and to the size of the polymer coils, which may be very large, up to a few hundreds of Ångstroms.

There are two ways of attaching polymer chains onto a surface, by natural adsorption or by chemically grafting the chains by one end. The structure of the polymer layers and the strength of the forces between surfaces are different in both cases.

II. STABILIZATION AND FLOCCULATION BY ADSORBED POLYMERS

The simplest and, probably, the first historically known[2] way to make a protective layer on the surface of colloidal particles is to adsorb polymer onto the surface.

A. POLYMER MOLECULE CONFORMATION IN A SOLUTION

It is well known[5,6] that due to polymer chain flexibility an individual macromolecule in the bulk of the solution acquires, as a result of thermal motion, a random coil conformation. If the solvent is good for polymer, i.e., monomer-monomer interaction has a character of excluded volume repulsion, the coil is swollen and, from a statistical point of view, is equivalent to a self-avoiding random walk. If the solvent is a θ-solvent for polymer, weak attraction between monomers compensates the excluded volume effect and the chain has the conformation of a Gaussian coil.

If the polymer chain consists of N freely joined monomers (segments) each of length a, the characteristic coil size (end-to-end distance or radius of gyration) scales like $R_f \approx N^{3/5}v^{1/5}a$ ("Flory-radius") or $R_0 \approx N^{1/2}a$ under the conditions of good or θ-solvent, respectively; here a^3v is the excluded volume parameter or second virial coefficient of monomer-monomer interaction and we have omitted numerical coefficients of the order of unity. The second virial coefficient, a^3v, characterizes the solvent quality for the polymer: $v > 0$ under conditions of good solvent (repulsive binary monomer-monomer interaction), $v = 0$ at the θ-point and $v < 0$ under the conditions of poor solvent (attractive binary monomer interaction). As monomers interact in the solvent medium, the variation of the solvent quality can be caused by the variation of the solvent composition (addition of precipitant), or by the variation of temperature.

Under the conditions of poor solvent (below the θ-point) individual polymer molecules collapse and form globules with the density $\approx |a^3v|$ and radius $R_g \approx N^{1/3}a|a^3v|^{-1/3}$. As individual globules tend to aggregate, macrophase separation of polymer solution below the θ-point occurs.

The interaction of polymer coils in the solution ("steric interaction") is of repulsive character under the conditions of good or θ-solvent (v ≥0), while polymer globules attract each other due to positive surface tension at the globule-solvent interface. Consequently, steric stabilization by attached polymers can be provided only if the disperse medium is a good or θ-solvent of the polymer; below the θ-point steric repulsion of protective polymer layers is altered by attraction which provokes dispersion flocculation.

B. POLYMER ADSORPTION

If an attraction between monomer units of the chains and particle surfaces takes place, an adsorption of polymer coils from the solution onto the particle surfaces may occur. This attraction can be of different origin, e.g., van der Waals interaction, hydrogen bond formation, etc. The monomer surface interaction energy depends, of course, on the solvent composition. A certain level of understanding of the main relationships of polymer adsorption has been achieved (see References 7 to 14 and an excellent review in Reference 15).

The adsorption of macromolecules onto the surface occurs only if the free energy gain per adsorbed monomer, ε, exceeds a certain critical value, ε_c, which is of the order of kT and is determined by local polymer and surface structure. Equivalently, one can define the critical adsorption temperature, T_c, below which polymer is adsorbed onto the surface.

At $\varepsilon < \varepsilon_c$ (or $T > T_c$) the entropy losses related to steric restrictions imposed on the polymer coil by an impermeable surface are not compensated by the attraction energy so that the polymer coil avoids the surface. At $\varepsilon > \varepsilon_c$ (or $T < T_c$) a polymer coil can be adsorbed at the surface as a whole, forming an alternating succession of trains (adsorbed sequences) and exposing into the solution numerous loops and two terminal tails.

Note that under the conditions of weak adsorption ($\varepsilon - \varepsilon_c \ll$ kT) most of the monomers belong to the loops and tails and only a small fraction of the monomers is in direct contact with the surface. However, due to the large value of N, the total free energy gain related to the chain adsorption, $\Delta F \approx (\varepsilon - \varepsilon_c)^{5/3}$ in good solvent condition, can be sufficiently large to provide strong and almost irreversible bonding of the macromolecule to the surface.

The characteristic thickness of the layer H, formed by individual adsorbed macromolecules (loop height) varies with ε as $H \approx |\varepsilon - \varepsilon_c|^{-1}$. However, if the bulk of the solution is not very diluted, the surface concentration of adsorbed macromolecules is sufficiently high for their overlapping even at small adsorption energy, $|\varepsilon - \varepsilon_c| \ll$ kT. The competition between macromolecules for adsorption area results in the formation of long loops, thus increasing the thickness of the adsorbed polymer layer[14] which becomes of the order R_f.

C. INTERACTION BETWEEN SURFACES COATED BY
ADSORBED POLYMER

The nature of the interaction between two surfaces coated by adsorbed polymer depends strongly on the amount adsorbed. If the surface coverage is sufficiently high and even saturated, so that both interacting surfaces are covered by a continuous and rather thick layer of adsorbed polymer loops, the approach of the surfaces gives rise to strong steric repulsion.

Thus, under this condition of high surface coverage, adsorbed polymer layers can, in principle, prevent particle aggregation, provided that the thickness of the protective layers exceeds the range of van der Waals attraction. The latter condition can be fulfilled with sufficiently long polymer chains ($R_f > 10^2$ Å) and if the polymer concentration in the bulk of the solution is sufficiently high to provide strong interchain competition for the adsorption area.

The picture is different when the adsorption sites of the two surfaces are far from being saturated, i.e., when sparsely adsorbed macromolecules cover only a small fraction of the total surface area. In this case a polymer chain can adsorb on the two surfaces at the same time if they are not too far apart and bridge them. This bridging effect provides an attractive contribution to the interparticle interaction at the separation $h \approx |\varepsilon - \varepsilon_c|^{-1}$. It is only when the interparticle separation is small, $h \ll |\varepsilon - \varepsilon_c|^{-1}$, that the entropy losses and the excluded volume interaction come into play and result in an increase of the interaction potential V_p. Thus, in the case of low surface coverage the potential curve of interaction of two surfaces covered by adsorbed polymer exhibits a minimum at $h \approx |\varepsilon - \varepsilon_c|^{-1}$. Of course, the bridging effect occurs also at high coverage, but in the latter case it is weaker due to a deficiency of vacant adsorbing sites on the opposite surfaces and is masked by a stronger repulsive interaction.

As at extremely low surface coverage the attraction due to bridging is also weak, the dependence of the attractive force between two polymer-covered surfaces on the amount of adsorbed polymer is non-monotonic.[16] It is clear that the attraction is stronger when approximately half the surface area is covered by adsorbed polymer. This condition is well known as the optimal condition of flocculation of colloidal dispersion by adsorbing polymer, i.e., for the purpose opposite to that of interest for us.

It thus appears that stabilization by adsorbed polymers is difficult to control. Adding polymers to a colloidal suspension may lead to a result opposite to the one desired, especially if one takes into account kinetic effects: collisions between particles in the initial stages of adsorption may lead to irreversible aggregation because the particles are not completely covered even if the amount of polymer in solution is in principle sufficiently high to provide stability. Another drawback arises from the structure of the adsorbed layer. The monomer concentration in the adsorbed layer decreases rapidly as a power function of the distance from the surface x.[11,31,33] Most of the monomers are located in the short loops of the proximal region closed to the wall and the steric repulsion between overlapping flat adsorbed layers is rather weak at large separation; it decreases with the same power law as the van der Waals attraction, i.e., x^{-3}. The balance between steric and van der Waals forces depends then on the precise value of the Hamaker constant. For these reasons, stabilization of colloidal suspensions by adsorbed homopolymers is not used very much in practice. More efficient results are obtained with grafted polymers, as we will describe now.

III. COLLOID STABILIZATION BY GRAFTED POLYMERS

Monolayers of polymers grafted by one end on a surface can be obtained in two ways: by chemical reaction of end functionalized macromolecules with reactive groups of the surface and by strong adsorption of the nonsoluble moiety of diblock copolymers, which anchors the soluble nonadsorbing block dangling into the solution.

The structure and stabilizing properties of grafted polymer monolayers have been studied intensively during the last two decades both theoretically[17,23] and experimentally by surface force measurements,[30] neutron scattering,[34] and reflectometry.[32]

A. MUSHROOMS AND BRUSHES

The conformation of end-grafted polymers depends strongly on the grafting density inversely proportional to the area occupied by one chain. If the chains are sparsely grafted so that the distance between neighboring chains is larger than the individual coil size, the grafted chains do not interact with each other and retain in first approximation their unperturbed bulk conformation. This is the so-called "mushroom" regime. Correspondingly, both components of the chain dimensions in the direction normal to the surface H and in the lateral direction $H_{//}$ scale like the unperturbed radius of gyration of the chain in good solvent $R_f \approx N^{3/5} v^{1/5} a$.

The opposite case of high grafting density and strong overlapping of neighboring coils corresponds to the "brush" regime. The name arises from the strong stretching of the chains in the direction normal to the surface, first described by Alexander[17] and de Gennes.[18]

This equilibrium stretching of the chains in a brush is determined by the competition between an osmotic force, due to the monomer-monomer interactions and an elastic force associated to the conformational entropy of the chains. As a result, the layer thickness H is proportional to the degree of polymerization N and increases with the grafting density and solvent strength:

$$H \approx Na \, (\sigma/a)^{-1/3} v^{1/3}$$

σ is the surface area per grafted chain. The average monomer concentration in the layer is then given by $\Phi = Na^3/\sigma H$.

More accurate analysis based on the self-consistent field approach[22,23] shows that the monomer concentration profile $\Phi(x)$ in grafted monolayers is not constant but varies parabolically with the distance from the surface x:

$$\Phi(x) = (3/2) \, [\Phi(1 - x^2/H^2)]$$

This behavior is related to the fact that the free ends of the chains are distributed throughout the layer thickness, the local stretching of every chain is stronger near the surface and decreases to zero at the free end.

This picture, established for a planar surface, remains valid in general if the radius of curvature of the surface is much larger than the layer thickness. In the opposite case, when polymer chains are grafted onto small particles of size b << H, the overall stretching is much weaker than in flat polymer brushes of the same grafting density. The overall picture is that of a star-like polymer described by Daoud and Cotton.[24] Balancing osmotic and elastic forces, one obtains that the layer thickness is proportional to the chain's radius of gyration.

$$H \approx N^{3/5} v^{1/5} a \cdot f^{1/5}$$

$f \approx b^2/\sigma$ is the number of star arms, i.e., the total number of chains grafted onto one particle.

The local monomer density (and the local chain stretching) decreases in the radial direction according to the following power law:

$$\Phi(r) \approx f^{2/3} v^{-1/3} a^{4/3} r^{-4/3}$$

Thus, most of the monomers are located near the periphery of the layer, where the local concentration is the lowest; this minimizes the repulsive monomer interactions.

B. INTERACTION BETWEEN GRAFTED POLYMER LAYERS AND COLLOID STABILITY

When two surfaces covered by grafted polymer approach each other, the layers are deformed and this gives rise to a steric repulsive force, which can be measured, for example, by the surface force apparatus.[30]

In the "mushroom" regime, the repulsive force is primarily of entropic origin and related to the configuration changes of the coils. The contribution of the repulsive monomer-monomer interactions ("osmotic forces") becomes significant only at small interparticle separation, when squeezed polymer coils ("pancakes") begin to overlap.

In the opposite "brush" regime, the strong repulsive interaction between opposite brushes, which occurs as soon as the distance between the surfaces 2h is smaller than 2H, is essentially due to the increase in osmotic forces caused by the monomer excluded volume interactions. In contrast to what happens in the "mushroom" regime, entropic contributions are negligible.

A weak compression of the brushes is even entropically favorable, because this compression decreases the stretching of the grafted chains in the two layers, which do not penetrate each other in first approximation.

Let us examine these two cases in more detail.

1. Repulsion Between Sparsely Grafted Polymer Layers

The increase in conformational free energy of a squeezed polymer coil is well known from the theory of polymer solution.[6] This free energy per unit area is given for a planar surface in the case of interest h < H by

$$F^p_{conf}(h) = V^p_p(h) \approx (2kT/\sigma a)Nv^{1/.5}(a/h)^{5/3}$$

The equivalent contribution to the interaction potential of two large spherical particles of radius b can be obtained from the preceding expression with the help of the Derjaguin approximation[4]

$$F^s_{conf} = V^s_p(h) \approx (2\pi kT/\sigma)Nv^{1/3}(a/h)^{2/3}$$

in the limit h < H << b.

We observe that in contrast to the case of electrostatic repulsion by a screened Coulomb potential, the free energy of steric repulsion behaves at a small distance h as a power law h^{-m}. The total potential energy of interaction including the contribution of van der Waals forces then presents a larger variety of behavior, determined by the combination of parameters $\gamma = (A/24\pi kT)(\sigma/Na^2)$.

At $\gamma < 1$ the total potential energy $V(h) = V_A(h) + V_p(h)$ decreases monotonically in the range a < h < H, the steric repulsion thus provides thermodynamic stability of the dispersion. In the opposite case, $\gamma > 1$, the potential curve has a deep primary minimum at h ≈ a and exhibits a maximum of height $V_{max} \approx kT(N/\sigma\gamma^2)bv$ at $h = h_{max} \approx \gamma^3/v$. Thus, at $\gamma > 1$ the system can be kinetically stabilized if $V_{max} >> kT$, but a decrease in the solvent strength (i.e., in v) leads to coagulation in the primary minimum somewhat above the θ-point where v vanishes.

Note that as the range of the repulsive steric interaction is restricted by the condition h < H < R_f, the total potential curve should exhibit a secondary minimum of depth $V'_{min} \approx Ab/12R_f$ at h ≈ H. Thus, even if the dispersion is stabilized with respect to coagulation in the primary minimum, aggregation in the secondary minimum can occur if $V_{min} > kT$.

Now we can summarize the stability conditions. A layer of grafted "mushrooms" provides:

1. Dispersion stability with respect to strong irreversible coagulation in the primary minimum if $\gamma < 1$, i.e., if the surface coverage is sufficiently high: $(Na^2/\sigma) > A/24\pi kT$
2. Dispersion stability with respect to weak aggregation in the secondary minimum, if the size of the grafted coils is of the order of, or larger than, the particle size, $R_f > b$, at least for organic dispersion where A ≈ kT

Note that in practice the condition $\gamma < 1$ may not correspond to a mushroom regime but rather to a brush regime, particularly if A >> kT.

2. Steric Interaction Between Polymer Brushes

One can find a detailed analysis of the interaction between two brushes, based on self-consistent field calculations, in Reference 27. We shall discuss here only the main features.

As already stressed, the chains belonging to unconstrained brushes are stretched perpendicularly to the surface. Therefore, approaching two brushes at a distance h smaller than their total thickness 2H decreases the stretching of the chains provided that the brushes do not interpenetrate each other significantly. This is indeed the case if h is larger than $2R_0 = 2N^{1/2}a$, then the width ζ of the interpenetration zone remains much smaller than the total brush thickness. The potential of steric interaction is thus the sum of two terms, an entropic term related to the stretching of the chains and the usual osmotic term related to the excluded volume interactions.

The contribution of the chain's conformational entropy is significant only at weak compression and is very sensitive to the details of the layer structure. The most precise description yields a very soft repulsive potential varying at h close to 2H as

$$V_p(h) \approx (2H - h)^3$$

in good solvent.

In the range of strong compression of the brushes, $h \ll 2H$, the entropic contribution as well as the details of the layer structure become unessential. The steric repulsive potential is determined only by the osmotic force, depending only on the average monomer concentration between the surfaces $\Phi(h) = 2N/\sigma h$.

$$V_p^p(h) \approx (2Na^2/\sigma)[v\Phi(h) + w\Phi^2(h)]$$

This expression is valid for a unit area of the two surfaces, assumed to be flat. For completeness we have included the contribution of the three body interactions to the osmotic pressure (w is the third virial coefficient of the monomers) so that the above expression can be used close to the θ-point or below (v <0).

Using the Derjaguin approximation we can also derive the potential of steric interaction between two spherical surfaces of radius b, covered by a dense layer of grafted polymers (H < b). In the range of weak compression, h < 2H, the potential is softer than for the planar case, $V_p^s(h) \approx (2H - h)^4$ (in a good solvent). In the range of strong compression ($h \ll 2H$), we get

$$V_p^s(h) \approx \pi b[-v(2N/\sigma)^2 \ln(h) + w(2N/\sigma)^3/h]$$

The complete analysis of the total potential of interaction V(h) including the van der Waals and steric contributions shows that three types of potential curves can be observed depending on the surface coverage and solvent strength. The dimensionless parameter characterizing the surface coverage in the brush regime is $g = (A/96\pi kTw)(\sigma/Na^2)^3$, which differs from the one defined in the previous section.

1. At high surface coverage or weak van der Waals attraction, g <1, the potential V(h) decreases monotonically in the range h < 2H and has a shallow secondary minimum at h = 2H both under the conditions of good and θ solvent. Consequently the thermodynamic stability of the dispersion is provided even by weakly overlapping grafted chains in the case of organic dispersions. Stabilization of mineral dispersions (with a larger A) would require a stronger overlap of the layers. Note that this thermodynamic stability must be lost just below the θ-point (v <0), because of the positive surface tension at the polymer solvent interface not included in our discussion (see Reference 27).
2. At lower surface coverage, or stronger van der Waals attraction, g >1, the potential curves V(h) exhibit a deep primary minimum and a potential barrier separating the primary and the secondary minima if $v > v^* \approx s^{-1/2}w^{3/4}(g - 1)$. Due to the presence of the potential barrrier, the dispersion can be kinetically stabilized against strong coagulation in a good solvent, while a decrease in solvent quality leads to a decrease of the height of the potential barrier and results in a slow coagulation at $v > v^*$ and in a fast coagulation if $v < v^*$, i.e., when the barrier disappears.

3. Even if the dispersion is thermodynamically (at g <1) or kinetically (at g >1, v > v*) stabilized against coagulation in the primary minimum, long range aggregation in the secondary minimum may occur if its depth $V'_{min} \approx Ab/24kT$ is larger than kT. For preventing this, the layer thickness H should be sufficiently large, H > b for organic dispersions and H >> b for mineral dispersions. This is of course better achieved at fixed surface coverage Na^2/σ, if the chains are very long. Note that if aggregation in the secondary minimum has occurred at v >0, an increase of v, leading to an increase of H and a decrease of V'_{min}, makes possible the redispersion of the aggregates.

IV. STERIC STABILIZATION OF SMALL COLLOIDAL PARTICLES

The stabilization of suspensions of small colloidal particles covered by grafted polymers whose size H is larger than the particle radius b deserves a special discussion because the potential of steric interaction cannot be obtained from the study of the planar case by using the Derjaguin approximation. This case, which can be important in practice if the Hamaker constant is large, has been studied by using the scaling theory of polymer solution.[29] The potential of steric interaction between two star-like spheres varies logarithmically with the distance h between the spheres, in the range b < h < H:

$$V_p^s(h) \approx f^{3/2}kT \ln(H/h)$$

H is the radius of the polymer corona discussed in Section III.A. The corresponding van der Waals potential of attraction in the same range of distance is

$$V_A^s(h) \approx A(b/h)^6$$

The total potential of interaction, $V^s(h)$, exhibits a deep primary minimum and a shallow secondary one separated by a potential barrier at $h \approx h_{max} \approx b(A/kTf^{3/2})^{1/6}$. The height of the potential barrier $V_{max} \approx kTf^{3/2}\ln(H/h_{max})$ is sufficiently high to provide kinetic stability if the total number of chains per particle f is large, f >> 1, and if the layer thickness is large, H > $b(A/kTf^{3/2})^{1/6}$. It is easy to see that the same conditions also provide the stability of the dispersion with respect to long-range aggregation in the secondary minimum at $h \approx H$.

V. CONCLUSION

The discussion we have presented concerns mainly the stabilization of colloidal suspensions of solid particles covered by irreversibly attached polymer chains. This case is rather well understood theoretically and we think that it provides a good starting point to describe what happens when polymer chains are attached to flexible deformable surfaces like the membranes of vesicles. This last case raises new problems, however. First one has to understand how the shape and elastic properties of the deformable membranes react to the presence of the polymer chains. Secondly one has to take into account that polymer chains may not be irreversibly attached to the surface, but may be simply in a reversible equilibrium of adsorption. This may occur in Stealth® liposomes if the grafted chains can drag their lipid anchor in the solution. The polymer surface coverage then may depend on the distance between the membranes; this may reduce the stability of the system. Third, the van der Waals forces between thin membranes are weaker than between bulk surfaces. The criteria of stability may therefore also change for this reason. Works along these lines are in progress.

ACKNOWLEDGMENTS

I thank Loïc Auvray, Philippe Auroy, and Jean Christophe Castaing for their hospitality at Laboratoire Léon Brillouin and for useful discussions.

REFERENCES

1. **Verwey, E.J.W. and Overbeek, J.Th.G.,** *Theory of Stability of Lyophobic Colloids,* Elsevier, Amsterdam, 1948.
2. **Napper D.H.,** *Polymeric Stabilization of Colloidal Dispersions,* Academic Press, London, 1983.
3. **Hamaker, H.C.,** *Physica,* 4, 1058, 1937.
4. **Israelachvili, J.N.,** *Intermolecular and Surface Forces,* Academic Press, London, 1985.
5. **Flory P.J.,** *Principle of Polymer Chemistry,* Cornell University Press, Ithaca, NY, 1953.
6. **de Gennes, P.G.,** *Scaling Concepts in Polymer Physics,* Cornell University Press, Ithaca, NY, 1979.
7. **di Marzio, E.A. and McCrakin, F.L.J.,** *Chem. Phys.,* 43, 539, 1965.
8. **Rubin, R.J.J.,** *Chem. Phys.,* 43, 2392, 1965.
9. **de Gennes, P.G.,** *Rep. Prog. Phys.,* 32, 187, 1969; *J. Phys.,* 37, 1445, 1976.
10. **Eisenriegler, E., Kremer, K., and Binder, K.,** *J. Chem. Phys.,* 77, 6296, 1982.
11. **de Gennes, P.G. and Pincus, P.,** *J. Phys. Lett.,* 44, 241, 1983.
12. **Birshtein ,T.M.,** *Macromolecules,* 12, 715, 1979; *Macromolecules,* 16, 45, 1983.
13. **Birshtein, T.M. and Borisov, O.V.,** *Polymer,* 32, 916, 1991.
14. **Bouchaud, E. and Daoud M.,** *J. Phys.,* 48, 1991, 1987.
15. **Binder, K.,** in *Phase Transitions and Critical Phenomena,* Vol. 8, Academic Press, London, 1983.
16. **Ji, H., Hone, D., Pincus, P.A., and Rossi, G.,** *Macromolecules,* 23, 698, 1990.
17. **Alexander, S.,** *J. Phys.,* 38, 983, 1977.
18. **de Gennes, P.G.,** *Macromolecules,* 13, 1069, 1980.
19. **Birshtein, T.M. and Zhulina, E.B.,** *Polym. Sci. USSR,* 25, 2165, 1983.
20. **Zhulina, E.B.,** *Polym. Sci. USSR,* 26, 885, 1984; **Birshtein, T.M. and Zhulina, E.B.,** *Polymer,* 25, 1453, 1984.
21. **Borisov, O.V., Zhulina, E.B., and Birshtein, T.M.,** *Polym. Sci. USSR,* 30, 772, 1988.
22. **Skvortsov, A.M., Gorbunov, A.A., Pavlushkov, I.V., Zhulina, E.B., Borisov, O.V., and Pryamitsyn, V.A.,** *Polym. Sci. USSR,* 30, 1706, 1988; **Zhulina, E.B., Pryamitsyn, V.A., and Borisov, O.V.,** *Polym. Sci. USSR,* 31, 205, 1989; **Zhulina, E.B., Borisov, O.V., and Pryamitsyn, V.A.,** *Macromolecules,* 24, 140, 1991.
23. **Milner, S.T., Witten, T.A., and Cates, M.E.,** *Europhys. Lett.,* 5, 413, 1988; *Macromolecules,* 21, 2610, 1988; **Milner S.T.,** *Europhys. Lett.,* 7, 695, 1988.
24. **Daoud, M. and Cotton, J.P.,** *J. Phys.,* 43, 531, 1982.
25. **Dolan, A.K. and Edwards, S.F.,** *Proc. Soc. London Ser. A,* 337, 509, 1974; *Proc. R. Soc. London Ser. A,* 343, 427, 1975.
26. **Gerber, P.R. and Moore, M.A.,** *Macromolecules,* 10, 476, 1977.
27. **Zhulina, E.B., Borisov, O.V., and Pryamitsyn, V.A.,** *J. Colloid Interface Sci.,* 137, 495, 1990.
28. **de Gennes, P.G.,** *C.R. Acad. Sci. (Paris),* 300, 839, 1985; *Adv. Colloid Interface Sci.,* 27, 189, 1987.
29. **Witten, T.A. and Pincus, P.A.,** *Macromolecules,* 19, 2509, 1986.
30. **Klein, J. and Luckham, P.F.,** *Macromolecules,* 17, 1041, 1984; *Macromolecules,* 18, 721, 1985; *Macromolecules,* 19 2007, 1986.
31. **Guiselin, O., Lee, L.T., Farnoux, B., and Lapp, A.,** *J. Chem. Phys.,* 95, 4632, 1991.
32. **Kent, M.S., Lee, L.T., Farnoux, B., and Rondelez, F.,** *Macromolecules,* 25, 6240, 1992.
33. **Auvray, L. and Cotton, J.P.,** *Macromolecules,* 20, 202, 1987.
34. **Auroy, P., Auvray, L., and Léger L.,** *Phys. Rev. Lett.,* 66, 719, 1991; **Auroy, P., Mir, Y., and Auvray, L.,** *Phys. Rev. Lett.,* 69, 93, 1992.

Chapter 5

PHYSICAL PROPERTIES OF POLYMER-GRAFTED BILAYERS

Kalina Hristova and David Needham

TABLE OF CONTENTS

I. INTRODUCTION

The behavior of solvated polymers that are adsorbed or grafted at the interface between biofluids and biomaterials is of interest to polymer physicists and bioengineers alike because of the unique repulsive properties that these polymers possess. For example, a strategy based on the conformation of water-soluble polymers grafted to lipid bilayers is leading to a more effective intravenous liposome drug delivery system.[1-7] As described elsewhere in this volume, drugs and conventional liposomes are rapidly cleared from the bloodstream by the reticuloendothelial system and other nonspecific mechanisms.[8] However, this clearance of liposomes can be significantly decreased by incorporating polyethylene glycol-linked lipids (molecular weights 2000 and 5000 Da). This leads to a substantial increase in the blood circulation time of these so-called "Stealth" liposomes.* It is believed that the mechanism of stabilization is a physical one:[1,2,4,5] the polymer creates a steric barrier to enhance the repulsive

---

* "Stealth" is a trademark of Liposome Technology, Inc.

properties of the liposomal surface, stabilizing the lipid bilayer against close approach of other macromolecules and cells. The application has driven recent interest in the basic physical properties of these interfacial structures.

For *in vivo* applications, the polymer-grafted liposomes are formed by spontaneous rehydration of a polymer-lipid/lipid mixture (for instance, distearoyl phosphatidyl-ethanola-mine-polyethyleneglycol/distearoyl phosphatidylcholine) in which the concentrations of polymer-lipids in the lipid mixture are less than 15 *M%*.[9,10] To characterize these so-formed polymer-grafted bilayers we use several different experimental methods that give information about the physical properties of the polymer-lipid/lipid system (i.e., X-ray diffraction, micropipette manipulations, nuclear magnetic resonance (NMR), calorimetry, electron and optical microscopy.[4,9,10] Of these the most powerful so far has been X-ray diffraction (see Chapter 7, McIntosh et al.). As discussed in Reference 4 and Chapter 7, the X-ray diffraction method gives information about: (1) the extension length of the polymer away from the lipid surface, (2) the repulsive pressure with which the grafted polymer opposes mutual surface compression, and (3) structural information about the lipid phase and indication of phase separation if it occurs.

The micropipette technique allows us to measure: (1) bilayer elastic constants, (2) mechanical stability of the bilayer, and (3) interbilayer adhesion energies. Together with these two techniques, the other spectroscopic and microscopic methods allow us to confirm the phase behavior of the lipid/polymer-lipid/water system.

With this experimental data we are now in the process of testing theoretical models (existing and new) that describe the combined influence of the two components: the lipid bilayer and the polymer-grafted layer.

There is already a significant literature regarding both components of the system as separate entities. The physical properties of lipid bilayers have been studied and different experimental techniques have been developed to test existing theories.[11-22] Also, there have been recent developments in polymer physics, concerning the properties of polymers adsorbed to,[23-26] depleted from,[24,25,27] or grafted on solid surfaces.[28-34] Some of these theories have been tested and partially confirmed.[26,35-40] The repulsive features have mostly concerned polymers adsorbed or grafted to solid surfaces.

However, to fully understand the influence of polymer-grafted lipids on the physical properties of lipid bilayers, we have begun to develop a theory that coherently merges membrane and polymer physics. This theory should prove useful in the further understanding of the properties of biosurfaces and the design of systems that will interact with them.

Here we describe theoretical predictions for the influence of the polymer-bearing lipids on the interactive and structural properties of lipid bilayers. For two bilayers in close proximity, we model the interbilayer steric interactive forces for different polymer-lipid concentrations. These predictions are relevant for polymer-lipid concentrations in the range 0 to 10 *M%*, when the bilayer structure is the stable phase. At the other concentration extreme of the phase diagram, 100% polymer-lipid in water forms micelles.[9] This implies that, at intermediate compositions, the lipid/PEG-lipid/water system has a complex phase behavior and so the conditions under which bilayers (i.e., liposomes) are formed have to be determined. Special attention therefore has to be paid to the self-assembling properties of the polymer-lipid/lipid system.

II. BACKGROUND ON PHYSICAL PROPERTIES OF LIPID BILAYERS THAT ARE IMPORTANT TO LIPOSOME DESIGN

Amphiphilic molecules spontaneously self-assemble in aqueous media into aggregates such as bilayers and micelles. This occurs when the concentration exceeds the critical micelle or critical bilayer concentration, respectively (CMC or CBC) (for instance, for distearoyl phosphatidylcholine, DSPC, the CBC is on the order of 10^{-10} *M*). Depending on the structure of the molecules and their concentration these aggregates can have different shapes (spherical, cylindrical, flat) and sizes. The common feature of the aggregates is that the polar (or hydrophilic) parts of the molecules are exposed to water and the apolar (or hydrophobic) parts are packed together and hidden inside the aggregate. The driving force for this self-assembly is an entropy-driven effect, called the hydrophobic effect.[41] It stems from the fact that immersing a hydrophobic object into a highly polar medium (such as water) leads to a different arrangement of the water molecules around the object than that of pure water and thus to a decrease in the entropy of the system. Thus, an aggregate exists mainly due to the fact that it is excluded from the solvent, since its molecules are not strongly bound — molecule-molecule interactions are of the weak van der Waals type. This is quantitatively manifested in the low elastic coefficients: (1) bending rigidity k_c and (2) area expansion modulus K_A in relation to other, more traditional engineering materials.[42]

For liquid bilayers k_c is of the order of 10^{-19} J. Typical values for the area compressibility modulus of fluid bilayers are $K_A \sim 0.2$ Nm^{-1} or 200 dyn/cm.[12] If we convert this surface compressibility into an equivalent bulk modulus by dividing by the membrane thickness ~5 nm, we get a compressibility ~10^7 Nm^{-2}, which is somewhere between that of an ordinary liquid and a gas. So we can view the membrane as a two-dimensional liquid that can be about 100 times more compressible than its embedding fluid.

Changing the cholesterol content (up to 50%) increases bilayer cohesion and dramatically changes bilayer physical properties (for instance, the incorporation of 50 *M*% cholesterol in SOPC membranes increases K_A by an order of magnitude and increases their strength by about six times).[15] Thus, varying the amount of cholesterol in the bilayer provides us with a simple means to vary the mechanical parameters (elastic constants and failure) of bilayers.

At a certain applied lateral isotropic tension τ_s (the tensile strength) a sudden and unavoid-able membrane rupture occurs. At the point of rupture the membrane is characterized by its maximal area change. Expansions bigger than this one cannot be achieved. The tensile strength τ_s of the bilayer and the maximum area change have been measured in micropipette experiments[15,22] as a function of cholesterol content. Thus, tensile strengths for purely lipid bilayers are ~1 to 6 dyn/cm. Addition of cholesterol, especially to saturated lipids, increases τ_s to ~50 dyn/cm.[42] Critical areal strains are 1 to 5% and reflect complete lack of "ductility" in these two-dimensional materials. For gel-phase bilayers the mechanism of rupture will depend on the presence of defects and is not so well understood.

Bilayers can also undergo substantial changes in physical state with temperature. Above a certain transition temperature the bilayer exists as a two-dimensional liquid. Below this temperature the bilayer is a two-dimensional solid and the lipid molecules are tightly packed in a crystal lattice. Compared to their liquid crystalline state, K_A for gel-phase bilayers is about five times higher.[22] This gel-liquid crystalline transition temperature varies with the chemical composition and hydration level of the membrane and so is another way of controlling the physical state of a liposome capsule.

The rehydration of a mixture of polymer-lipids and lipids will lead to their spontaneous assembly into bilayer or micellar aggregates. If the hydrocarbon tails of the lipids and polymer-lipids are identical, one could expect the two types of lipids to mix well. In these aggregates, the hydrocarbon chains of the polymer-lipids will be packed together and the polar heads with their long hydrophilic polymer chains will be exposed to water. Since polyethylene glycol (PEG) does not adsorb on lipid surfaces,[26] it will remain attached to the bilayer at only one point (i.e., grafted, but not adsorbed to the bilayer). The chain will extend into the aqueous medium, thus maximizing its interaction with the solvent.

If the polymer is grafted on a *solid* surface the lateral repulsion between the polymer chains is unlikely to have any effect on the structure of the solid since the building units of solids are held together by strong chemical forces. In contrast, when the "substrate" is a lipid bilayer polymer chain repulsion may create membrane curvatures, area expansion (increased area per molecule), or even changes of phase (bilayer to micelle) due to the low elastic coefficients and weak bonds between the lipid molecules. To understand the influence of the polymer on the bilayer we have to therefore also characterize the polymer system itself.

The behavior of polymer chains grafted on solid surfaces was first studied by de Gennes.[28] According to him, two concentration regimes can be identified for the polymer chains at the surface, and the physical characteristics of the polymer system are different in these two regimes. When the grafting density is low, the chains form separate "mushrooms", each with a size $R_F \cong aN^{3/5}$, where N is the degree of polymerization and *a* is the monomer size.* Such a situation is shown for a polymer-lipid bilayer system in Figure 1a. When the grafting level is high, the chains overlap laterally to form a continuous "brush" and the polymer system can now be considered a semidilute solution. The brush regime is shown in Figure 1b. Brushes have been described with scaling and mean-field theories. In the scaling scheme[28-30,43,44] it is supposed that the monomer concentration is constant throughout the polymer layer. In the mean-field scheme Milner et al.[31-34] include the possibility that a chain may not be uniformly stretched so that the monomer concentration varies throughout the polymer layer. They show that the self-similar concentration profile is parabolic and the brush is "softer" upon compression than predicted by scaling arguments.

We can apply these theoretical studies of polymers grafted on solid surfaces to the bilayer system. However, due to the characteristics of the bilayer as a self-assembling system a new phenomenon occurs: the polymer perturbs the surface it is grafted to, which in turn affects the characteristics of the polymer.

We will now consider the consequences of this mutual interaction between the grafted polymer and the lipid bilayer. These considerations will include the nature of the steric barrier that the polymer layer provides when two such bilayers interact, and the phase behavior of the polymer-lipid/lipid/water system. The influence of the polymer on the mechanical properties of the bilayer is considered in detail elsewhere.[45]

III. POLYMER STERIC INTERACTIONS: POLYMER EXTENSION LENGTH AND POLYMER COMPRESSIBILITY

When two polymer-grafted bilayers are opposed and made to approach each other under applied compressive osmotic forces (as in X-ray diffraction experiments,[4,9] the nature of the pressure vs. distance relation that characterizes polymer compressibility is dependent on polymer surface concentration. As discussed, two regimes can be identified: dilute mushrooms and dense brushes. Expressions for polymer extensions away from the surface in these two regimes have been defined in scaling and mean-field approaches. In the scaling scheme, the extension of the mushroom away from the surface is of the order of the Flory radius:

$$L_{mush} = R_F \cong aN^{3/5} \tag{1}$$

where N is the degree of polymerization (N = 46 for PEG-2000) and *a* is the monomer size (*a* = 3.5 Å for PEG-2000). According to Equation 1, L_{mush} for PEG-2000 is about 35 Å.

The extension of the brush away from the surface depends on the concentration of the polymer. In the scaling scheme it is given by:

$$L_{sc} \cong N \frac{a^{5/3}}{D^{2/3}} \tag{2}$$

---

* In polymer theories the term "monomer" is used for the repeat unit in the polymerized chain. Materials science terminology distinguishes "monomer" as the original chemical species that is polymerized to form a polymer made of up to N repeating "mer" units.

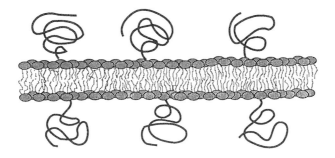

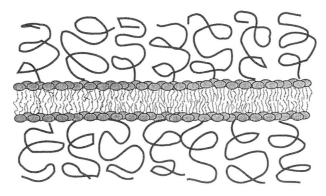

FIGURE 1. (a) Schematic diagram showing a polymer-grafted bilayer at low grafting concentration (mushrooms). (b) Schematic diagram showing a polymer-grafted bilayer at high grafting concentration (brush).

where D, the distance between points of grafting, is a measure of the polymer surface coverage.

In the mean-field scheme proposed by Milner et al., extension is given by:

$$L_{MWC} = \left(\frac{12}{\pi^2}\right)^{1/3} N \frac{a^{5/3}}{D^{2/3}} \qquad (3)$$

A. MUSHROOMS

In the X-ray diffraction experiment two mushroom làyers oppose each other. To introduce the scaling concepts for compression let us discuss the behavior of a single mushroom, compressed between two parallel solid plates. The question was first considered by de Gennes[23] and Daoud and de Gennes.[44] For strong compression (i.e., when the distance between the plates h is small compared to the random coil size of the polymer) they obtain for the free energy of the compressed polymer:

$$F_{conf}(h) \cong kTN\left(\frac{a}{h}\right)^{5/3} \qquad (4)$$

The corresponding pressure (force per unit area) is

$$P_{conf}(h) = -\frac{1}{D^2}\frac{dF_{conf}}{d(h)} \cong \frac{kTN}{D^2 a}\left(\frac{a}{h}\right)^{8/3} \qquad (5)$$

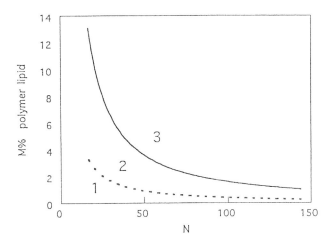

FIGURE 2. Regimes of polymer behavior: interdigitated mushroom, mushroom, and brush. (**1**) D > 2R_F — interdigitated mushrooms; (**2**) 2R_F > D > R_F — mushrooms; (**3**) D < R_F — brushes.

This formula describes the behavior of a strongly compressed mushroom. However, to interpret the full range of X-ray diffraction data from a fraction of an atmosphere applied pressure to pressure of several atmospheres it is important to know the polymer confinement energy when the distance between plates is close to the random coil size (i.e., small compressions). We suggest that in this case Flory's mean-field arguments are more appropriate.

Flory proposed that the energy of a random polymer coil in solution (for athermal solvent) has the form:

$$F = kT \frac{R_F^2}{Na^2} + kTa^3c^2R_F^3 \tag{6}$$

where c = N/R_F^3 is the monomer concentration inside the coil. The first term describes the entropy of the chain, the second accounts for the interactions between the monomers.

We assume that although the chain end is attached to the surface, the free energy of the polymer is given by Equation 6. The free energy has its minimum for $R_F = aN^{3/5}$, the Flory radius. If we compress the mushroom by a small amount with respect to its equilibrium size, it will respond with pressure which scales as:

$$P_{mf}(h) \cong \frac{kTN^{1/5}}{D^2} \left(\frac{R_F^3}{(h)^4} - \frac{h}{R_F^2} \right) \tag{7}$$

Now consider the experimental case of two compressed mushroom-covered surfaces in close proximity. For this case, depending on the distance D between points of grafting we define two subregimes of polymer behavior (see Figure 2):

1. 2R_F > D > R_F — mushrooms at low and high compression (region 2).
2. D > 2R_F — interdigitated mushrooms at low and high compression (region 1).

We now consider the functional form of the steric pressure in each of these subregimes and under a range of compressive loads.

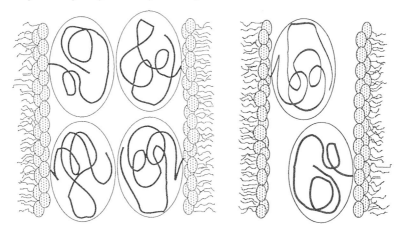

FIGURE 3. (a) Schematic diagram showing compressed mushrooms; (b) schematic diagram showing compressed interdigitated mushrooms.

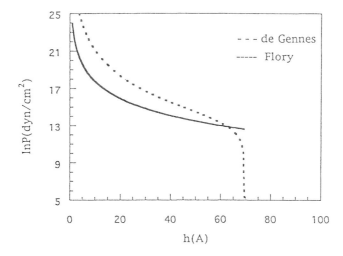

FIGURE 4. Theoretical pressure vs. interbilayer distance curves for polymer mushrooms: 1.2 $M\%$ polymer lipid, N = 46, a = 3.5 Å.

1. Mushrooms, Low Compression

If $2R_F > D > R_F$ the mushrooms on the opposite surfaces are dense enough so that they are squeezed one against the other (Figure 3a). We model this as compressing of each mushroom between two plates at separation h/2, where h is the separation between the bilayer surfaces — Equation 7 for P_{mf}(h/2) holds. It is plotted in Figure 4 for N = 46, a = 3.5 Å, 1.2 $M\%$ polymer lipid for the whole range of compression.

2. Mushrooms, High Compression

The elastic energy of the highly compressed mushroom is described by P_{conf} (h/2), where P_{conf} is given by de Gennes' relationship (Equation 5) and is presented in Figure 4 for the whole range of compression for N = 46, a = 3.5 Å, 1.2 $M\%$ polymer lipid.

If we confine just one chain between two surfaces under the assumption that the local average monomer density does not change, the elastic energy is the only energy that opposes compression. A result of this compression will be the spreading of the mushroom (i.e., its size in the lateral direction $R_{\parallel}$ will increase). de Gennes computes this size to depend on the degree of compression as

$$R_{\parallel} \cong aN^{3/4}\left(\frac{a}{h/2}\right)^{1/4} \tag{8}$$

If the mushrooms are grafted on a surface at a given concentration, then at a certain h_{cr}, for which

$$R_{\parallel} \cong aN^{3/4}\left(\frac{a}{h/2}\right)^{1/4} = D \tag{9}$$

the squeezed mushrooms begin to overlap. Then, to compress the system further, we also need to overcome the osmotic repulsion in the surface polymer solution, which can now be considered semidilute. From Equation 9 we find h_{cr} to scale as:

$$\frac{h_{cr}}{2} = \frac{a^5 N^3}{D^4} \tag{10}$$

The osmotic force per unit area for $h < h_{cr}$ is[44]

$$P_{osm}(h/2) = \frac{kT}{(h/2)^2 D} N^3 \left(\frac{a}{D}\right)^5 \tag{11}$$

and the overall repulsion P_{sc} due to the compressed mushrooms for $h < h_{cr}$ will therefore be:

$$P_{sc} = P_{conf}(h/2) + P_{osm}(h/2) \tag{12}$$

where P_{conf} and P_{osm} are given by Equations 5 and 11, respectively.

3. Interdigitated Mushrooms, Low Compression

If $D > 2R_F$ the bilayers are sparsely covered by polymer. When two surfaces approach it would appear that a mushroom grafted to one of them can be compressed directly against the other surface; i.e., the mushrooms from opposite bilayers can interdigitate (Figure 3b). In the case of 100% interdigitation, each mushroom is compressed between two plates of separation h. If we assume that the degree of interdigitation depends on the surface coverage, two mushrooms will interdigitate with probability $(D - R_F)/D$ (note that D is greater than $2R_F$). Then the average force will be:

$$P(h) = \frac{D - R_F}{D} P_{mf}(h) + \frac{R_F}{D} P_{mf}(h/2) \tag{13}$$

Note that the force between mushrooms is steeper than interdigitated mushrooms. The major assumption here is that the mushrooms do not move on the surface. This is true if lipid diffusion in the plane of the bilayer does not occur. In the gel-phase lipid system the diffusion is slow, but still present. The diffusion coefficient is on the order of 10^{-10} cm²/s. Let us calculate the average distance that a lipid moves in the course of t = 12 h, which is a characteristic time for the equilibration of the multilamellar liposomal system used in the X-ray diffraction equipment. This distance is given by the formula:

$$\bar{x} = \sqrt{2Dt} \tag{14}$$

where D is the diffusion coefficient. We find $\bar{x}$ to be of the order of 30 μm so in the course of these 12 h lipids will mix well. Since there is compression applied, this compression may

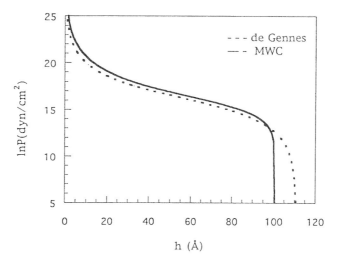

FIGURE 5. Theoretical pressure vs. interbilayer distance curves for polymer brushes: 12 *M*% polymer lipid, N = 46, *a* = 3.5 Å.

lead to lipid rearrangement such that a 100% interdigitation is achieved. For low compression, however, this effect might not be significant. In the case of a 100% interdigitation the average force is given by $P_{mf}(h)$ (Equation 7).

4. Interdigitated Mushrooms, High Compression

In this case, we assume that due to high compression polymer-lipids rearrange such that a 100% interdigitation occurs. The repulsive force will be given by Equation 5.

B. BRUSHES

de Gennes[28] and Alexander[29] consider the physical response of the polymer when two brush-covered surfaces are brought in close contact. The repulsive pressure is found to depend on the distance between the two surfaces h as:

$$p_{sc}(h) \cong \frac{kT}{D^3}\left[\left(\frac{2L_{sc}}{h}\right)^{9/4} - \left(\frac{h}{2L_{sc}}\right)^{3/4}\right] \tag{15}$$

Milner and coworkers[31-34] calculate the repulsive force in terms of their mean-field theory to be:

$$P_{MWC}(h) = \frac{5}{9D^2}F_{MWC}\left(\frac{L_{MWC}}{(h/2)^2} - \frac{h}{L_{MWC}^2} + \frac{(h/2)^4}{L_{MWC}^5}\right) \tag{16}$$

where F_{MWC} is given by:

$$F_{MWC} = \frac{9}{10}kT\left(\frac{\pi^2}{12}\right)^{1/3}N\frac{a^{4/3}}{D^{4/3}} \tag{17}$$

The two theoretical predictions are plotted in Figure 5 for N = 46, *a* = 3.5 Å, 12 *M*% polymer lipid.

These theoretical schemes are now being tested. The predictions are being compared to X-ray data. The comparison between theory and experiment is to be published elsewhere.

IV. PHASE BEHAVIOR OF THE POLYMER-LIPID/LIPID MIXTURE IN AQUEOUS MEDIUM AND MAXIMUM CONCENTRATION OF POLYMER-LIPIDS IN THE BILAYER

As suggested by the presented theory and experimental studies,[9] increasing the concentration of grafted polymer, as well as the molecular weight of the polymer, further improves the repulsive properties of the lipid bilayer surfaces, creating a denser, larger brush. The brush, however, will eventually perturb the surface that is grafted to such that at high concentrations of polymer-lipid a transition must occur from a bilayer to a micellar phase.[9] We propose two possible scenarios for this transition: (1) the transition can be due to reaching critical material parameters of the bilayer, or (2) it can be determined by the thermodynamics of the self-assembling amphiphilic system. For each scenario we define a critical concentration: (1) n_{sat} (which we will call the saturation limit of polymer-lipid in the bilayer) is determined by the material characteristics of the bilayer, and (2) n_{tr} (which will be called the thermodynamic crossover) is the polymer-lipid concentration for which micelle formation becomes energetically favorable.

We will now discuss the two possible schemes for micelle formation. First, we note that since by definition there are no lateral interactions between polymers in the mushroom regime the phase transition to micelles which is determined by polymer lateral interactions should not occur until polymer concentrations are within the brush regime.

A. PHASE TRANSITION, DETERMINED BY THE MATERIAL PROPERTIES OF THE BILAYER

In the brush regime there is energy stored in the brush which increases with the molecular weight of the polymer and the grafting concentration. This stored free energy is expressed as a lateral tension between the brushes. If the bilayer structure is stable, the lateral pressure in the two polymer layers (see Figures 1b and 6) will be felt as an isotropic tension in the bilayer. Just as in micropipette experiments[15,42] where applied tension leads to area change, the polymer will cause area expansion of the bilayer, which will be opposed by the cohesive forces in the bilayer (Figure 6). This tension will increase with the molecular weight and the grafting concentration and eventually reach the critical tension that the bilayer can support as a material. The maximum concentration will therefore be the one at which the critical material parameters are attained; concentrations higher than this one cannot be achieved. So the material properties define a saturation limit n_{sat}. Increasing the polymer-lipid concentration above n_{sat} will lead to a transition from bilayer into a mixed bilayer and micellar phase. The excess polymer-lipids are expected to segregate to form micelles. These micelles have been observed in an electron micrograph for a 100% polymer-lipid.[9]

To calculate the maximal concentration of polymer-lipid in the bilayer we developed a simple lateral force model.[45] We suppose that the maximum concentration of polymer-lipid in the bilayer is the one at which the lateral steric repulsion between the polymer chains equals the bilayer tensile strength τ_s, measured by micropipette experiments.[15] The prediction is plotted in Figure 7 as a function of cholesterol content in the SOPC bilayer.

Based on this prediction we expect then that we can control and manipulate this maximum concentration by changing both the polymer molecular weight and the lipid composition (e.g., incorporating cholesterol that is known to increase the tensile strength τ_s of the bilayer[15] or using gel-phase lipids). Thus, we propose that increasing the bilayer cohesion is a unique means of achieving a high density of grafting in these self-assembling lipid systems.

The saturation concentration due to these material properties gives the absolute limit of polymer concentration in the bilayer. If the lipid molecules formed only bilayers then this would be the only scheme for the saturation limit. In fact, lipid molecules are polymorphic. (Lipid polymorphism is the ability of the lipid molecules to form different types of aggregates under different conditions.) As pointed out by Israelachvili,[46] fully hydrated lipid molecules

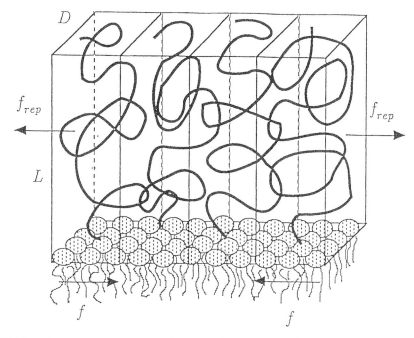

FIGURE 6. Schematic diagram showing that the brush lateral pressure is balanced by bilayer cohesion.

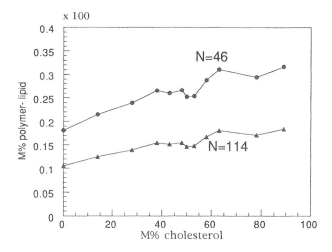

FIGURE 7. Maximum concentration of polymer-lipid that can be incorporated in the SOPC bilayer as a function of cholesterol content.

such as DSPC, egg PC, etc., which have a predominantly cylindrical shape (the area per head equals the area per tail) will form bilayers. But changes in the temperature, pH, or, as discussed here, high concentration of incorporated polymer-lipid which leads to enhanced repulsions between the lipid headgroup, can induce a phase transition to more stable micellar or cubic phases.

B. PHASE TRANSITION DETERMINED BY THE THERMODYNAMICS OF A SELF-ASSEMBLING POLYMER/LIPID/LIPID SYSTEM (THE MINIMUM ENERGY REQUIREMENT)

Curving the grafting surface will relax the lateral tension in the polymer layer. The higher the curvature, the bigger the relaxation. The highest curvature will be obtained if the lipids

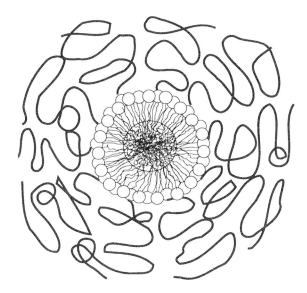

FIGURE 8.   Schematic diagram showing a polymer-grafted micelle.

pack into micelles (cylindrical or spherical) (Figure 8). This will cost additional energy due to "unfavorable packing" of the hydrocarbon chains. While the latter energy does not depend on the molecular weight and grafting density, the energy stored in the polymer brush will depend crucially on them: the bigger the molecular weight of the polymer, the bigger the relaxation when a polymer-lipid is transferred from a bilayer to a micelle. That is why, for high molecular weight, the brush energy decrease due to curving of the surface will be bigger than the hydrocarbon energy increase due to micelle packing. A phase transition from bilayers to micelles will therefore occur. The concentration n_{tr} (the thermodynamic crossover) at which this occurs will depend on the molecular weight and the grafting density of the polymer and on the structure of the hydrocarbon tails: for instance, number of CH_2 groups and number of double bonds, which determines how "difficult" it is to pack the chains in a micelle.

To understand the mechanism of this phase transition we study the changes in the energy of the system that occur upon an increase in the concentration of the polymer-lipids in the lipid mixture. We obtain that, upon an increase in the concentration of polymer-lipid in the sample, the energy minimum gradually moves from a bilayer into a micellar phase.

For concentrations lower than n_{tr} the dominant phase is the bilayer. Increasing the concentration above the thermodynamic crossover n_{tr} has the effect of decreasing the probability for formation of bilayers. Above n_{tr} a mixed phase exists — bilayers and micelles have the same energy and coexist. So the transition from bilayers to micelles is not a first-order transition.

Polymer-lipid concentration in the bilayer is equal to the system concentration (i.e., concentration in the original anhydrous mixture) for concentrations less than n_{tr}. Above n_{tr} it is equal to n_{tr}, as dictated by the behavior of a mixed-phase region.

In summary, in the first scheme, the phase transition from bilayers to micelles is determined by the critical parameters of the bilayer as a material. In the second one, the minimum energy requirement is the driving force for the phase transition. Both schemes predict micelle formation and the existence of a mixed phase above a certain critical polymer-lipid concentration. Also, in both schemes the polymer-lipid concentration in the bilayer increases up to this critical concentration and then levels off. In the first scheme this concentration is the saturation limit of the polymer-lipids in the bilayer n_{sat} and is determined by the material parameters of the bilayer. In the second scheme the critical concentration is the thermodynamic crossover n_{tr}.

Both schemes are possible. The one that occurs first (i.e., has a lower critical concentration) will determine the phase behavior of the system. This will depend on the material parameters and micelle-forming properties of the lipids that make up the bilayer. For phospholipids such as DSPC, SOPC, and their mixtures with cholesterol we obtain theoretically that n_{tr} is much lower than n_{sat}. (The results will not be presented here in detail, they are obtained numerically and are to be published elsewhere.) For molecular weight 2000 Da in gel-phase bilayers the concentration n_{tr} is about 15 *M%* and for liquid bilayers it is about 8 *M%*. Thus, the proposed analysis predicts that the driving force for the phase transition is the minimum energy requirement.

V. DISCUSSION

The theoretical analyses presented here and elsewhere[45] can provide valuable information that is directly related to the Stealth liposomes and their expected performance in service:

1. The concentration of polymer-lipid in the liposomes is not necessarily equal to the concentration in the anhydrous mixture. Bilayer concentrations are only equal to the concentration in the anhydrous mixture for concentrations less than n_{sat} or n_{tr}, whichever is less. Bilayers in the two phase region (bilayers and micelles) cannot have higher concentrations than these limits.
2. Increasing the polymer-lipid concentration above n_{tr} will induce micelle formation. Thus, in Stealth liposome preparations there is no point increasing the polymer-lipid concentration above n_{sat} or n_{tr}, whichever is less. (In our theoretical study we obtain that for molecular weights 2000 and 5000 Da $n_{sat} > n_{tr}$.) For molecular weight 2000 Da for gel-phase bilayers the concentration n_{tr} is about 15 *M%* and for liquid bilayers it is about 8 *M%*. Increasing the polymer-lipid concentrations above these values will not improve the repulsive properties of the bilayers and hence will not affect their circulation times.
3. The theoretical predictions show that the grafted polymer layer changes the elastic constants and the tensile strength of the bilayer.[45] These changes occur within the brush regime and make the bilayer weaker. It may appear that for the sake of stability it is better if the polymer chains barely touch and interact: i.e., at the very borderline between mushrooms and brushes, which is about 5% in liquid bilayers for PEG-2000.
4. The nature of the lipids in the bilayer appears to be of importance. The higher the cohesion, the more stable the system. This suggests that the mixture of saturated (gel-phase) lipids such as DSPC with cholesterol may be the optimum material for Stealth liposomes preparation.
5. For concentrations less than n_{sat} or n_{tr} the theory allows us to predict the extension of the polymer from the surface. It is given by Equation 1 for the mushroom regime and by Equation 3 for the brush regime. At these distances from the lipid surface the polymer steric interactions are "switched on" and their magnitude grows almost exponentially towards the surface, dominating over the electrostatic and van der Waals forces. For distances larger than L this steric interaction is zero. If $L \leq \lambda_D$, the Debye length (a condition achievable in low salt) there will be a long electrostatic tail to the repulsion.[47] If $L \gg \lambda_D$ the electrostatic contribution will be negligible. Then, if L is within the range of the van der Waals forces (we can expect this for molecular weights of 2000 and smaller at low surface coverage) some weak attraction between the polymer-covered bilayers may exist.[48] The extension L should also give an indication of the maximum size of PEG that can be incorporated in a targetable liposome such that a protein molecule attached to the surface can still be accessible for binding to its target molecule.

6. When considering the mechanism of repulsion between two identical polymer-covered bilayers, we showed that the polymer acts as a good repulsive barrier against the close approach of another surface. As some preliminary experiments suggest, however, the polymer layer is not an effective barrier against small proteins and surfactant molecules.[49] It appears that these small molecules can easily diffuse through the polymer layer. Thus, the liposome in circulation may well get opsonized, but the steric barrier that the polymer creates does not allow for the close approach of macrophages and other cells with phagocytic activity to the opsonized liposome surface. Thus, its long circulation time in the bloodstream is still maintained.

REFERENCES

1. **Blume, G. and Cvec, G.,** Liposomes for the sustained drug release in vivo, *Biochim. Biophys. Acta,* 1029, 91, 1990.
2. **Klibanov, A.L., Maruyama, K., Torchilin, V.P., and Huang, L.,** Amphipathic polyethylene glycols effectively prolong the circulation time of liposome, *FEBS Lett.,* 268, 235, 1990.
3. **Mori, A., Klibanov, A.L., Torchilin, V.P., and Huang, L.,** Influence of the steric barrier activity of amphipathic poly(ethylene glycol) and ganglioside gm1 on the circulation time of liposomes and on the target binding of immunoliposomes in vivo, *FEBS Lett.,* 284, 263, 1991.
4. **Needham, D., Hristova, K., McIntosh, T.J., Dewhirst, M., Wu, N., and Lasic, D.,** Polymer-grafted liposomes: physical basis for the "stealth" property, *J. Liposome Res.,* 2, 411, 1992.
5. **Papahadjopoulos, D., Allen, T., Gabizion, A., Mayhew, E., Matthay, K., Huang, S.K., Lee, K., Woodle, M.C., Lasic, D.D., Redemann, C., and Martin, F.J.,** Sterically stabilized liposomes: pronounced improvements in blood clearance, tissue disposition, and therapeutic index of encapsulated drugs against implanted tumors, *Proc. Natl. Acad. Sci. U.S.A.,* 88, 11460, 1991.
6. **Senior, J., Delgado, C., Fisher, D., Tilcock, C., and Gregoriadis, G.,** Influence of surface hydrophilicity of liposomes on their interaction with plasma protein and clearance from the circulation: studies with poly(ethylene glycol)-coated vesicles, *Biochim. Biophys. Acta,* 1062, 77, 1991.
7. **Mayhew, E., Lasic, D.D., Babbar, S., and Martin, F.J.,** Pharmacokinetics and antitumor activity of epirubicin encapsulated in long circulating liposomes incorporating a polyethylene glycol-derivatized phospholipid, *Int. J. Cancer,* 51, 1, 1992.
8. **Chonn, A., Semple, S.C., and Cullis, P.R.,** Association of blood proteins with unilamellar liposomes in vivo. Relation to circulation halftimes, *J. Biol. Chem.,* 267, 18759, 1992.
9. **Needham, D., McIntosh, T.J., and Lasic, D.,** Repulsive interactions and mechanical stability of polymer-grafted lipid membranes, *Biochim. Biophys. Acta,* 1108, 40, 1992.
10. **Kenworthy, A.K., McIntosh, T.J., Needham, D., and Hristova, K.,** Steric interactions between bilayers containing lipids with covalently attached polyethylene glycol, *Biophys. J.,* 64, A348, 1993.
11. **Cevc, G. and Marsh, D.,** *Phospholipid Bilayers,* John Wiley & Sons, New York, 1987.
12. **Evans, E. and Skalak, R.,** *Mechanics and Thermodynamics of Biomembranes,* CRC Press, Boca Raton, FL, 1980.
13. **Bloom, M., Evans, E., and Mouritsen, O.G.,** Physical properties of the fluid lipid bilayer component of cell membranes: a perspective, *Q. Rev. Biophys.,* 24, 293, 1991.
14. **Bloom, M.,** The physics of soft, natural materials, *Phys. Can.,* 7, 531, 1992.
15. **Needham, D. and Nunn, R.S.,** Elastic deformation and failure of lipid bilayer membranes containing cholesterol, *Biophys. J.,* 58, 997, 1990.
16. **Bivas, I., Hanusse, P., Bothorel, P., Lalanne, J., and Aguerre-Chariol, O.,** An application of optical microscopy to the determination of the curvature elastic modulus of biological and model membranes, *J. Phys. (Paris),* 48, 855, 1987.
17. **Helfrich, W.,** Steric interactions of fluid membranes in multilayer systems, *Z. Naturforsch.,* 33a, 306, 1977.
18. **Faucon, J.F., Mitov, M.D., Meleard, P., Bivas, I., and Bothorel, P.,** Bending elasticity and thermal fluctuations of lipid membranes, *J. Phys. (Paris),* 50, 2389, 1989.
19. **Faucon, J.F., Meleard, P., Mitov, M.D., Bivas, I., and Bothorel, P.,** Thermal fluctuations of giant vesicles and elastic properties of bilayer lipid membranes, *Prog. Colloid Polym. Sci.,* 79, 11, 1989.
20. **Parsegian, V.A. and Ninham, B.W.,** van der Waals forces in many-layered structures: generalization of the lifshitz result for two semi-infinite media, *J. Theor. Biol.,* 38, 101, 1973.
21. **LeNeveu, D.M., Rand, R.P., Parsegian, V.A., and Gingell, D.,** Measurement and modification of forces between lecithin bilayers, *Biophys. J.,* 18, 209, 1977.
22. **Evans, E. and Needham, D.,** Physical properties of surfactant bilayer membranes: thermal transitions, elasticity, rigidity, cohesion, and colloidal interactions, *J. Phys. Chem.,* 91, 4219, 1987.

23. **de Gennes, P.G.,** *Scaling Concepts in Polymer Physics,* Cornell University Press, Ithaca, NY, 1985.
24. **de Gennes, P.G.,** Polymer solutions near an interface. I. Adsorption and depletion layers, *Macromolecules,* 14, 1637, 1981.
25. **de Gennes, P.G.,** Polymers at an interface. II. Interaction between two plates carrying adsorbed polymer layers, *Macromolecules,* 15, 492, 1982.
26. **Evans, E. and Needham, D.,** Attraction between lipid bilayer membranes in concentrated solutio , of nonadsorbing polymers: comparison of mean-field theory with measurements of adhesion energy, *M omolecules,* 21, 1822, 1988.
27. **Joanny, J.F., Leibler, L., and de Gennes, P.G.,** Effects of polymer solutions on colloidal stability, *J. Polym. Sci. (Phys.),* 17, 1073, 1979.
28. **de Gennes, P.G.,** Conformation of polymers attached to an interface, *Macromolecules,* 13, 1069, 1980.
29. **Alexander, S.,** Adsorption of chain molecules with a polar head. A scaling description, *J. Phys. (Paris),* 38, 983, 1977.
30. **de Gennes, P.G.,** Model polymers at interfaces, in *Physical Basis of Cell-Cell Adhesion,* Bongrand, P., Ed., CRC Press, Boca Raton, FL, 1988.
31. **Milner, S.T., Witten, T.A., and Cates, M.E.,** A parabolic density profile for grafted polymers, *Europhys. Lett.,* 5, 413, 1988.
32. **Milner, S.T.,** Compressing polymer brushes: a quantitative comparison of theory and experiment, *Europhys. Lett.,* 7, 695, 1988.
33. **Milner, S.T., Witten, T.A., and Cates, M.E.,** Theory of the grafted polymer brush, *Macromolecules,* 21, 2610, 1988.
34. **Milner, S.T., Witten, T.A., and Cates, M.E.,** Effects of polydispersity in the end-grafted polymer brush, *Macromolecules,* 22, 853, 1989.
35. **Hadziioannou, G., Patel, S., Granick, S., and Tirrell, M.,** Forces between surfaces of block copolymers adsorbed on mica, *J. Am. Chem. Soc.,* 108, 2869, 1986.
36. **Patel, S., Tirrell, M., and Hadziioannou, G.,** A simple model for forces between surfaces bearing grafted polymers applied to data on adsorbed copolymers, *Colloids Surfaces,* 31, 157, 1988.
37. **Personage, E., Tirrell, M., Watanabe, A., and Nuzzo, R.G.,** Adsorption of poly(2-vinylpyridine)-poly(sterene) block copolymers from toluene solution, *Macromolecules,* 24, 1987, 1991.
38. **Tirrell, M., Personage, E., Watanabe, H., and Dhoot, S.,** Adsorbed block copolymer layers: assembly and tailoring of polymer brushes, *Polym. J.,* 23, 641, 1991.
39. **Argillier, J.F. and Tirrell, M.,** Adsorption of water soluble ionic/hydrophobic diblock copolymer on a hydrophobic surface, *Theor. Chim. Acta,* 82, 343, 1992.
40. **Marques, C., Joanny, J.F., and Leibler, L.,** Adsorption of block copolymers in selective solvents, *Macromolecules,* 21, 1051, 1988.
41. **Tanford, C.,** *The Hydrophobic Effect,* John Wiley & Sons, New York, 1980.
42. **Needham, D.,** Cohesion and permeability of lipid bilayer vesicles, in *Permeability and Stability of Lipid Bilayers,* Bongrand, P., Ed., CRC Press, Boca Raton, FL, 1994,
43. **de Gennes, P.G.,** Scaling theory of polymer adsorption, *J. Phys. (Paris),* 37, 1445, 1976.
44. **Daoud, M. and de Gennes, P.G.,** Statistics of macromolecular solutions trapped in small pores, *J. Phys. (Paris),* 38, 85, 1977.
45. **Hristova, K. and Needham, D.,** Influence of grafted polymer on the physical properties of lipid bilayers, *J. Surface Interface Sci.,* submitted.
46. **Israelachvili, J.N.,** *Intermolecular and Surface Forces,* Academic Press, San Diego, CA, 1985.
47. **Kuhl, T., Leckband, D.E., Lasic, D.D., and Israelachvili, J.N.,** Modulation of interaction forces between lipid bilayers exposing short-chained ethylene oxide headgroups, *Biophys. J.,* submitted.
48. **Yoshioka, H.,** Surface modification of haemoglobin-containing liposomes with polyethylene glycol prevents liposome aggregation in blood plasma, *Biomaterials,* 12, 861, 1991.
49. **Needham, D.,** unpublished results.

Chapter 6

MOLECULAR MECHANISM OF LIPOSOME AND IMMUNOLIPOSOME STERIC PROTECTION WITH POLY(ETHYLENE GLYCOL): THEORETICAL AND EXPERIMENTAL PROOFS OF THE ROLE OF POLYMER CHAIN FLEXIBILITY

V.P. Torchilin, M.I. Papisov, A.A. Bogdanov, V.S. Trubetskoy, and V.G. Omelyanenko

TABLE OF CONTENTS

I. INTRODUCTION

The evident drawback of liposomal preparations as potential carriers for biologically active compounds[1] is their fast elimination from the circulation and capture by the cells of the reticuloendothelial system (RES). The recognition and capture of liposomes occurs, primarily, in liver and spleen, and is believed to be the result of fast opsonization of liposomes with blood opsonins.[2] To make liposomes capable of delivering pharmaceutical agents to targets other than the RES, attempts have been made to prolong their lifetime in the circulation, such as by variation of the liposome size,[3] coating liposomes with some plasma proteins,[4] and RES blockade by presaturation with empty liposomes before the application of "therapeutical" ones.[5]

The really important breakthrough in this area, however, was achieved with the discovery of so-called long circulating liposomes.[6-9] Originally,[10] such liposomes have been prepared by the incorporation of ganglioside GM1 into the liposomal membrane. Another approach deals with liposome coating with poly(ethylene glycol) or PEG,[11-16] reviewed in Reference 17. The protective effect of GM1 on liposomes has already been clarified,[18] whereas the molecular mechanism of PEG action remains obscure. The explanations of the phenomenon involve the participation of PEG in the repulsive interactions between PEG-grafted membranes and other particles,[19] the role of surface charge and hydrophilicity of PEG-coated liposomes,[20] and, generally speaking, the decreased rate of plasma protein (opsonins) adsorption on the hydrophilic surface of pegylated liposomes.[21]

In an attempt to explain what PEG properties underlie its ability to prevent liposome opsonization, we hypothesized that the molecular mechanism of PEG protective action is determined by the behavior of a polymer molecule in a solvent, and includes the formation of an impermeable polymeric "cloud" over the liposome surface, even at relatively low polymer concentration.[22] Here we would like to discuss the further development of the model and submit some experimental support for it.[23,24] Moreover, our investigations have led us to the conclusion that the protection of liposome with PEG can be effectively combined with antibody coupling to the liposome surface, yielding long circulating immunoliposomes.[14,22] The coexistence of PEG and antibody on the liposome surface can be successfully described within our model, and optimal PEG concentration for the preparation of long circulating liposomes and immunoliposomes can be predicted as well as optimal PEG molecular weight.

II. IMPORTANCE OF POLYMER FLEXIBILITY AND DENSITY OF "STATISTICAL" CLOUD

As we have already mentioned, liposome elimination from the blood proceeds mainly via liposome recognition by phagocytic cells, mediated by plasma protein (opsonins) adsorption onto the liposome surface. The most evident approach to slow down the liposome clearance is, therefore, to prevent protein contacts with the membrane. To perform our analysis of this approach, we are assuming that (1) protecting polymer contacts with plasma proteins does not result in the opsonization, and (2) the polymer itself does not contain any cell-specific moieties. These assumptions exclude from our consideration hydrophobic and/or significantly charged polymers because of their obvious potential to bind proteins via hydrophobic and/or electrostatic interactions. Evidently, the protective layer of a polymer over the liposome surface has to combine abilities to escape opsonization and recognition by cells (the best for it is to look like water from outside) and to prevent the penetration of opsonizing proteins to the liposome surface. How can these properties "work" simultaneously?

Before discussing it, let us introduce some quantitative criteria for the characterization of liposome protection with a polymer.[22] Considering the diffusional movement of a protein molecule from the blood towards the liposome surface as the initial step of a protein-to-liposome interaction, we can express the degree of liposome protection as the probability for the protein to collide with a polymer (P_{pol}) instead of liposome (P_{lip}). Thus, if the protein is unable to reach the liposome surface and $P_{lip} = 0$, so $P_{pol} = 1 - P_{lip} = 1$. When $P_{pol} = 0$, no interaction between protein and polymer occurs and no protection is achieved. If we will consider the interaction between liposome and opsonin in terms of chemical kinetics, P_{lip} should represent the "steric factor" in the equation of the reaction rate:

$$\frac{d[LO]}{dt} = k \cdot [L][O] = P_{lip} \cdot k' \cdot [L][O]$$

where [LO] is the concentration of liposome/opsonin complex, k' is the rate constant, [L] and [O] are the concentrations of lipid (outer liposome monolayer only) and opsonin, respectively.

To forecast possible P values, we can describe the behavior of a liposome-grafted polymer molecule in terms of statistical physics, e.g., applying a simplified model of a polymer solution.[25] We can consider it, for example, as a three-dimensional network, in which each cell may be occupied either with a polymer unit or with a solvent (water) molecule. From this point of view, the more flexible the polymer is, i.e., the more independent is the motion of any polymeric unit relative to the neighboring one, the larger is the total number of its possible conformations and the higher is the transition rate from one conformation to another. It means that water-soluble flexible polymer statistically exists as a distribution ("cloud") of probable conformations. Figure 1 shows how this cloud is formed (under some simple assumptions) and its density and uniformity increase with the increase in the number of possible conformations. The polymer flexibility correlates with its ability to occupy with high frequency many cells in solution, temporarily squeezing water molecules out of them (i.e., making them imperme-

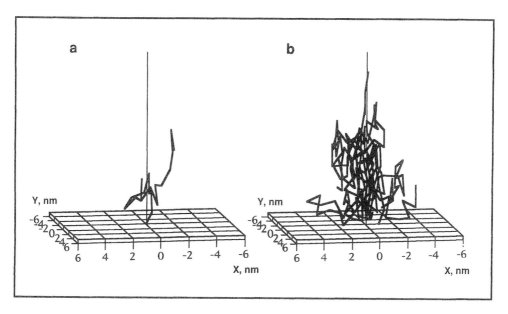

FIGURE 1. Computer simulation of conformational "cloud" formation by surface-attached polymer. Polymer molecule is conditionally assumed to consist of 20 segments, 1 nm each. Produced by random flight simulation; unrestricted segment motion assumed. Space restriction: $Z > 0$, (a) one conformation; (b) eleven conformations. (From Torchilin, V.P., et al., *Biochim Biophys. Acta,* 1195, 11, 1994. With permission.)

able for other solutes). To reach the liposome surface, protein molecules have to penetrate the whole cloud, formed by the liposome-attached polymer molecules. From the computer analysis it follows (see the data in Figure 2) that a flexible polymer forms the conformational cloud with very high density in its central part, whereas a rigid polymer of the same length (its segment was conditionally assumed to be five times longer than for the flexible polymer) forms a broad, but loose and thus permeable, cloud. Thus, a relatively small number of water-soluble and very flexible polymer molecules can create sufficient numbers of high-density conformational "clouds" over the liposome surface, protecting the latter from being opsonized and recognized by RES cells. These molecules form protective "umbrellas" on the liposome surface, and the P_{lip} value depends on their amount (N_P), effective square (S_P), and "reliability" (P^*), expressed as

$$P^* = \frac{1}{S_P} \int_S \frac{dP_{pol}}{dS_P}$$

where P^* is the average P_{pol} within the "umbrella" volume (see Figure 3). It is easy to see that the average P_{pol} value for the entire liposome is

$$P_{pol} = \frac{N_P S_P}{S_{lip}} P^* \quad \text{or} \quad P_{lip} = 1 - \frac{N_P S_P}{S_{lip}} P^*$$

where S_{lip} is the liposome surface square. The last equation can be transformed into:

$$P_{pol} = \gamma \frac{S_P}{S_l} P^*$$

where γ is the polymer/lipid molar ratio in the outer monolayer, and S_l is the average area occupied by a single lipid molecule (for the given liposome size and composition). At high γ values, polymer will be "stretched" out of the liposome forming dense "brush".[26] Therefore,

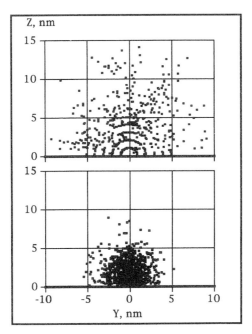

FIGURE 2. Distribution of polymer conformations in space; slice X = 0 ± 0.25 nm. Produced by random flight simulation (Z >0, polymer length 20 nm, 440 conformations). *Upper panel*, segment length is 5 nm (rigid polymer); *lower panel*, segment length is 1 nm (flexible polymer). Calculations for the rigid polymer have been done under the assumption that only every fifth 1-nm segment may change the direction. (From Torchilin, V.P., et al., *Biochim. Biophys. Acta*, 1195, 11, 1994. With permission.)

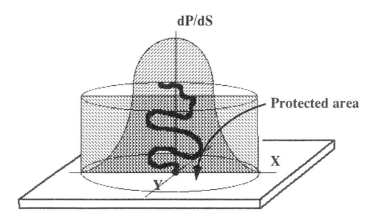

FIGURE 3. The distribution of P_{pol} density in the vicinity of the polymer molecule attached to the liposome surface (bell-shaped curve) and the average P_{pol} value within the "umbrella" volume. (From Torchilin, V.P. and Papisov, M.I., *J. Liposome Res.*, 4, 725, 1994. With permission.)

γ (S_P/S_1) is always <1. The maximal protection can be achieved when $\gamma \cdot S_P \approx S_1$ (the polymer "clouds" are practically fused) and P^* value is close to 1.

As we can see for a rigid chain polymer (where unit motion is hindered), even good water solubility and hydrophilicity may not provide sufficient protection for the liposome surface. The number of possible conformations for such polymers is lower; besides, the conformational transitions proceed with a slower rate than those of a flexible polymer (compare Figures 2a and 2b). It means that the density of the conformational "cloud" for a rigid polymer will be very uneven during a single collision act, and the number of water molecules disturbed much smaller. In terms of the "cellular" model, it appears that there should exist a sufficient

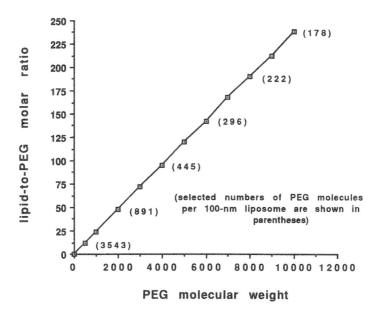

FIGURE 4. The dependence of the number of PEG molecules required for complete protection of a 100-nm liposome on PEG molecular weight.

water space through which the normal diffusion of plasma proteins toward the liposome surface is still possible. Thus, to protect the liposome one has to bind a much larger number of rigid polymer molecules on the liposome surface. This difference between flexible and rigid polymers may only increase with the polymer molecular weight. Good water solubility and hydrophilicity alone cannot provide sufficient protective effect.[27] Only when the same polymer combines both hydrophilicity and flexibility, can it serve as an effective liposome protector even at relatively low concentration of the surface-immobilized macromolecules (the most obvious example of such polymers is PEG). Other possible candidates are, for example, poly(acrylamide) and poly(vinylpyrrolidone), provided that these polymers do not have any specific affinity towards cells or plasma proteins.

The next question to answer is how can we estimate the size of the area on the liposome surface protected with a single polymer molecule of a given molecular weight? Or, how many polymer molecules do we need to protect the liposome of a given size? Such parameters as the average end-to-end distance of a polymer random coil in solution, $R_{0,sol}$, give us certain insight on the "cloud" density (assuming that PEG forms a random coil in solution).[22] A polymer molecule is located mainly in the volume "between the ends" (inside the appropriate sphere for the molecule in solution and inside the hemisphere for the surface-immobilized polymer). So, we can assume that the $R_{0,sol}$ value nearly corresponds to the radius of the "dense cloud" (R), which can hardly be penetrated with a protein molecule. The end-to-end distance for the surface-attached molecule might be about two times longer than $R_{0,sol}$ (for a flexible polymer, the fixation of one end does not influence molecular mobility significantly). Thus, we can assume the radius of the protected area being from $R_{0,sol}$ to $2R_{0,sol}$, and within this area $P^* \approx 1$. Using this simple approach and $R_{0,sol}$ values for PEG of different molecular weight published in Reference 28, we can estimate the area of the liposome surface that can be protected by a single PEG molecule. Assuming about $4.25 \cdot 10^4$ lipid molecules in the outer monolayer of a single bilayer 100-nm liposome,[29,30] we can calculate the molar ratio PEG-to-lipid required for 100% protection of the liposome surface (surface area, S, $3.14 \cdot 10^4$ nm²), see Figure 4. One should mention that the figures calculated match well with published experimental data,[14,31] thus the end-to-end distance may be used as an estimate for the

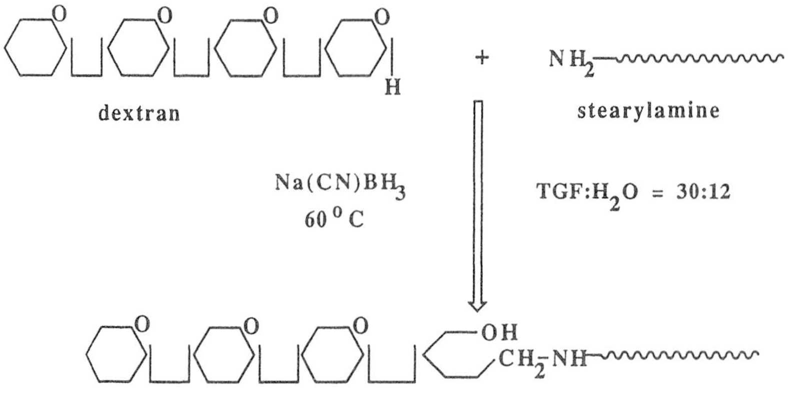

FIGURE 5. Synthesis of hydrophobic dextran derivative for incorporation into the liposomal membrane.

protected square radius. As clearly follows from Figure 4, the size of the protected area depends linearly on the polymer molecular weight.

To prove our hypothesis experimentally we have investigated the efficacy of the fluorescence quenching of the liposome-incorporated fluorescent phospholipid *N*-[7-nitrobenz-2-oxa-1,3-diazol-4-yl]-dioleoyl phosphatidyl ethanolamine (NBD-PE) with soluble rhodamine-modified ovalbumin (Rh-OVA), depending on the type and quantity of the liposome-attached polymer.[24] Both fluorescent labels were obtained from Avanti Polar Lipids. Rh conjugation with OAB was performed as in Reference 32. The preparation of phospholipid-modified PEG was described in Reference 11; fatty acid "tail"-containing dextran (Figure 5) was prepared as in Reference 33, with slight modifications. Liposomes were prepared by the detergent (octyl glycoside) dialysis method from phosphatidyl choline (PC) and cholesterol (Ch) in 7:3 molar ratio with the addition of 1 mol% of NBD-PE and different quantities of PEG-PE (molecular weight 5000) or dextran-stearylamine (molecular weight 6000). The kinetics of NBD fluorescence quenching with increasing quantities of Rh-OAB from the solution was registered spectrofluorimetrically (see Figure 6). The increase in the liposomal PEG-PE concentration resulted in decrease of the liposomal NBD quenching. As far as the whole process is limited only by Rh-OAB diffusion from the solution to the liposome surface, it is evident that the presence of PEG on the surface (even at such low concentrations as 0.2 mol%) creates diffusional hindrances for this process. At PEG concentration of about 1 mol% these difficulties are more pronounced (compare with the data from Figure 4). At the same time, similar quantities of liposome-incorporated dextran-stearylamine practically did not influence NBD quenching with Rh-OAB, which well supports our hypothesis.

A similar experiment has been performed[24] using liposome surface-incorporated fluorescein and anti-fluorescein antibody able to quench antigen fluorescence upon forming antigen-antibody complex. In this case, the presence of 1 mol% PEG, but not dextran, on the liposome surface noticeably decreased both the rate of fluorescein quenching and the total quantity of fluorescein accessible for the interaction with antibody. The use of small PEG concentration (0.2 mol%) revealed the existence of two different fluorescein pools on the liposome surface. One of them was quenched with the same kinetics as fluorescein on PEG-free liposomes, whereas the quenching kinetics for another was close to that for fluorescein on PEG-liposomes with high PEG content. Figure 7 helps to elucidate the phenomenon observed. At low PEG concentration we can find two types of reactive centers on the liposome surface: (1) those that are located within the PEG cloud, and (2) those that are located on the nonprotected part of the surface. The kinetic parameters of the reactive site depend on its location on the liposome surface when the quantity of PEG on the liposome is not sufficient to form an even protective cloud over the liposome surface. The data obtained support our hypothesis well.

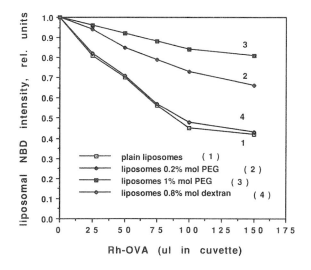

FIGURE 6. Quenching of liposomal NBD fluorescence with Rh-OVA from the solution at different Rh-OVA concentrations: (**1**) plain liposomes; (**2**) liposomes with 0.2 mol% PEG-PE; (**3**) liposomes with 1 mol% PEG-PE; (**4**) liposomes with 0.8 mol% dextran-stearylamine. (Modified from Torchilin, V.P., et al., *Biochim. Biophys. Acta,* 1195, 11, 1994.)

III. LONG CIRCULATING IMMUNOLIPOSOMES — COEXISTENCE OF ANTIBODY AND POLYMER ON THE LIPOSOME SURFACE

The reasonable doubt which arises when PEG is considered as a protection for *immuno*liposome is that it can create steric hindrances for normal antibody-target interaction.[13] Within our model the picture does not look so bad. We can consider three cases for the coimmobilization of antibody and water-soluble flexible polymer on the same liposome (for the constant polymer molecular weight), see Figure 8. In the first case (Figure 8A), the polymer does not form a complete cloud around the liposome. Antibody molecules should be excluded from the volume occupied with PEG (the mechanism of exclusion is similar to that for polymer protective action, yet involves lateral diffusion), and two separate zones appear on the surface — "umbrellas" of PEG and antibodies in between. This structure should successfully interact with target antigens, so liposome binding still occurs. At the same time the clearance rate (or, more exactly, the opsonization rate) depends on the surface area available for the opsonin-to-membrane interaction. Taking into account that opsonization is often a cascade-type process, and the binding of, e.g., a single C3 molecule can be amplified fast, the clearance time of such liposomes may be close to that for "normal" immunoliposomes.

In the second (optimal) case (Figure 8B) the liposome surface is completely coated with overlapping "umbrellas" of PEG probable conformations, but still some areas of more loose conformational density are available, into which free antibodies (capable of the lateral diffusion) can be squeezed from the dense PEG areas. In this case opsonins have very limited opportunity for interaction with the liposome, yet antibodies are still able to recognize and to bind the target. When high-molecular-weight PEG is used, some decrease in binding ability can be achieved.[13,14]

The third case (Figure 8C) requires very high-molecular-weight PEG or a very high degree of surface modification, in such a way that PEG is forming conformationally stretched "brushes".[26] It is evident that in this case surface-immobilized antibody cannot overcome steric hindrances and bind to the target. On the other hand, opsonins also cannot interact with the liposome surface, which allows very long circulation times to such liposomes; half-

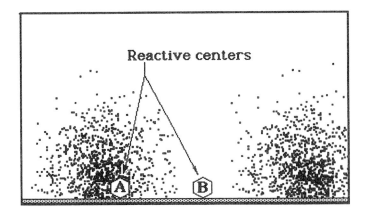

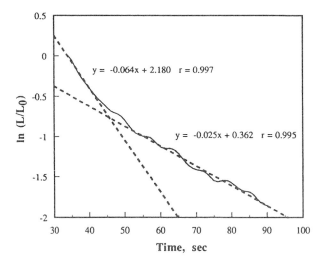

FIGURE 7. *Top panel,* the scheme of reactive site location on the liposome surface at low PEG concentration. The site can be either sterically hindered by polymer "cloud" (A), or exposed and readily available for a variety of interactions (B). Kinetic parameters of chemically identical but differently located sites may differ. *Bottom panel,* linearization of liposomal fluorescein quenching with anti-fluorescein antibody in liposomes with 0.2 mol% PEG-PE. L_0 and L, initial and current fluorescence, respectively. Two phases on the kinetic curve can be seen: initial fast phase, reflecting the quenching of exposed antigen, and subsequent slow phase, reflecting the quenching of PEG-protected antigen. (From Torchilin, V.P., et al., *Biochim. Biophys. Acta,* 1195, 11, 1994. With permission.)

clearance time in rabbits for immunoliposomes containing 10 mol% of PEG was more than 10 h.[14] Brushes of this type, which should provide extremely long circulation times to liposomes, can probably be used for the long circulating immunoliposome preparation, if antibodies will be immobilized on the liposome surface via the long spacer group or even directly on termini of some PEG molecules.

The approach developed opens also the opportunity to predict the optimal size (molecular weight) of the protective polymer molecule in order to provide a desirable combination of liposome longevity and targetability (see Figure 9). The assumption that a certain level of polymer cloud density (L) provides sufficient protection to the liposome surface suggests the existence of a polymer length optimum (Figure 9B). Below the optimum, the polymer cloud will not have enough density even near the attachment point (Figure 9A), whereas very long polymers provide density much above the necessary one (Figure 9C).

The experimental confirmation of the possibility of PEG and antibody coimmobilization on the liposome surface resulting in long circulating immunoliposomes, is described in Chapter 19.

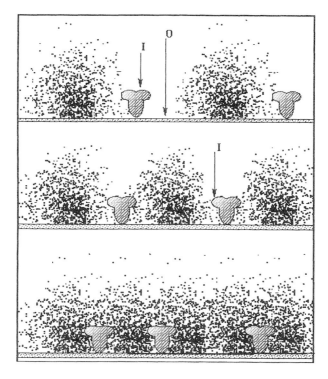

FIGURE 8. Possible location of immunoglobulin (I) on the liposome surface at different PEG concentration. (A) Low PEG concentration; partially nonprotected liposome surface can be reached by opsonins (O). (B) Intermediate (optimal) PEG concentration; no free surface available for opsonins, whereas immunoglobulin still can interact with antigen. (C) High PEG concentration; polymer forms dense "brush" sterically hindering antibody.

IV. WHAT IF PROTECTIVE POLYMERS THEMSELVES ARE OPSONIZED?

Opsonization of a polymer complicates the real liposome behavior considerably. In this case we have to talk not only about the possibility of liposome surface opsonization, but also about the "productive" contacts between plasma opsonins and protective polymers. In terms of the model proposed it can be described as transformation of P_{pol} into $(P_{pol} + P_{pol/opsonin})$ and, thus,

$$\frac{d[LO]}{dt} = \left(P_{lip} \cdot k' \cdot [L][O]\right) + \left(P_{pol/opsonin} \cdot k'' \cdot [L][O]\right)$$

where k' and k'' are the appropriate rate constants.

If the polymer opsonization is taken into account, some brief comments have to be made on the nature of the polymer used. It is evident that protective polymer preferably should not contain OH-groups (like polysaccharides), which are targets for C3; NH_2-groups (like poly-lysine), which are targets for C4; and any other nucleophiles capable of reacting with cyclothioesters of complement proteins (the presence of anionic groups on the liposome coated with OH-containing polymer can, however, diminish the opsonization rate).[34] Some polymer moieties (for example, carbohydrates) can be recognized by plasma lectins as well as by cells directly; for this reason one should avoid the presence of mannose, galactose, and other "signal" groups.

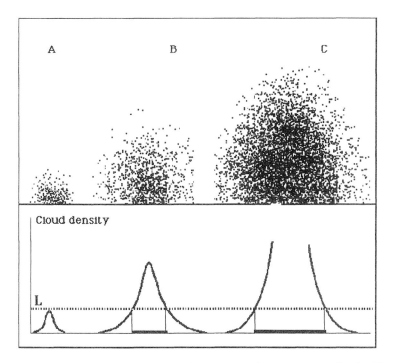

FIGURE 9. Polymers of different length grafted onto the liposome surface and corresponding densities of polymer "clouds" (nonquantitative computer simulation). L, level of polymer "cloud" density providing sufficient protection for the surface; A, short-chain polymer cannot create sufficiently dense "cloud", even if it is highly flexible; B, polymer with optimal chain length provides sufficient protection and still does not exclude the possibility of immobilized protein to interact with its substrate; C, polymer with excessive molecular weight provides unnecessary high "cloud" density and sterically hinders any coimmobilized protein (antibody). (From Torchilin, V.P., et al., *Biochim. Biophys. Acta,* 1195, 11, 1994. With permission.)

V. CONCLUSION

We realize that our model is a simplified one and does not consider the possible role of water tightly bound to polymers (which should, to some extent, increase the polymer rigidity), the possibility of PEG-immunoliposome opsonization via partially oxidized PEG, and probably some other more or less important events. Other aspects of polymer-coated liposome behavior in the organism involve different molecular and cellular mechanisms of the higher level of hierarchy. Nevertheless, in our opinion the model suggested can explain some peculiarities of PEG-coated liposomes on the molecular level that result in their long circulation *in vivo,* and allows design of long circulating immunoliposomes.

REFERENCES

1. **Gregoriadis, G., Ed.,** *Liposomes as Drug Carriers,* John Wiley & Sons, New York, 1988.
2. **Senior, J.H.,** Fate and behavior of liposomes in vivo: a review of controlling factors, *Crit. Rev. Ther. Drug Carriers Syst.,* 3, 123, 1987.
3. **Abra, R.M. and Hunt, C.A.,** Liposome disposition in vivo. III. Dose and vesicle-size effects, *Biochim. Biophys. Acta,* 666, 493, 1981.
4. **Torchilin, V.P., Berdichevsky, V.R., Barsukov, A.A., and Smirnov, V.N.,** Coating liposomes with proteins decreases their capture by macrophages, *FEBS Lett.,* 111, 184, 1980.
5. **Ellens, H., Mayhew, E., and Rustum, Y.M.,** Reversible depression of the reticuloendothelial system by liposomes, *Biochim. Biophys. Acta,* 714, 479, 1982.
6. **Allen, T.M.,** Stealth™ liposomes: avoiding reticuloendothelial uptake in liposomes, in *Liposomes in the Therapy of Infectious Diseases and Cancer,* Lopez-Berestein, G. and Fidler, I.J., Eds., Alan R. Liss, New York, 1989, 405.

7. **Papahadjopoulos, D., Allen, T.M., Gabizon, A., Mayhew, E., Huang, S.K., Lee, K.D., Woodle, M.C., Lasic, D.D., Redemann, C., and Martin, F.J.,** Sterically stabilized liposomes — improvements in pharmacokinetics and antitumor therapeutic efficacy, *Proc. Natl. Acad. Sci. U.S.A.*, 88, 11460, 1991.

8. **Blume, G. and Cevc, G.,** Liposomes for sustained drug release in vivo, *Biochim. Biophys. Acta*, 1029, 91, 1990.

9. **Namba, Y., Sakakibara, T., Masada, M., Ito, F., and Oku, N.,** Glucuronate-modified liposomes with prolonged circulation time, *Chem. Pharm. Bull.*, 38, 1663, 1990.

10. **Allen, T.M. and Chonn, A.,** Large unilamellar liposomes with low uptake into the reticuloendothelial system, *FEBS Lett.*, 223, 42, 1987.

11. **Klibanov, A.L., Maruyama, K., Torchilin, V.P., and Huang, L.,** Amphipathic polyethyleneglycols effectively prolong the circulation time of liposomes, *FEBS Lett.*, 268, 235, 1990.

12. **Mori, A., Klibanov, A.L., Torchilin, V.P., and Huang, L.,** Influence of steric barrier activity of amphipathic poly(ethyleneglycol) and ganglioside GM1 on the circulation time of liposomes and on the target binding of immunoliposomes in vivo, *FEBS Lett.*, 284, 263, 1991.

13. **Klibanov, A.L., Maruyama, K., Beckerleg, A.M., Torchilin, V.P., and Huang, L.,** Activity of amphipathic poly(ethylene glycol) 5000 to prolong the circulation time of liposomes depends on the liposome size and is unfavorable for immunoliposome binding to target, *Biochim. Biophys. Acta*, 1062, 142, 1991.

14. **Torchilin, V.P., Klibanov, A.L., Huang, L., O'Donnell, S., Nossiff, N.D., and Khaw, B.A.,** Targeted accumulation of polyethylene glycol-coated immunoliposomes in infarcted rabbit myocardium, *FASEB J.*, 6, 2716, 1992.

15. **Senior, J., Delgado, C., Fisher, D., Tilcock, C., and Gregoriadis, G.,** Influence of surface hydrophilicity of liposomes on their interaction with plasma proteins and clearance from the circulation: studies with poly(ethylene glycol)-coated vesicles, *Biochim. Biophys. Acta*, 1062, 77, 1991.

16. **Woodle, M.C., Matthay, K.K., Newman, M.S., Hidayat, J.E., Collins, L.R., Redemann, C., Martin, F.J., and Papahadjopoulos, D.,** Versatility in lipid compositions showing prolonged circulation with sterically-stabilized liposomes, *Biochim. Biophys. Acta*, 1105, 193, 1992.

17. **Woodle, M.C. and Lasic, D.D.,** Sterically stabilized liposomes, *Biochim. Biophys. Acta*, 1113, 171, 1992.

18. **Allen, T.M., Austin, G.A., Chonn, A., Lin, L., and Lee, K.C.,** Uptake of liposomes by cultured mouse bone marrow macrophages: influence of liposome composition and size, *Biochim. Biophys. Acta*, 1061, 56, 1991.

19. **Needham, D., McIntosh, T.J., and Lasic, D.D.,** Repulsive interactions and mechanical stability of polymer-grafted lipid membranes, *Biochim. Biophys. Acta*, 1108, 40, 1992.

20. **Gabizon, A. and Papahadjopoulos, D.,** The role of surface charge and hydrophilic groups on liposome clearance in vivo, *Biochim. Biophys. Acta*, 1103, 94, 1992.

21. **Lasic, D.D., Martin, F.G., Gabizon, A., Huang, S.K., and Papahadjopoulos, D.,** Sterically stabilized liposomes — a hypothesis on the molecular origin of the extended circulation times, *Biochim. Biophys. Acta*, 1070, 187, 1991.

22. **Torchilin V.P. and Papisov M.I.,** Why do polyethylene glycol-coated liposomes circulate so long?, *J. Liposome Res.*, 4, 725, 1994.

23. **Torchilin V.P., Trubetskoy, V.S., Milshteyn, A.M., Canillo, J., Wolf, G.L., Papisov, M.I., Bogdanov, A.A., Jr., Narula, J., Khaw, B.A., and Omelyanenko, V.G.,** Targeted delivery of diagnostic agents by surface-modified liposomes, *J. Contr. Release*, 28, 45, 1993.

24. **Torchilin V.P., Omelyanenko, V.G., Papisov, M.I., Bogdanov, A.A., Jr., Trubetskoy, V.S., Herron J.N., and Gentry, C.A.,** Poly(ethylene glycol) on the liposome surface: on the mechanism of polymer-coated liposome longevity, *Biochim. Biophys. Acta*, 1195, 11, 1994.

25. **des Cloizeaux, J. and Jannink, G.,** *Polymers in Solution. Their Modelling and Structure*, Clarendon Press, Oxford, 1990, pp. 63, 280, 539.

26. **Milner, S.T.,** Polymer brushes, *Science*, 251, 905, 1991.

27. **Blume, G. and Cevc, G.,** Bilayer modification and longevity of liposomes, in *Abstracts of Conference 2nd Liposome Research Days*, Leiden University, The Netherlands, 1992, SC19.

28. **Kurata, M. and Tsunashima, Y.,** Viscosity — molecular weight relationships and unperturbed dimensions of linear chain molecules, in *Polymer Handbook*, Brandup, J. and Himmelgut, E.H., Eds., John Wiley & Sons, New York, 1989, VII/1.

29. **Huang, C. and Mason, J.T.,** Geometric packing constraints in egg phosphatidylcholine vesicles, *Proc. Natl. Acad. Sci. U.S.A.*, 75, 308, 1978.

30. **Enoch, H.G. and Strittmatter, P.,** Formation and properties of 1000-A-diameter, single-bilayer phospholipid vesicles, *Proc. Natl. Acad. Sci. U.S.A.*, 76, 145, 1979.

31. **Allen, T.M. and Hansen, C.,** Pharmacokinetics of Stealth versus conventional liposomes: effect of dose, *Biochim. Biophys. Acta*, 1068, 133, 1991.

32. **Wilchek, M., Spiegel, S., and Spiegel, Y.,** Fluorescent reagents for the labeling of glycoconjugates in solution and on cell surfaces, *Biochem. Biophys. Res. Commun.*, 92, 125, 1980.

33. **Wood, C. and Kabat, E.A.,** Immunochemical studies of conjugates of isomaltosyl oligosaccharides to lipid. I. Antigenicity of the glycolipids and the production of specific antibodies in rabbits, *J. Exp. Med.*, 154, 432, 1981.
34. **Carreno, M., Labarre, D., Morillet, F., Jozefowicz, M., and Kazatchkine, M.,** Regulation of the human alternative complement pathway: formation of ternary complex between factor H, surface-bound C3B and chemical groups on non-activating surfaces, *Eur. J. Immunol.*, 19, 2145, 1989.

Chapter 7

MEASUREMENT OF THE RANGE AND MAGNITUDE OF THE REPULSIVE PRESSURE BETWEEN PEG-COATED LIPOSOMES

T.J. McIntosh, A.K. Kenworthy, and D. Needham

TABLE OF CONTENTS

I. INTRODUCTION

As described previously[1-4] and elsewhere in this book, liposomes containing phospholipids with poly(ethylene glycol) covalently attached to their polar headgroups (PEG-lipids) are currently being developed for *in vivo* delivery of drugs to tumors and other sites in the body. A key feature of these PEG-liposomes is that, when injected into the bloodstream, they have a greatly increased circulation time compared to conventional liposomes.[1,5-10] Whereas conventional liposomes are quickly removed from the blood circulation by macrophages primarily located in the spleen and liver, PEG-liposomes tend to remain in the bloodstream for extended periods of time. This means that PEG-liposomes can be used to deliver drugs by being targeted to specific sites in the body.[2,3] For example, since the microvasculature in some solid tumors is relatively leaky, PEG-liposomes extravasate and accumulate in those tumors.[4,9,11-16] The mechanism by which PEG-liposomes avoid macrophages and stay in the blood circulation is thought to involve a "steric barrier" formed around the liposome by the attached PEG molecules.[7,9,16-19] It has been argued that such a barrier could sterically prevent contact between liposomes and opsonins, proteolipids, or macrophage cell surfaces.[10,17] The effectiveness of this barrier appears to depend on the incorporation of a sufficient concentration of PEG-lipids, with appropriately sized PEGs, into liposomes.

In this chapter we describe our ongoing work to measure the steric barrier caused by the incorporation of various PEG-lipids into liposomes. X-ray diffraction methods are used to measure the range and magnitude of the steric barrier around liposomes as a function of: (1) the size of the PEG covalently attached to the phospholipid headgroup, (2) the concentration of the PEG-lipid in the liposome, and (3) the structure (gel or liquid crystalline) of the lipid matrix in the liposome. These are three design features that are expected to influence the *in vivo* performance of drug-laden liposomes. The results of our experiments are strongly correlated with the observed blood circulation times of PEG-liposomes.

II. MATERIALS AND METHODS

PEG-lipids were a kind gift from Liposome Technology, Inc., Menlo Park, California. These PEG-lipids contained polyethylene glycol (PEG) covalently attached to the amine group of distearoylphosphatidylethanolamine (*N*-(carbamyl-poly(ethylene glycol) methyl ether)-1,2-distearoyl-*sn*-glycero-3-phosphoethanolamine, sodium salt). In these studies we used PEG-lipids with PEG chains of molecular weights of 750, 2000, and 5000. These PEG-lipids are abbreviated here as PEG-750, PEG-2000, and PEG-5000, respectively. Stearoyloleoyl-phosphatidylcholine (SOPC) and distearoylphosphatidylcholine (DSPC) were purchased from Avanti Polar Lipids and cholesterol, dextran, and poly(vinylpyrrolidone) (PVP) were purchased from Sigma Chemical Co.

To measure the range and magnitude of the repulsive pressure between bilayer surfaces, an "osmotic stress" technique[20,21] was used. In this method known osmotic pressures are applied to multibilayer systems and the distance between apposing bilayers at each applied pressure is measured by X-ray diffraction analysis. At equilibrium, the total repulsive pressure between bilayers is balanced by the total attractive pressure, which is the sum of the van der Waals attractive pressure and the applied osmotic pressure. Since under the conditions of our experiments the applied pressure is significantly larger than the van der Waals pressure, the total repulsive pressure between bilayers can be equated to the applied osmotic pressure.[20,21] Therefore, results of this method provide the total repulsive pressure between bilayers as a function of distance between bilayers.

For the X-ray diffraction experiments, it is essential to study multilamellar bilayer systems. Two types of lipid systems were examined, unoriented suspensions of multilamellar vesicles (MLVs) and oriented multibilayers. The first step in making either type of these systems was to codissolve in chloroform the appropriate molar ratios of phospholipid and PEG-lipid. To make unoriented liposomes (MLVs), the chloroform was removed by rotary evaporation, aqueous polymer solutions (see below) were added to the dry lipid, and the suspensions were allowed to equilibrate for several hours at temperatures above the lipid's main phase transition temperature. Oriented multilayers were formed by standard techniques,[22] by placing a small drop of the lipid in chloroform onto a curved glass substrate and allowing the chloroform to evaporate under a gentle stream of nitrogen. Known osmotic pressures were applied to these systems by the following published procedures.[20-23] Osmotic stress was applied to the liposomes by incubating them in aqueous solutions of large neutral polymers, either dextran or PVP. Since these polymers are too large to enter the lipid lattice, they compete for water with the lipid multilayers, thereby applying an osmotic pressure.[20,21] Osmotic pressures for the dextran and PVP solutions have been previously measured.[21,24,25] Pressure was applied to the oriented multibilayers by incubating them in constant relative humidity atmospheres maintained with saturated salt solutions, as detailed previously.[22,26]

The multilamellar lipid vesicles were sealed in quartz capillary tubes and mounted on a point collimation X-ray camera, whereas the oriented multilayers on the glass substrate were mounted in a controlled humidity chamber on a single-mirror (line-focused) X-ray camera such that the X-ray beam was oriented at a grazing angle relative to the multilayers.[22,27] For both oriented and unoriented specimens, X-ray diffraction patterns were recorded on X-ray films which were densitometered with a Joyce-Loebl microdensitometer. After background subtraction, integrated intensities, I(h), were obtained for each order h by measuring the area under each diffraction peak. Depending on the specimen geometry, standard correction factors were applied to obtain X-ray structure factors. These structure factors were used to calculate one-dimensional electron density profiles across the bilayer at a resolution of about 7 Å.[22,25,27,28]

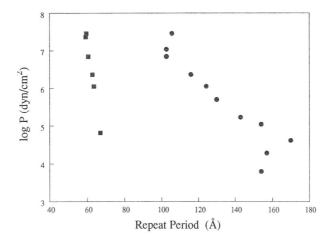

FIGURE 1. Plot of logarithm of applied pressure (log P) vs. lamellar repeat period for 2:1 SOPC:cholesterol bilayers in the absence (squares) and presence (circles) of 5 mol% DSPE:PEG-2000. Osmotic stress experiments were performed on oriented multibilayers in relative humidity atmospheres (log P = 7.5) and on unoriented multilamellar liposomes in PVP or dextran solutions containing 100 mM NaCl (log P <7.5). (From Needham, D., McIntosh, T. J., and Lasic, D. D., *Biochim. Biophys. Acta,* 1108, 40, 1992. With permission of Elsevier Science Publishers.)

III. RESULTS

Typical experimental data from these osmotic stress experiments are presented in Figure 1, which shows the logarithm of applied osmotic pressure (log P) plotted vs. the lamellar repeat period obtained from X-ray diffraction for liposomes containing a 2:1 molar ratio of SOPC:cholesterol in the presence and absence of 5 mol% PEG-2000.[19] These are similar lipid compositions to those used in experiments measuring the effect of PEG-2000 on blood circulation time.[9,17] The data in Figure 1 show that the presence of this concentration of PEG-2000 has a large effect on the lamellar repeat period at each value of applied pressure. For example, at an applied pressure P = 6×10^4 dyn/cm^2 (log P = 4.8) the repeat period for 2:1 SOPC:cholesterol bilayers is increased by about 100 Å by the incorporation of 5 mol% PEG-2000.

The lamellar repeat periods in Figure 1 represent a single unit cell containing two components, a fluid space separating apposed bilayers, and the bilayer itself. These repeat periods are therefore the sum of both bilayer thickness and interbilayer fluid space. The question is, how do PEG-lipids influence lipid bilayer structure, interbilayer separation, or both? Electron density profiles, calculated from the diffraction data, are used to determine both the effects that the incorporation of PEG-lipid has on bilayer organization and to estimate the distance between adjacent bilayers at each applied pressure.[19] Figure 2 shows electron density profiles for 2:1 SOPC:cholesterol bilayers in the presence and absence of PEG-2000 at the same applied pressure (2.8×10^7 dyn/cm^2). Each profile shows two unit cells, containing adjacent bilayers and the fluid space between them. For each profile, the bilayer on the left is centered at 0 Å. For this bilayer, the low-density trough at 0 Å corresponds to the terminal methyl groups of the lipid hydrocarbon chains, the high-density peaks located at about ± 25 Å correspond to the high-density phospholipid headgroups, and the medium-density regions between the headgroup peaks and the terminal methyl trough correspond to the methylene groups of the hydrocarbon chains. "Stick figure" representations of the lipids are included in this figure to identify these regions of the bilayer. The medium-density region between headgroup peaks from adjacent bilayers corresponds to the fluid space between adjacent bilayers, where the PEG is presumably localized. Figure 2 shows two major points. First, the electron density distribution across the bilayer, and therefore the bilayer structure, is not

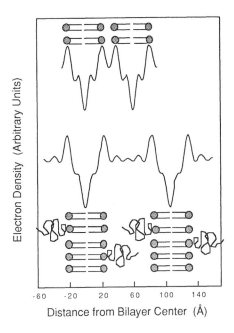

FIGURE 2.   Electron density profiles for 2:1 SOPC:cholesterol bilayers in the absence (top) and presence (bottom) of 5 mol% DSPE:PEG-2000, both at an applied pressure of 3.2×10^7 dyn/cm². For each profile, two unit cells are shown. Stick figure models indicate the positions of the lipid molecules in each profile. In these stick figures, the circles represent the phospholipid headgroups, the straight lines represent the hydrocarbon chains, and the curvy lines represent the covalently attached PEG. (From Needham, D., McIntosh, T. J., and Lasic, D. D., *Biochim. Biophys. Acta,* 1108, 40, 1992. With permission of Elsevier Science Publishers.)

appreciably changed by the incorporation of 5 mol% PEG-2000. Second, the distance between apposing bilayers is markedly increased by the incorporation of this concentration of PEG-2000.

Profiles such as those shown in Figure 2 can be used to estimate the distance between apposing bilayers at each applied pressure. As noted previously,[19,22,27] the definition of the lipid/water interface is somewhat arbitrary, because the bilayer is not smooth, the lipid headgroups are mobile, and water penetrates into the bilayer headgroup region. We operationally define the bilayer width as the total thickness of the SOPC:cholesterol bilayers. At the resolution of the profiles shown in Figure 2, the physical edge of the bilayers lies about 5 Å outward from the center of the high-density peaks in the electron density profiles.[19,22,27] Thus, we estimate the total thickness of the SOPC:cholesterol bilayer to be the distance between the headgroup peaks across the electron density profile plus 10 Å. With this definition of bilayer thickness, the X-ray diffraction data are used to determine the distance between adjacent bilayers for each applied pressure by subtracting this bilayer thickness from the observed lamellar repeat period.

Figure 3 shows a plot of the logarithm of applied pressure vs. the distance between bilayers for 2:1 SOPC:cholesterol bilayers in the presence and absence of 5 mol% PEG-2000 with the aqueous solution containing 100 m*M* NaCl (solid symbols). After subtracting the bilayer thickness from the lamellar repeat periods given in Figure 1, it now becomes clear that for all applied pressures the incorporation of PEG-lipid markedly increases the fluid space between apposing bilayers. For example, at the lowest applied pressure (4×10^4 dyn/cm²) the fluid separation is increased by about 100 Å and at the highest applied pressure (3×10^7 dyn/cm²) the bilayers are still separated by a distance of about 50 Å. This shows that both the interbilayer separation and the total repulsive pressure between bilayers are tremendously increased by the incorporation of PEG-2000.

Although we hypothesized that the physical basis of the strong repulsive pressure provided by the PEG-lipid arose from a steric pressure caused by the extension of the PEG from the

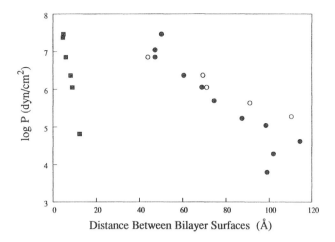

FIGURE 3. Plot of logarithm of applied pressure (log P) vs. distance between the surfaces of adjacent bilayers composed of 2:1 SOPC:cholesterol bilayers in the absence (solid squares) and presence (solid and open circles) of 5 mol% DSPE:PEG-2000. The solid and open circles represent data obtained with stressing solutions containing 100 mM and 1 mM NaCl, respectively. (From Needham, D., McIntosh, T. J., and Lasic, D. D., *Biochim. Biophys. Acta*, 1108, 40, 1992. With permission of Elsevier Science Publishers.)

bilayer surface, there are two other long-range repulsive pressures that are known to operate out to 50 Å from each bilayer surface, namely, electrostatic repulsion caused by charge on the bilayer surface,[29] and the undulation pressure caused by thermally driven bilayer undulations or fluctuations.[30,31] Thus, before further investigation of the nature of steric interactions between the PEG-lipid bilayers, experiments were performed to provide information on the contribution of these other potentially significant pressures in the system.

First, we checked whether electrostatic repulsion, arising from the negatively charged PEG-lipid incorporated at 5 mol%, was making a significant contribution to the total repulsive pressure by performing experiments with different ionic strengths. According to double layer theory, both the range and magnitude of electrostatic repulsion should depend on ionic strength.[29] Figure 3 shows that the pressure-distance data for 2:1 SOPC:cholesterol liposomes are nearly the same in stressing solutions containing either 1 mM or 100 mM NaCl. The invariance of the pressure-distance relations under these two different ionic strengths indicates that electrostatic repulsion is not a significant factor in determining the total repulsive pressure between bilayers containing PEG-2000 for the range of separations shown in Figure 3.

Next, experiments were performed to determine the relative contribution to the total pressure of thermally induced undulations. Such undulations or fluctuations in the bilayer are thought to give rise to a long-range repulsive interaction that depends strongly on the bending stiffness of the bilayer.[30,31] Figure 4 shows pressure-distance relations for 2:1 SOPC:cholesterol bilayers and DSPC bilayers containing 5 mol% PEG-2000. At the temperature at which these experiments were performed (20°C), our wide-angle diffraction patterns show that 2:1 SOPC:cholesterol/PEG-2000 bilayers are in the liquid crystalline phase, whereas DSPC/PEG-2000 bilayers are in the much less bendable gel phase. The similarity of the pressure distance data for 2:1 SOPC:cholesterol bilayers and DSPC bilayers containing 5 mol% PEG-2000 provides strong evidence that thermally driven undulations are a small factor in determining the total repulsive pressure for the systems shown in Figure 4.

Since the experiments presented in Figures 3 and 4 rule out significant contributions to the total repulsive pressure by electrostatics or bilayer undulations, we can now argue that this large, long-ranged repulsive force must arise from a steric barrier created by the presence of the covalently attached PEG at the bilayer surface.

Modern theories of steric repulsion caused by grafted polymers predict that the range and magnitude of the repulsion depend on both the grafting density of the polymer (the number

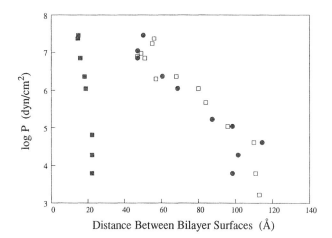

FIGURE 4.   Plot of logarithm of applied pressure (log P) vs. distance between the surfaces of adjacent bilayers composed of 2:1 SOPC:cholesterol bilayers in the absence (solid squares) and presence of 5 mol% PEG-2000 (solid circles) and DSPC in the presence of 5 mol% PEG-2000 (open squares). All stressing solutions contained 100 mM NaCl. SOPC:cholesterol data are taken from Needham, D., McIntosh, T. J., and Lasic, D. D., *Biochim. Biophys. Acta,* 1108, 40, 1992.

of polymer chains per unit surface area of the substrate) and the length of the polymer.[32-35] To test these theories, we[36,37] are now in the process of measuring pressure-distance relations between bilayers containing PEG-lipids as a function of both PEG grafting density and polymer length. Figure 5 shows some preliminary data from these experiments. In this figure, the distance between apposing bilayers at a constant applied pressure of 1×10^5 dyn/cm^2 is shown for DSPC:PEG-lipid suspensions for varying concentrations of three different PEG-lipids (PEG-750, PEG-2000, and PEG-5000). For DSPC suspensions containing any of the three PEG-lipids, the distance between bilayer surfaces increases with increasing PEG-lipid concentration up to about 10 mol%, and then levels off despite further increases in PEG-lipid concentration. However, for the same bilayer concentration of a given PEG-lipid, the maximum distance between apposing bilayer surfaces is a strong function of the size of the grafted PEG chain, being about 50 Å for PEG-750, about 100 Å for PEG-2000, and about 200 Å for PEG-5000.

If we assume that the PEG molecules interact at the mid-plane between apposing bilayers, then the extension length of a given PEG from the bilayer surface can estimated by dividing the interbilayer separation distance by 2 (Figure 5). The PEG extension lengths vary as a function of both molecular size and surface concentration, as compiled in Table 1. Thus, the approximate thickness of the polymer layer adjacent to the PEG-liposome is obtained from the X-ray measurements. This thickness is important for at least two reasons: (1) it gives the approximate distance that a large macromolecule, cell, or other surface can approach the PEG-liposome, and (2) it provides a basis for the design of targetable PEG-liposomes containing antibodies, antibody fragments, or other receptor moieties whose active groups must extend beyond the PEG-layer in order to be available for binding.

IV. RELEVANCE TO DRUG DELIVERY

The results of our osmotic stress/X-ray experiments can be correlated with the observed blood circulation times for liposomes containing PEG-lipids. As noted above, it has been argued that the long circulation times of PEG-liposomes are due to the steric barrier provided by the PEG chains.[7,9,17,18] Our experiments support this idea, and provide quantitative data that help to explain several of the observed phenomena in blood circulation trials with PEG-liposomes.

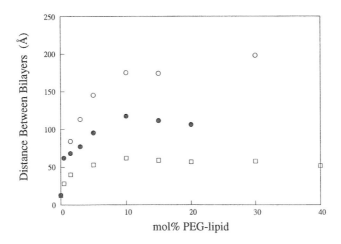

FIGURE 5. Plot of distance between apposing DSPC:PEG-lipid bilayers as a function of PEG-lipid concentration for PEG-750 (open squares), PEG-2000 (solid circles), and PEG-5000 (open circles). All data were recorded at an applied osmotic pressure of 1×10^5 dyn/cm^2.

TABLE 1
PEG Extension Length (Å) at an
Applied Pressure of 1×10^5 dyn/cm^2

PEG-Lipid	PEG-Lipid Concentration (mol%)		
	1.5	**5**	**10**
PEG-750	20	27	31
PEG-2000	34	48	59
PEG-5000	42	73	88

Three specific observations on the blood circulation time of liposomes can be rationalized in terms of our X-ray measurements. First, it has been observed that PEG-liposome blood circulation time depends on the length and concentration of the PEG-lipid, as the length of the circulation times measurably increases as PEG size is increased from 750 to 2000.[7,38] (The effects on blood circulation time of increasing the PEG size from 2000 to 5000 are not clear at this time, as one group finds an increased circulation time with PEG-5000 as compared to PEG-2000,[7] whereas another group finds a similar circulation time for liposomes containing comparable concentrations of PEG-2000 and PEG-5000.[38]) These results can be rationalized in terms of a steric barrier model, since we observe that the range of the steric barrier increases with increasing PEG size (Figure 5). Or, put another way, our experiments (Figures 3, 4, and 5) provide an estimate for the minimum range and magnitude of the steric barrier necessary to substantially increase liposome circulation times. Second, although there are few data available, it appears that the effect of PEG-lipid on blood circulation depends on the concentration of PEG-lipid in the liposome, reaching a maximum effect near 10 mol% PEG-lipid.[39] We find that the range of the steric barrier provided by the incorporation of PEG-lipid also reaches a maximum near 10 mol% (Figure 5). The addition of higher concentrations of PEG-lipid does not further increase the magnitude of the steric barrier, and, in the cases of PEG-2000 and PEG-5000, larger concentrations of PEG-lipid are even detrimental to the liposome drug delivery system, since they break up the liposomes and form micelles.[37] Third, two recent studies find that the blood circulation time of liposomes containing PEG-lipids is similar for bilayers in the gel and liquid crystalline phases.[39,40] This is in accord with our findings that the steric barrier provided by PEG-2000 is nearly identical in gel and liquid crystalline bilayers (Figure 4).

V. CONCLUSIONS

X-ray diffraction analysis of osmotically stressed multibilayers containing PEG-lipids provides direct information on the steric barrier produced by PEG-lipids in liposomes. The range and magnitude of this steric repulsion depend on both the concentration of PEG-lipid in the bilayer and the size of the PEG chain. The measured steric barrier is large and long-ranged for the concentrations of PEG-lipids that appreciably increase the blood circulation time of liposomes used in drug delivery.

ACKNOWLEDGMENTS

This work was supported by a grant to TJM from the National Institutes of Health, GM27278.

REFERENCES

1. **Allen, T. M., Hansen, C., Martin, F., Redemann, C., and Yau-Young, A.,** Liposomes containing synthetic lipid derivatives of poly(ethyleneglycol) show prolonged circulation half-lives in vivo, *Biochim. Biophys. Acta,* 1066, 29, 1991.
2. **Ahmad, I., Longenecker, M., Samuel, J., and Allen, T. M.,** Antibody-targeted delivery of doxorubicin entrapped in sterically stabilized liposomes can eradicate lung cancer in mice, *Cancer Res.,* 53, 1484, 1993.
3. **Blume, G., Cevc, G., Crommelin, M. D. J. A., Bakker-Woudenberg, I. A. J. M., Kluft, C., and Storm, G.,** Specific targeting with poly(ethylene glycol)-modified liposomes: coupling of homing devices to the ends of the polymeric chains combines effective target binding with long circulation times, *Biochim. Biophys. Acta,* 1149, 180, 1993.
4. **Mayhew, E. G., Lasic, D., Babbar, S., and Martin, F. J.,** Pharmacokinetics and antitumor activity of epirubicin encapsulated in long-circulating liposomes incorporating a polyethylene glycol-derivatized phospholipid, *Int. J. Cancer,* 51, 302, 1992.
5. **Blume, G. and Cevc, G.,** Liposomes for the sustained drug release in vivo, *Biochim. Biophys. Acta,* 1029, 91, 1990.
6. **Allen, T. M. and Hansen, C.,** Pharmokinetics of Stealth versus conventional liposomes: effect of dose, *Biochim. Biophys. Acta,* 1068, 133, 1991.
7. **Mori, A., Klibanov, A. L., Torchilin, V. P., and Huang, L.,** Influence of the steric barrier activity of amphipathic poly(ethylene glycol) and ganglioside GM1 on the circulation time of liposomes and on the target binding of immunoliposomes in vivo, *FEBS Lett.,* 284, 263, 1991.
8. **Klibanov, A. L., Maruyama, K., Torchilin, V. P., and Huang, L.,** Amphipathic polyethylene glycols effectively prolong the circulation time of liposomes, *FEBS Lett.,* 268, 235, 1990.
9. **Papahadjopoulos, D., Allen, T. M., Gabizon, A., Mayhew, E., Matthay, K., Huang, S. K., Lee, K.-D., Woodle, M. C., Lasic, D. D., Redemann, C., and Martin, F. J.,** Sterically stabilized liposomes: improvements in pharmacokinetics and antitumor therapeutic efficacy, *Proc. Natl. Acad. Sci. U.S.A.,* 88, 11460, 1991.
10. **Senior, J., Delgado, C., Fisher, D., Tilcock, C., and Gregoriadis, G.,** Influence of surface hydrophilicity of liposomes on their interaction with plasma protein and clearance from the circulation: studies with poly(ethylene glycol)-coated vesicles, *Biochim. Biophys. Acta,* 1062, 77, 1991.
11. **Williams, S. S., Alosco, T. R., Mayhew, E., Lasic, D. D., Martin, F. J., and Bankert, R. B.,** Arrest of human lung tumor xenograft growth in severe combined immunodeficient mice using doxorubicin encapsulated in sterically stabilized liposomes, *Cancer Res.,* 53, 3964, 1993.
12. **Vaage, J., Donovan, D., Mayhew, E., Uster, P., and Woodle, M.,** Therapy of mouse mammary carcinomas with vincristine and doxorubicin encapsulated in sterically stabilized liposomes, *Int. J. Cancer,* 54, 959, 1993.
13. **Wu, N. Z., Da, D., L., R. T., Needham, D., Whorton, A. R., and Dewhirst, M. W.,** Increased microvascular permeability contributes to preferential accumulation of Stealth liposomes in tumor tissue, *J. Cancer Res.,* in press, 1993.
14. **Allen, T. M., Agrawal, A. K., Ahmad, I., Hansen, C. B., and Zalipsky, S.,** Antibody-mediated targeting of long-circulating (Stealth) liposomes, *J. Liposome Res.,* in press, 1993.
15. **Huang, S. K., Lee, K.-D., Hong, K., Friend, D. S., and Papahadjopoulos, D.,** Microscopic localization of sterically stabilized liposomes in colon carcinoma-bearing mice, *Cancer Res.,* 52, 5135, 1992.
16. **Needham, D., Hristova, K., McIntosh, T. J., Dewhirst, M., Wu, N., and Lasic, D. D.,** Polymer-grafted liposomes: physical basis for the "Stealth" property, *J. Liposome Res.,* 2, 411, 1992.

17. **Lasic, D. D., Martin, F. J., Gabizon, A., Huang, S. K., and Papahadjopoulos, D.,** Sterically stabilized liposomes: a hypothesis on the molecular origin of the extended circulation times, *Biochim. Biophys. Acta,* 1070, 187, 1991.

18. **Woodle, M. C., Collins, L. R., Sponsler, E., Kossovsky, N., Papahadjopoulos, D., and Martin, F. J.,** Sterically stabilized liposomes. Reduction in electrophoretic mobility but not electrostatic surface potential, *Biophys. J.,* 61, 902, 1992.

19. **Needham, D., McIntosh, T. J., and Lasic, D. D.,** Repulsive interactions and mechanical stability of polymer-grafted lipid membranes, *Biochim. Biophys. Acta,* 1108, 40, 1992.

20. **LeNeveu, D. M., Rand, R. P., and Parsegian, V. A.,** Measurement of forces between lecithin bilayers, *Nature,* 259, 601, 1976.

21. **Parsegian, V. A., Rand, R. P., Fuller, N. L., and Rau, R. C.,** Osmotic stress for the direct measurement of intermolecular forces, *Methods Enzymol.,* 127, 400, 1986.

22. **McIntosh, T. J., Magid, A. D., and Simon, S. A.,** Steric repulsion between phosphatidylcholine bilayers, *Biochemistry,* 26, 7325, 1987.

23. **McIntosh, T. J., Simon, S. A., Needham, D., and Huang, C.-H.,** Interbilayer interactions between sphingomyelin and sphingomyelin:cholesterol bilayers, *Biochemistry,* 31, 2020, 1992.

24. **Vink, H.,** Precision measurements of osmotic pressure in concentrated polymer solutions, *Eur. Polymer J.,* 7, 1411, 1971.

25. **McIntosh, T. J., Magid, A. D., and Simon, S. A.,** Range of the solvation pressure between lipid membranes: dependence on the packing density of solvent molecules, *Biochemistry,* 28, 7904, 1989.

26. **Parsegian, V. A., Fuller, N., and Rand, R. P.,** Measured work of deformation and repulsion of lecithin bilayers, *Proc. Natl. Acad. Sci. U.S.A.,* 76, 2750, 1979.

27. **McIntosh, T. J., Magid, A. D., and Simon, S. A.,** Cholesterol modifies the short-range repulsive interactions between phosphatidylcholine membranes, *Biochemistry,* 28, 17, 1989.

28. **McIntosh, T. J. and Simon, S. A.,** The hydration force and bilayer deformation: a reevaluation, *Biochemistry,* 25, 4058, 1986.

29. **Israelachvili, J. N.,** *Intermolecular and Surface Forces,* Academic Press, London, 1991, chap. 12.

30. **Evans, E. A. and Parsegian, V. A.,** Thermal-mechanical fluctuations enhance repulsion between bimolecular layers, *Proc. Natl. Acad. Sci. U.S.A.,* 83, 7132, 1986.

31. **Harbich, W. and Helfrich, W.,** The swelling of egg lecithin in water, *Chem. Phys. Lipids,* 36, 39, 1984.

32. **DeGennes, P. G.,** Polymers at an interface: a simplified view, *Adv. Colloid Interface Sci.,* 27, 189, 1987.

33. **Milner, S. T.,** Polymer brushes, *Science,* 251, 905, 1991.

34. **Milner, S. T., Witten, T. A., and Cates, M. E.,** Theory of the grafted polymer brush, *Macromolecules,* 21, 2610, 1988.

35. **Milner, S. T., Witten, T. A., and Cates, M. E.,** A parabolic density profile for grafted polymers, *Europhys. Lett.,* 5, 413, 1988.

36. **Kenworthy, A. K., McIntosh, T. J., Hristova, K., and Needham, D.,** Steric interactions between gel and liquid crystalline phosphatidylcholine bilayers containing lipids with covalently attached polyethyleneglycol, *Biophys. J.,* 66(Abstr.), A287, 1994.

37. **Kenworthy, A. K., Simon, S. A., and McIntosh, T. J.,** Effects of lipids with covalently attached polyethyleneglycol on distearoylphosphatidylcholine bilayer structure, *Biophys. J.,* 66(Abstr.), A287, 1994.

38. **Woodle, M. C., Matthay, K. K., Newman, M. S., Hidayat, J. E., Collins, L. R., Redemann, C., Martin, F. J., and Papahadjopoulos, D.,** Versatility in lipid compositions showing prolonged circulation with sterically stabilized liposomes, *Biochim. Biophys. Acta,* 1105, 193, 1992.

39. **Woodle, M. C. and Lasic, D. D.,** Sterically stabilized liposomes, *Biochim. Biophys. Acta,* 1113, 171, 1992.

40. **Blume, G. and Cevc, G.,** Molecular mechanism of the lipid vesicle longevity in vivo, *Biochim. Biophys. Acta,* 1146, 157, 1993.

Chapter 8

MODULATION AND MODELING OF INTERACTION FORCES BETWEEN LIPID BILAYERS EXPOSING TERMINALLY GRAFTED POLYMER CHAINS

T. L. Kuhl, D. E. Leckband, D. D. Lasic, and J. N. Israelachvili

TABLE OF CONTENTS

I. INTRODUCTION

The presence of terminally grafted polymers at the solid-solution interface of colloidal particles, liposomes, biological cells, and other biosurfaces is currently receiving considerable attention, especially regarding sterically stabilized liposomes used for drug delivery. Previously, the use of liposomes as drug delivery vesicles was limited by their rapid clearance from circulation by the mononuclear phagocytotic system. Recent studies have found that circulation times can be greatly enhanced, from hours to days, by incorporating into the liposomes a small amount of modified lipids whose headgroups are derivatized with polyethylene oxide (PEO) creating what are now referred to as "Stealth"® liposomes.[1-5] The enhanced circulation time is believed to be due to steric stabilization of the liposomes by the grafted polymer, preventing their close approach to cellular surfaces in the body, which is necessary to prevent recognition interactions, adhesion binding, and subsequent destruction of the liposomes by the immune system. The special properties of PEO: chemical inertness, biocompatibility, water solubility (hydrophilicity), insensitivity to changes in solution ionic conditions due to its nonionic character, and low protein adsorption have made it a particularly suitable polymer for use as a steric stabilizer in aqueous media, including body fluids.[6-9]

Although the exact nature of the stealth property of such liposomes *in vivo* is still not fully understood due to the complexities of living organisms, model systems have been studied.[10] Using X-ray diffraction, Needham and colleagues (1992) recently reported that at moderate surface coverages the polymer moiety acts as a steric barrier, increasing the repulsion between oriented multilayers and unoriented multiwalled liposomes.[11,12] We have extended this work and present here the results of detailed measurements of the interaction forces, F, as a function of distance, D, between supported lipid bilayers of double-chained zwitterionic lipids in aqueous solution as modified by the attachment of PEO chains to the headgroups.[13] In addition, the lateral interactions (pressure-area or Π-A curves) of headgroup-attached PEO chains in monolayers at the air-water interface were also measured. The results of the force-distance (F-D) measurements and corresponding Π-A curves were compared with each other and with various theories which attempt to model the steric interactions of grafted polymer chains.

Unlike most of the surface grafted polymer systems that have been investigated to date, we used ethylene oxide (EO) chains covalently bound to the headgroup of distearoyl phosphatidyl ethanolamine (DSPE), thereby forming the surfactant-polymer $DSPE-EO_{45}$. As polymers, these are very short chains; however, the modified $DSPE-EO_{45}$ molecules may also be considered as lipids or surfactants having an extra large, hydrophilic headgroup. One aim of looking at low-molecular-weight polymers or short chains is to test the limits of polymer scaling theories. Another is to establish a link between the interactions of simple surfactant or lipid molecules and those having many flexible segments in their headgroups, e.g., gangliosides, glycolipids, sugars, polysaccharides, etc. On the practical side, the effective steric stabilization of many colloidal particles, especially vesicles and liposomes, is more likely to be accomplished with short chains than with highly hysteretic and less economical long-chain polymers.

II. EXPERIMENTAL METHODS

A. MONOLAYERS AT THE AIR-WATER INTERFACE

PEO chains were covalently bound to the terminal amine group of DSPE, thereby forming the modified lipid $DSPE-EO_{45}$. The PEO chains had a mean molecular weight of 2000 (45 monomers), $R_F = 35$ Å, and a polydispersity of 1.1.[7,10]

Monolayers were composed of a mixture of DSPE and 1 to 10% $DSPE-EO_{45}$. Because the polymer is covalently bound to the DSPE headgroup, the surface coverage of polymer chains can be quantitatively controlled in two ways: (1) varying the mole fraction of the components ($DSPE-EO_{45}$ to DSPE) and (2) varying the surface area per molecule at the air-water interface. Surface pressure measurements were made with a standard Whilhelmy balance.[14]

B. BILAYERS

The model "Stealth" liposome studied was composed of two interacting supported bilayers prepared by Langmuir-Blodgett (LB) deposition. The support consists of thin, molecularly smooth, back silvered mica sheets which are glued onto cylindrical fused silica discs. The

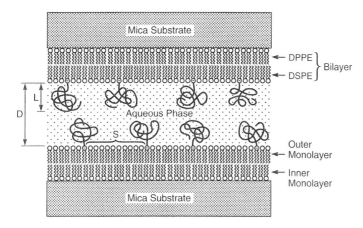

FIGURE 1.   Schematic representation of bilayer-coated mica surfaces at the nonoverlapping surface density of 1.3% DSPE-EO$_{45}$ including definition of distances, where contact is defined by D = 0, L is the thickness of the polymer layer, and s is the average distance between polymer chains.

inner layer of the bilayer was DPPE which binds electrostatically to the mica surface. The outer monolayer of the bilayer was composed of a mixture of unmodified DSPE (the matrix lipid) doped with 1.3, 4.5, or 9.0 mol% DSPE-EO$_{45}$. The DSPE molecular area was 43 Å^2 in all cases. At the lower concentration of 1.3% DSPE-EO$_{45}$, the polymer chains are not perturbed by their neighbors and are in the "mushroom" regime. The 4.5% concentration gives rise to a weakly interacting or "weak overlap" regime, whereas at 9.0% the polymer chains are strongly interacting and in the "brush" regime. Hence, with these three concentrations we could follow how the interaction varied as a function of polymer surface coverage (grafting density) from the nonoverlapping (mushroom) to the strongly overlapping (brush) regimes. Figure 1 shows a schematic representation of the model system containing 1.3% DSPE-EO$_{45}$.

C. SURFACE FORCE MEASUREMENTS

The surface forces apparatus (SFA) technique has been previously described (Figure 2).[15,16] Briefly, both the attractive and repulsive interaction forces are measured between two bilayer-coated mica sheets, which are back silvered and glued onto cylindrical silica discs. The separation between the treated mica surfaces was measured with an optical technique based on multiple beam interferometry producing fringes of equal chromatic order (FECO). From the position of the FECO fringes in a spectrometer, the distance between the two surfaces could be measured with a resolution of 1 Å. The distance between the surfaces is controlled by a series of coarse and fine micrometers. Additional control of the separation was achieved with a piezoelectric crystal attached to the upper disc. The force between the bilayer surfaces is measured from the deflection of a variable cantilever spring supporting the lower surface with a resolution of 100 nN.

This technique enabled us to measure the force-distance (F-D) profiles between bilayer-coated mica sheets and to investigate the thickness, compressibility, and relaxation processes occurring in this polymer-modified bilayer system as a function of the terminally grafted polymer surface coverage.

III. THEORETICAL PREDICTIONS FOR TERMINALLY GRAFTED POLYMER CHAINS

A. ALEXANDER-DE GENNES (A-dG) POLYMER BRUSH THEORY (1981)

Alexander was the first to propose a theoretical scaling description for the structure of terminally grafted polymer chains.[17] de Gennes later extended this scaling analysis to the interaction of two opposing surfaces in solution, each with a symmetrical layer of grafted

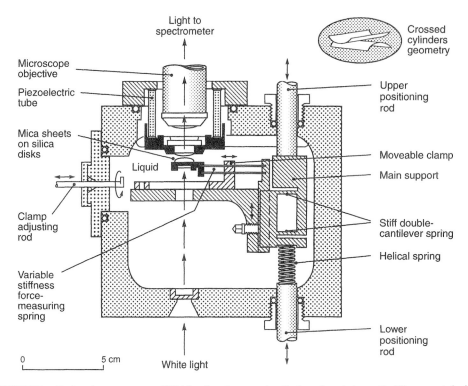

FIGURE 2. Surface forces apparatus (SFA) for directly measuring the force laws between the bilayer-coated mica surfaces in solution at the angstrom resolution level. Forces are measured from the deflection of the force-measuring spring.

polymer.[18,19] The controlling parameter in both theories is the surface density of grafted polymer chains, σ. If $\sigma < \pi R_F^2$, where $R_F = a\, N^{3/5}$ is the Flory radius of the coil in a good solvent, the neighboring grafted chains will begin to significantly overlap. In a good solvent environment, this will lead to an osmotic pressure within the polymer layer or brush. In order to reduce the resulting osmotic pressure, the polymer chains will extend away from the surface into the solvent phase. This swelling involves the stretching of the chains. Alexander and de Gennes balanced the energetics of the osmotic pressure and elastic resistance to stretching by using the simplest representations of each effect. The osmotic pressure was calculated assuming a constant density of segments in the polymer layer, which yields the following step function:

$$\Pi(\sigma, L) \cong \frac{kT}{a^3}\left(\frac{Na^3}{L\sigma}\right)^{9/4} \tag{1}$$

where a is the length of a segment, N is the number of segments per chain, and L is the brush thickness. The elastic energy per chain was given by the energetics of a spring distorted from its mean extension:

$$E(\sigma, L) \cong \frac{kTL}{Na^2}\left(\frac{Na^3}{L\sigma}\right)^{1/4} \tag{2}$$

Hence, the free energy per chain in a high density brush can be written as

$$E(\sigma, L) \cong kTN\left(\frac{Na^3}{L\sigma}\right)^{5/4} + \frac{kTL^2}{Na^2}\left(\frac{Na^3}{L\sigma}\right)^{1/4} \tag{3}$$

The brush thickness or extension from the surface, L, is given by minimizing the free energy:

$$L = \frac{Na^{5/3}}{\sigma^{1/3}} = \frac{Na^{5/3}}{s^{2/3}} \tag{4}$$

where $s = \sqrt{\sigma/\pi}$ is the average distance between grafting points on the surface.

When two flat surfaces bearing grafted chains approach to less than twice the brush thickness (D < 2L), the chains become compressed. Assuming constant s, with increasing polymer concentration the osmotic pressure inside each brush increases more than the elastic restoring forces decrease, thereby giving rise to a net repulsive pressure between the two surfaces. Assuming that the brushes do not interdigitate, the above equations can be used to compute this repulsive pressure yielding:[18,19]

$$P(D) \cong \frac{kT}{s^3}\left[\left(\frac{2L}{D}\right)^{9/4} - \left(\frac{D}{2L}\right)^{3/4}\right] \quad \text{for } D < 2L \tag{5}$$

Integration of Equation 5 over the separation distance D yields the interaction energy, E(D), between the two plates. The force as a function of separation distance between two curved cylindrical surfaces of radius R, as used in the SFA, can be obtained using the Derjaguin approximation[20]

$$\frac{F(D)}{R} = 2\pi E(D) = 2\pi \int P(D)dD = \frac{16\pi kTL}{35s^3}\left[7\left(\frac{2L}{D}\right)^{5/4} + 5\left(\frac{D}{2L}\right)^{7/4} - 12\right] \tag{6}$$

We note that all the above scaling equations are missing a numerical prefactor.

B. MILNER, WITTEN, AND CATES (MWC) MEAN FIELD CALCULATION FOR POLYMER BRUSHES (1988)

Using a mean field analysis, Milner, Witten, and Cates (1988) also studied the regime of high-density polymer brushes.[21] They calculated the density profile and free energy of a strongly stretched grafted polymer brush assuming binary interactions between segments. They found that the density profile of the brush layer is parabolic, not the step function assumed by A-dG. Moreover, the free ends of the chains were distributed throughout the entire brush layer with a maximum in the probability of finding an end at about 0.7L, rather than confined to the outer extremity of the brushes as in the A-dG approach. Even with these differences, the MWC results do not differ much from the simpler analysis of A-dG, thus providing similar scaling law exponents. The mean field energy of a strongly stretched brush was found to be:[22]

$$E(L) = \frac{1}{2}\left(\frac{\pi^2}{12}\right)^{1/3}\frac{NkT\,\omega^{2/3}\upsilon^{1/3}}{\sigma^{5/3}}\left[\frac{L}{D} + \left(\frac{D}{L}\right)^2 - \frac{1}{5}\left(\frac{D}{L}\right)^5\right] \tag{7}$$

where ω is the excluded volume parameter and υ is a concentration-dependent material parameter with dimensions of inverse length squared (Å^{-2}). The corresponding repulsive force between two curved surfaces bearing terminally grafted chains as a function of their separation is then:

$$\frac{F(D)}{R} = 4\pi\left[E(L) - E(L_o)\right] \tag{8}$$

where the equilibrium noncompressed brush thickness, L_o, is given by:

$$L_o(\sigma) = \left(\frac{12}{\pi^2}\right)^{1/3} \frac{N\omega^{1/3}\upsilon^{-1/3}}{\sigma^{1/3}} = \left(\frac{12}{\pi^2}\right)^{1/3} \frac{N\omega^{1/3}\upsilon^{-1/3}}{s^{2/3}} \tag{9}$$

C. DE GENNES THEORY FOR DILUTE MUSHROOMS (1985)

de Gennes extended the above scaling analysis (Section III.A) to the low-density case where the polymer coils on a surface (known as "mushrooms" for their assumed shape) do not interact or overlap with their neighbors, i.e., $\sigma > \pi R_F^2$.[23] The interaction between the chains on opposing surfaces begins once the separation is less than the extension of a chain, $D \leq L_m \approx R_F$, where L_m is the height of a mushroom. As the separation decreases below L_m, the chains will become compressed with a resulting repulsion due to their confinement. Under conditions of high compression, scaling arguments predict a confinement energy of:

$$E(D) \cong kT\left(\frac{L_m}{D}\right)^{5/3} = kTN\left(\frac{a}{D}\right)^{5/3} \quad \text{per chain} \quad \text{for } D < L_m \tag{10}$$

and the corresponding force as a function of separation distance between two curved cylindrical surfaces of radius R is

$$\frac{F(D)}{R} \cong \frac{2\pi kT}{\sigma}\left[\left(\frac{L_m}{D}\right)^{5/3} - 1\right] \tag{11}$$

Again, this scaling theory is missing a numerical prefactor of order unity.

D. DOLAN AND EDWARDS (D&E) THEORY FOR DILUTE
MUSHROOMS (1974)

An early self-consistent mean field calculation for the density profile and free energy of grafted polymer chains was carried out by Dolan and Edwards (1974).[24] Their approach is based on the configurational free energy of a random flight polymer chain and, as in the de Gennes theory, applies in the limit of moderate surface coverage where the grafted chains do not overlap or interact with their neighbors, i.e., $\sigma > \pi R_g^2$. Essentially, the theory assumes that each chain or mushroom interacts with an opposing surface independently of the other chains. Although the repulsive energy per unit area was found to be a complex series, analytical expressions were determined in the two limits of large and small separations.

In the limit of large separations, $D \geq a\sqrt{3N} = \sqrt{3aL_e}$, where $L_e = aN$ is the length of the fully extended polymer chain, the repulsive interaction energy as a function of the separation distance between two flat plates bearing nonoverlapping mushrooms is given by:

$$E(D) \cong \frac{2kTN}{\sigma}\exp\left(\frac{-3D^2}{2aL_e}\right) \quad \text{per unit area} \tag{12}$$

whereas, in the limit of small separations, $D \leq \sqrt{3aL_e}$,

$$E(D) = \frac{kTN}{\sigma}\left(\frac{\pi^2 aL_e}{6D^2} + \log\left(\frac{3D^2}{8\pi aL_e}\right)\right) \quad \text{per unit area} \tag{13}$$

Over the distance regime from $0.5R_g < D < 6R_g$, the full numerical solution for the repulsive force between two crossed cylinders can be approximated by an exponential expression:[25]

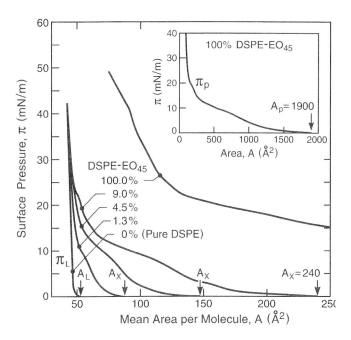

FIGURE 3. Monolayer compression (Π-A) isotherms of mixed DSPE-DSPE-EO$_{45}$ at $21 \pm 1°$C, plotted on a linear scale. The inset is the isotherm for pure 100 mol% DSPE-EO$_{45}$. The area, A, is the mean area per molecule at the air-water interface, where $A = XA_L + (1 - X)Ap$ (Equation 15). X is the mole fraction of DSPE and A_L and A_p are the areas per DSPE and DSPE-EO$_{45}$ molecule, respectively.

$$\frac{F(D)}{R} = \frac{72\pi kT}{\sigma} \exp\left(\frac{-D}{R_g}\right) \tag{14}$$

In the discussion of experimental results below, we demonstrate that the D&E equations can be used under good solvent conditions by substituting $R_g \approx R_F$.[13]

IV. RESULTS

A. LATERAL INTERACTIONS OF LIPID MONOLAYERS AT THE AIR-WATER INTERFACE

Figure 3 shows the surface pressure-area (Π-A) curves of unmodified DSPE doped with various concentrations of DSPE-EO$_{45}$ (0, 1.3, 4.5, 9.0, and 100 mol%) at $21 \pm 1°$C. The inset is an isotherm of pure (100%) DSPE-EO$_{45}$. As the concentration of DSPE-EO$_{45}$ increases, the area per molecule, A, at which a lateral pressure is measured steadily increases. This additional surface pressure is due to the lateral steric interaction between the PEO chains and lipids at the air-water interface.

The covalently bound polymer chains have four possible interactions with lipids and the air-water interface: (1) the polymer does not adsorb and remains totally in the water subphase, (2) the polymer adsorbs at the air-water interface mixing with the lipid, (3) the polymer adsorbs but segregates from the lipid at the air-water interface,[26] or (4) the polymer adsorbs to both. From previous studies it is known that PEO can be spread and compressed at the air-water interface.[27] Hence, at large areas per molecule the EO$_{45}$ chains, although covalently bound to the DSPE, should be in the "expanded state" on the air-water interface. From the pure 100% DSPE-EO$_{45}$ isotherm at large A, it is apparent that once the packing area per molecule is ~2000 Å^2 (Figure 3, inset), the polymer chains begin to interact laterally with each other.

This critical area per EO_{45} chain agrees well with a previous measurement of 2200 $Å^2$ for pure PEO chains (EO_{400} scaled to EO_{45}) at the air water interface.[27]

In order to assess whether the polymer mixes with or segregates from the lipid, a simple analysis of the Π-A data of the mixed systems can be made. For example, if the polymer mixes ideally with the lipid, as in case (2) above, we can assume ideal mixing equations of the form:

$$A_x = X A_L + (1 - X) A_p \tag{15}$$

$$\Pi_x = X \Pi_L + (1 - X) \Pi_p \tag{16}$$

where A_x is the mean area at the air-water interface, A_L is the area per DSPE lipid molecule, A_p is the area per EO_{45} chain, and X is the mole fraction of DSPE. The area per DSPE and DSPE-EO_{45} molecules calculated at the onset of a measurable surface pressure should correspond to the separate contributions of each of the components in the mixture. Thus, in the case of the 9.0% mixture, a surface pressure (of ~ 0.1 mN/m) is first measured at a packing area of $A_x \approx 240$ $Å^2$ per molecule (Figure 3). Assuming DSPE has a packing area of $A_L \approx 53$ $Å^2$ at this pressure (see Figures 3 and 4) and applying Equation 15, the DSPE-EO_{45} molecules require a packing area of $A_p \approx 1930$ $Å^2$ per molecule, in agreement with the value obtained in the pure DSPE-EO_{45} monolayer (Figure 3, inset). In fact, for all of the mixtures (1.3, 4.5, and 9.0%) the dilute low-pressure regime gives a packing area of 2000 ± 100 $Å^2$ per DSPE-EO_{45} molecule, which correlates well with the area per molecule found in the 100% DSPE-EO_{45} and the pure polymer system. Thus, at low pressures or low surface densities, the polymer is constrained at the interface and mixes with the lipid, as shown schematically in Figure 5A. Moreover, Equations 15 and 16 apply to the entire range of compositions and surface pressures, indicative of mixed behavior throughout, as shown by the dashed curves in Figure 4.

As the area decreases, the polymer chains are increasingly compressed on the surface and a corresponding increase in the surface pressure is measured. When the isotherms are plotted on a log plot (Figure 4), an inflection point at $\Pi_{pc} = 10 \pm 2$ mN/m becomes apparent. This inflection point probably corresponds to the critical pressure at which the compressed polymer chains cease to remain at the interface, and significantly enter the water subphase. In fact, when pure PEO polymer chains are compressed at the air-water interface, the surface pressure plateaus at a critical value of 10 mN/m.[27] This transition from adsorption to depletion behavior at the air-water interface for polymer/lipid systems was previously hypothesized by de Gennes (1990).[26]

Thus, above this critical pressure, it becomes energetically more favorable for the polymer chains to extend into solution rather than to remain at the air-water interface. We find that in all cases (1.3, 4.5, and 9.0% DSPE-EO_{45}), the critical pressure corresponds to a packing area per DSPE-EO_{45} chain of $\sigma = 500 \pm 20$ $Å^2$, or 11.1 $Å$ per segment, approximately the van der Waals radius of a segment. Hence, once the surface pressure of the mixture at the air-water interface surpasses this critical value, the polymer layer collapses and the chains enter the subphase. Thus, a simple excluded volume analysis of the experimental data suggests a transition from adsorption to depletion of the polymer chains at the critical pressure, $\Pi_{pc} = 10$ mN/m, above which the polymer monolayer collapses.

B. FORCE MEASUREMENTS OF LIPID BILAYER INTERACTIONS

In the absence of a grafted polymer layer, neutral DSPE bilayers adhere by entering an attractive van der Waals energy minimum at D ≈ 6 to 8 $Å$, below which repulsive hydration and protrusion forces oppose closer approach.[28] When DSPE bilayers contain bound PEO, the van der Waals attraction is overcome at all separations by the steric and electrostatic repulsion

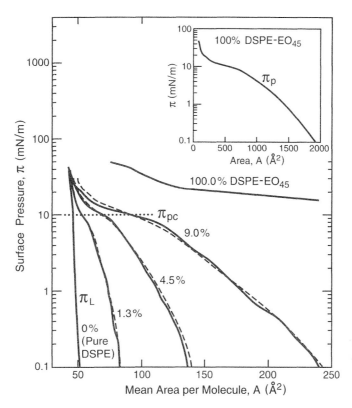

FIGURE 4. Monolayer compression (Π-A) isotherms of mixed DSPE-DSPE-EO$_{45}$ at 21 ± 1°C, plotted on a log scale. The inset is the isotherm for pure 100% DSPE-EO$_{45}$. The dashed curves are the predicted isotherms for the mixtures using Equations 15 and 16 and the pressure-area data for the pure components.

of the polymer chains (Figure 6). The electrostatic repulsion arises because DSPE-EO$_{45}$ is negatively charged at the phosphate group since the EO chain is covalently bound to the terminal amine group. This leads to an additional long-range electrostatic double-layer repulsion between the surfaces. An important consideration is whether the electrostatic repulsion or the steric repulsion is responsible for the enhanced circulation times of modified liposomes *in vivo*.

An analysis of the force distance data can be carried out by fitting the measured force curves to the theoretical double layer repulsion, and attributing the difference to the steric repulsion between the polymer layers. Theoretically and experimentally, both the electrostatic and steric forces decay roughly exponentially but with different decay lengths.[25] In the case of electrostatic forces, the decay is determined by the ionic strength (the Debye length), whereas for steric interactions, the decay is determined by the thickness of the polymer layer. Unless the two decay lengths happen to be similar, two different exponential dependencies are measured from which the electrostatic and steric contributions can be distinguished and accurately determined.

Figure 6 shows a log plot of the force-distance curves of the three polymer surface coverages studied at 21°C in dilute KNO$_3$ solutions. The longer-ranged electrostatic repulsion extends beyond D > 100 Å, while the steric repulsion is seen as a much steeper and stronger short-range repulsion below D ≈ 100 Å. In order to facilitate the analysis of the steric repulsion, the electrostatic portion of the force profiles must first be removed. The electrostatic component of the force profile was determined by fitting the long-range (large-distance) parts of the curves to the theoretical double layer repulsion with constant surface charge boundary

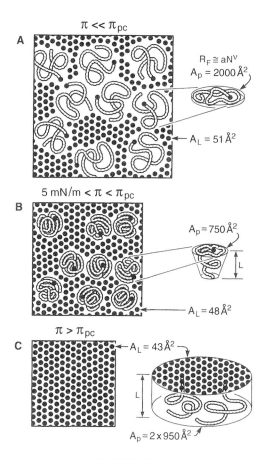

FIGURE 5. Schematic representation of the 4.5% DSPE-EO$_{45}$ mixture at the air-water interface as a function of the surface pressure, Π. As Π approaches the critical pressure, Π_{pc}, the polymer coils undergo a transition from adsorption (2-D polymer chains confined at the air-water interface) to depletion (3-D structure with the chains extending into the water subphase), where $A_p = \sigma$ is the area per DSPE-EO$_{45}$ molecule, A_L is the area per DSPE molecule, and L is the distance the polymer chains extend from the lipid surface.

conditions and assuming that the dielectric constant of the medium was that of pure water, i.e., $\varepsilon = 78.4$.[25] Based on these assumptions, the electrostatic repulsion was calculated by numerically solving the nonlinear Poisson-Boltzmann equation;[29] the fitted values are plotted as continuous lines in Figure 6 and the parameters are tabulated in Table 1.

An unusual feature of the results is the trend of decreasing surface charge density, σ_e, with increasing DSPE-EO$_{45}$ content in the bilayer. Since the DSPE-EO$_{45}$ headgroups are negatively charged, this trend appears to be counterintuitive. However, throughout the previous analysis we have assumed that the dielectric medium was that of pure water, neglecting the effect of the ethylene oxide chains at the bilayer surface. Pure polyethylene oxide has a dielectric constant, $\varepsilon = 11$, much lower than that of water.[30] Hence, as the concentration of DSPE-EO$_{45}$ increases, the dielectric constant of the medium surrounding the charged headgroups decreases. The overall effect of this local reduction in ε would be (1) a lower degree of dissociation, and (2) a more rapid decay of the double layer potential away from the surface. Both of these effects would account for the observed trend of decreasing surface charge with increasing DSPE-EO$_{45}$ content. A related phenomenon is the well-known reduction in the degree of headgroup dissociation of charged surfactants from approximately 100% when dissolved as monomers in solution, to approximately 25% when associated in micelles,[31,32] i.e., when surrounded by the low dielectric constant environment of the hydrocarbon chains.

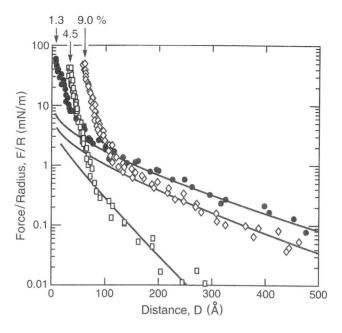

FIGURE 6. Measured force vs. distance on a log plot as a function of PEO surface coverage at 21°C. The longer-ranged electrostatic repulsion extends a few hundred angstroms, while the steric repulsion is seen as a much steeper, and stronger, short-range repulsion. (●) Forces between bilayers containing 1.3% DSPE-EO$_{45}$ in 0.5 mM KNO$_3$, where the electrostatic curve is fit using σ_e = 8 mC/m^2. (□) Forces between bilayers containing 4.5% DSPE-EO$_{45}$ in 4.2 mM KNO$_3$ with σ_e = 6.5 mC/m^2. (◇) Forces between bilayers containing 9.0% DSPE-EO$_{45}$ in 0.7 mM KNO$_3$ with σ_e = 5 mC/m^2. D = 0 is defined as the contact between nominally dehydrated DSPE bilayers without any polymer layer. For two cylindrically curved surfaces of radius R at a distance D apart, the force F(D) between them is related to the interaction energy per unit area E(D) between two planar surfaces by the Derjaguin approximation E(D) = F(D)/2πR.

TABLE 1
Electrostatic Double-Layer Parameters

Surface Coverage of DSPE-EO$_{45}$ (mol%)	KNO$_3$ Electrolyte Concentration (mM)	Measured Debye Length κ^{-1} (Å)	Surface Potential Ψ_0 (mV)	Surface Charge Density σ_e (mC/m^2)
1.3	0.51	130	105	8.0
4.5	4.2	45	50	6.5
9.0	0.69	115	70	5.0

V. ANALYSIS OF RESULTS

A. GENERAL CONSIDERATIONS

Figure 7 is a compilation of the steric forces alone after subtracting the electrostatic repulsion from the overall force curves of Figure 6. The steric force profiles are seen to be roughly exponential functions of separation and to depend on the polymer surface coverage — generally increasing with coverage. In the 1.3% "mushroom" regime the steric force extends ~35 Å per surface, which is the same as the theoretical Flory radius, $R_F = a N^{3/5} = 35$ Å. In the 9.0% "brush" regime the polymer chains extend significantly further from the bilayer surface, by about 65 to 70 Å per surface, which is twice R_F. It is interesting to compare the 1.3% DSPE-EO$_{45}$, noninteracting mushroom regime with the 4.5%, weakly overlapping

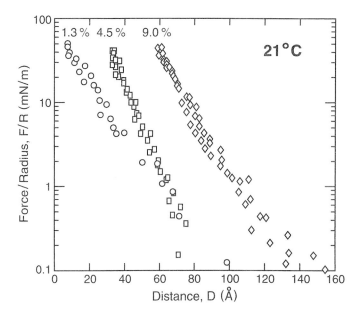

FIGURE 7. Steric force-distance profiles as a function of DSPE-EO$_{45}$ coverage. The steric forces were determined by subtracting the electrostatic double-layer repulsion assuming constant surface charge from the measured force profiles (Figure 6).

regime. In both cases, the polymer extends about 35 to 40 Å or one R$_F$ per surface, exactly the extension expected theoretically for a random coil in a good solvent.[23] However, closer in, the repulsion at 1.3% coverage is much less than at 4.5%. This is a result of the much larger free space available for accommodating the opposing chains on both surfaces at the nonoverlapping 1.3% coverage.

We would also note that for each of the DSPE-EO$_{45}$ grafting densities investigated here, the force curves were reversible and did not show hysteresis (curves not shown). In other surface force studies of adsorbed or physically grafted PEO chains on mica, the force profiles demonstrated significant hysteresis, where the measured repulsion was greater on approach than separation.[33-35]

The experimental values of the packing densities and polymer layer thickness deduced for each of the PEO surface concentrations shown in Figure 7 are listed in Table 2. From these values the projected area of a PEO chain on the bilayer surface appears to be about 900 Å^2 ($\sim$R$_F^2$) not 3600 Å^2 as would be deduced from A = πR$_F^2$. Based on this projected area of 900 Å^2 and an extension of 35 Å, the volume of a non- or weakly interacting coil is 31,500 Å^3 (area × extension). When the packing area in the 9.0% "brush" regime is used, the chains would need to extend 65 Å to attain the same coil volume and hence density. This is in very good agreement with the measured extension of 65 to 70 Å, indicative of internal consistency in the results.

B. COMPARISON OF MEASURED FORCE-DISTANCE PROFILES WITH THEORY

As described in Section III, a number of theories have attempted to model the interaction forces between surfaces bearing terminally bound sterically stabilizing polymers. One of the difficulties in applying these theories is that the controlling parameter σ, the surface grafting density of polymer chains, is usually unknown. This major drawback is successfully overcome in this study, where the surface coverage was accurately controlled and determined during the LB deposition. In this section we investigate the applicability of these theoretical treatments to model the interactions of short polymer chains grafted to a bilayer surface.

TABLE 2
Polymer Layer Specifications

Surface Coverage of DSPE-EO$_{45}$ (mol%)	PEO Chain[a] Extension L (Å)	Packing Area per Molecule (Å^2)	Packing Area per PEO Chain σ (Å^2)	Distance between Grafting Sites s (Å)
1.3	35	43	3,300	33
4.5	35–40	43	960	18
9.0	65–75	43	480	12

[a] Determined from range of forces in Figure 7. Theoretically, $R_F = 35$ Å, $R_g = 23$ Å.

1. Low Coverage (1.3%)

At a concentration of 1.3% DSPE-EO$_{45}$, the polymer surface coverage resembles that of dilute noninteracting mushrooms. The suitable theories for modeling this regime are the scaling analysis of de Gennes (dG, Equation 11) and the self-consistent mean field calculation of Dolan and Edwards (D&E, Equations 12 to 14).[17-19,24] Figure 8 displays the experimental steric force (see Figure 7) and the resulting curves from these theoretical equations. As seen, both the dG scaling theory and the approximation for D&E[14] more complex mean field theory are remarkably accurate at modeling the experimental force-profile for these short terminally grafted chains.

The fit shown for Equation 11 requires a prefactor of 1.6, since scaling arguments do not provide one. This is very close to the expected order unity.[23] Moreover, the theoretically determined polymer thickness agrees well with the value found experimentally, 36 ± 5 vs. 35 Å. On the other hand, although D&E analysis was based on theta solvent conditions, it is apparent that their equations can be applied to this good solvent system by substituting the Flory radius R_F for R_g. At short distances where the osmotic repulsion dominates, Equation 13 accurately represents the measured force profile and the fitted value (165 Å) for the fully extended chain length, L_e, agrees well with the actual value of 155 Å. In order to obtain a

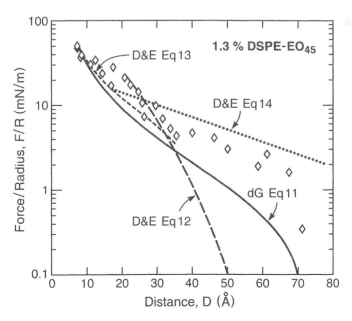

FIGURE 8.   Data of the "dilute mushroom" regime of 1.3% DSPE-EO$_{45}$ in Figure 7 fit to the theoretical equations of Section III. The equation parameters are listed in Table 3.

TABLE 3
Parameters for Theoretical Equations

Theoretical Equation	Model-Determined Thickness (Å)	Prefactor (if Used)	Fitted Theoretical Thickness (Å)	Correlation Coefficient
1.3% L_{exp} = 35 Å				
dG Equation 11	36	1.6 ± 0.5	36 ± 5	0.91
D&E (D > 3aL_e) Equation 12			27 (L_e)	0.77
D&E (D < 3aL_e) Equation 13		0.146	165 (L_e)	0.90
D&E Equation 14	35		29.5	0.84
4.5% L_{exp} = 34–40 Å				
dG Equation 11	36	6.14	31.9	0.94
D&E Equation 14	35		24	0.83
AdG Equation 6	36	1.65	33.3	0.96
MWC Equation 8	38	61.2[a]	36	0.97
9.0% L_{exp} = 65–70 Å				
AdG Equation 6	70	1.83	58.6	0.95
MWC Equation 8	70	66.8[a]	67	0.97

[a]   Prefactor for MWC is $\sigma^{-5/3} \omega^{2/3} \upsilon^{-1/3}$

reasonable fit, a prefactor was allowed to scale the force. At larger distances, Equation 12 was not able to fit the data very well even when a numerical prefactor was allowed to vary. The experimentally measured force decayed less rapidly than the distance squared dependence of this theoretical equation. Hence, this analytical approximation to the complex series solution is not adequate. However, the approximation to the full numerical solution (Equation 14) provides a reasonable fit to the data and the fitted polymer thickness and is within 15% of the experimental value. Both Equations 11 and 14 produce reasonable fits to the data and are in very good agreement for the polymer thickness. Table 3 lists the parameter values for each of the theoretical equations.

2. Moderate Coverage (4.5%)

At a concentration of 4.5% DSPE-EO$_{45}$, the polymer chains begin to laterally overlap and interact with their neighbors on the surface. Although this regime has not been specifically addressed, we can test the limits of both the nonoverlapping and strongly interacting brush theories to model this intermediate regime. Again, both of the scaling theories (Equations 6 and 11) are able to fit the experimental force curve and provide reasonable estimates of the polymer thickness (Figure 9), indicative of the applicability of scaling theories to model the entire range of polymer surface coverages. Although Equation 14 provides a reasonable fit to the data, it should be noted that the fitted thickness here is less than the previously determined thickness for the more dilute coverage of 1.3% DSPE-EO$_{45}$. Hence, this trend is the opposite of what would be expected if the theory was able to accurately model the measured force profiles. Lastly, the more complex self-consistent mean field analysis of Milner, Witten, and Cates fits the data very well, but as expected does not vary much from the simpler scaling approach.[21]

3. High Coverage (9.0%)

The experimentally determined polymer thickness at 9.0% DSPE-EO$_{45}$ is 70 to 75 Å. As shown in Figure 10, both the scaling theory of Alexander and de Gennes and the self-consistent mean field analysis of Milner, Witten, and Cates fit the experimental data very well and are virtually indistinguishable (Table 3). However, the parabolic density profile of the

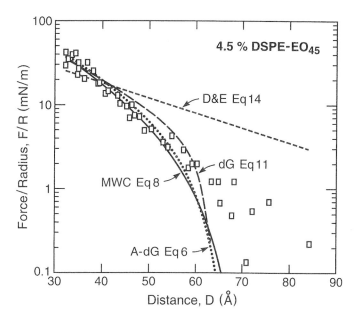

FIGURE 9. Data of the "overlapping mushroom" regime of 4.5% DSPE-EO$_{45}$ in Figure 7 fit to the theoretical equations of Section III.

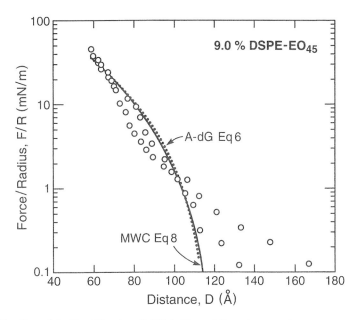

FIGURE 10. Data of the "brush" regime (9.0%) in Figure 7 fit to the theoretical equations of Section III.

MWC model results in a larger and more accurate estimate of the polymer thickness as compared to the experimental data than the simpler scaling approach. Although these theories provide reasonable fits to the experimental data, they both underestimate the longer-range portion of the force profile. More recent molecular dynamic and Monte Carlo simulations predict a smoothly decaying asymptotic tail of the polymer layer as found in these experiments, rather than the sharp cut-off in the profile assumed by the above theories.[36,37]

In summary, the experimental data are well fit in all three regimes by the simple scaling analysis by Alexander and de Gennes, providing reasonable estimates of the polymer thickness and the force profile when a prefactor of order unity is allowed to vary. Moreover, the self-consistent mean field analysis of Dolan and Edwards is able to provide reasonable fits at low surface coverages and can be extended to good solvent conditions. Lastly, the theory by Milner, Witten, and Cates for strongly stretched chains is able to fit the data in both the weakly overlapping regime (4.5%) as well as in the brush regime (9.0%), providing very reasonable estimates of the polymer thickness, although the results vary little from the simpler scaling arguments.

C. COMPARISON OF LATERAL INTERACTIONS WITH SCALING THEORY

In Section IV we demonstrated that the pressure-area isotherms for mixtures of DSPE and DSPE-EO$_{45}$ could be modeled by assuming ideal mixing between the pure components, where the polymer chains remained adsorbed to the air-water interface up to a critical pressure of 10 mN/m. We now analyze this lateral repulsion in terms of scaling theories on polymer interactions in two dimensions (2-D).

In 2-D, the surface pressure at low concentrations is that of an ideal gas of noninteracting chains;

$$\Pi \cong kTc \tag{17}$$

where c is the number of chains per unit area. This simple model breaks down when the chains begin to strongly interact. For concentrations above the overlap concentration, $c^* \approx 1/R_F^2$, scaling theory predicts that Π can still be described as an ideal gas, but of smaller units called blobs of size ξ. Hence, assuming the polymer chains do not interpenetrate, the surface pressure is given by:[23]

$$\Pi \cong \frac{kT}{\xi^2} \tag{18}$$

The scaling form of ξ with concentration is obtained from the requirement that $\xi = R_F$ at $c = c^*$ and that ξ is independent of N above c^*. Thus,[38]

$$\xi \cong R_F^2 \left(\frac{c}{c^*} \right)^{\nu/1-2\nu} \tag{19}$$

where the radius of gyration for an isolated chain is given by $R_F \approx aN^\nu$, and $\nu = 3/(d + 2)$ is the characteristic exponent which depends only on the spatial dimensionality, d.[23] Combining Equations 19 and 18 we find that

$$\Pi \cong \frac{kT \left(\dfrac{c}{c^*} \right)^{2\nu/2\nu-1}}{R_F^2} \approx c^{2\nu/2\nu-1} \sim \sigma^{-2\nu/2\nu-1} \tag{20}$$

In two dimensions, $\nu \approx 0.75$, whereas for theta solvents $\nu \approx 0.5$ for all dimensionalities. Hence, the surface pressure has a power law dependence on σ which can vary from σ^{-3} to $\sigma^{-\infty}$ as the solvent quality varies from a good to theta solvent.[38]

Applying Equation 20 to the pressure area data at concentrations above c^* for 100% DSPE-EO$_{45}$ (see inset Figure 3), we find $\Pi \sim \sigma^{-4}$ for $0 < \Pi \le 5$ mN/m. Thus, in the semidilute regime the air-water interface does not correspond to a theta or good solvent, but to one of intermediate quality. Here, we have assumed that the segments remain on the air-water interface and do not become partially submerged at low to intermediate surface densities. At higher surface pressures, this simple scaling analysis breaks down, since the segments begin to enter the water subphase.

VI. DISCUSSION AND CONCLUSIONS

The exact basis of the extended circulation times *in vivo* of liposomes containing lipids with bulky hydrophilic headgroups is still not fully understood. There is considerable debate as to whether the ethylene oxide chains merely provide a steric barrier around the liposome preventing the close approach of other protein and cellular surfaces in the body or if the particular biocompatibility and chemical inertness of PEO also play a significant role.[7,10]

Protein adsorption onto PEO-coated surfaces has been shown to be very low in a number of studies.[8,39-41] Further, adsorption of proteins decreases with increasing molecular weight of the PEO chains up to a limiting value of 1500, after which increasing the molecular weight did not further reduce the adsorption, presumably due to a decrease in the grafting density of PEO chains in the steric layer. It would appear that a minimally thick but dense layer of PEO is necessary to protect the liposome's lipid surface. In this way, proteins and other blood components will not have access to the lipid surface, and the lipid headgroups will be protected. Moreover, short-ranged van der Waals attraction and longer-ranged electrostatic interactions are dominated by the steric repulsion between the polymer chains as the layers become compressed.

As we have shown the steric barrier is easily modified by controlling the amount of DSPE-EO_{45} in the bilayer. Based on our LB studies, stable monolayers are formed at the air-water interface with concentrations ranging from about 1 to 10 mol% and they are easily transferred onto hydrophobic substrates (DPPE-coated mica) by LB deposition. The lateral interaction of the polymer chains is clearly indicated by the higher surface pressure readings at a given packing area compared to that of pure DSPE. This higher surface pressure manifests itself as a lateral stress on the bilayer. In the case of our supported bilayers, this lateral stress is not large enough to disrupt the bilayer structure. However, free liposomes or vesicles in solution are much more flexible and the same lateral stresses could increase the area per headgroup or cause more dramatic bulges and buckling of the bilayer surface. Each of these effects would lead to the exposure of hydrophobic groups on the surface, which are ideal sites for protein adsorption. Our data indicate that at concentrations above 9% DSPE-EO_{45} the PEO chains strongly repel each other, producing large destabilizing lateral stresses on the bilayers. Complementary work of Needham and Hristova[44] using the osmotic stress technique has also indicated that the bilayer structure becomes disrupted at high concentrations of PEO-modified lipids. Moreover, they recently calculated the theoretical maximum concentration of DSPE-EO_{45} to be ~15% before the lateral pressure would actually destabilize a lipid bilayer, by surpassing its tensile strength.[42,45]

In regards to modeling the interaction between short polymer chains or large lipid headgroups, we have shown that current theoretical models of randomly coiled, high-molecular-weight chains can be invoked to quantitatively fit the interaction forces between surprisingly short PEO chains "terminally grafted" to a bilayer interface. However, the long-range exponentially decaying portion of the steric force profile was not adequately represented at higher polymer surface densities. More recent molecular dynamic and Monte Carlo simulations predict a smoothly decaying asymptotic tail as found in these experiments.[36,37] The use of such theories may help predict the optimal steric barrier thickness and density in order to protect *in vivo* liposomes from degradation.

In conclusion, interbilayer interactions are greatly modified by the addition of a small concentration of DSPE-EO_{45}. The PE bilayers no longer adhere since the van der Waals attraction is swamped by the steric repulsion of the polymer chains. These short polymer chains provide a physical barrier or "force buffer" around the bilayer preventing the close approach of other surfaces, the strength and thickness of which was easily modulated by varying the concentration of the modified DSPE-EO$_{45}$ within the bilayer. This steric barrier correlates directly with the longer *in vivo* circulation time found for liposomes containing modified lipids with ethylene oxide headgroups. Moreover, the ineffectiveness of liposomes as drug carriers observed at higher concentrations of these bulky lipids correspond with the onset of large lateral stresses within the bilayer generated because the polymer chains begin to strongly overlap.[43,44] Lastly, theoretical models developed for long terminally grafted polymer chains were found to fit the experimental data.

ACKNOWLEDGMENTS

This work was supported by a grant from the National Institutes of Health (PHS GM47334). We thank Phil Pincus for illuminating discussions.

REFERENCES

1. **Klibanov, A., Maruyama, K., Torchilin, V., and Huang, L.,** Amphipathic polyethyleneglycols effectively prolong the circulation time of liposomes, *FEBS Lett.*, 268, 235, 1990.
2. **Lasic, D., Martin, F., Gabizon, A., Huang, S., and Papahadjopoulos, D.,** Sterically stabilized liposomes: a hypothesis on the molecular origin of the extended circulation times, *Biochim. Biophys. Acta*, 1070, 187, 1991.
3. **Allen, T., Hansen, C., Martin, F., Redemann, C., and Yau-Young, A.,** Liposomes containing synthetic lipid derivatives of poly(ethylene glycol) show prolonged circulation half-lives in vivo, *Biochim. Biophys. Acta*, 1066, 29, 1991.
4. **Allen, T., Austen, G., Chonn, A., Lin, L., and Lee, K.,** Uptake of liposomes by cultured mouse bone marrow macrophages: influence of liposome composition and size, *Biochim. Biophys. Acta*, 1061, 56, 1991.
5. **Papahadjopoulos, D., Allen, T., Gabizon, A., Mayhew, E., Matthay, K., Huang, S., Lee, K., Woodle, M., Lasic, D., Redemann, C., and Martin, F.,** Sterically stabilized liposomes: improvements in pharmacokinetics and antitumor therapeutic efficacy, *Proc. Natl. Acad. Sci. U.S.A.*, 88, 11460, 1991.
6. **Bailey, F. and Koleske, J.,** Configuration and hydrodynamic properties of the polyoxyethylene chain in solution, in *Nonionic Surfactants*, Vol. 23, Schick, M., Ed., Marcel Dekker, New York, 1990, 927.
7. **Woodle, M. and Lasic, D.,** Sterically stabilized liposomes, *Biochim. Biophys. Acta*, 1113, 171, 1992.
8. **Golander, C.-G., Herron, J., Lim, K., Claesson, P., Stenius, P., and Andrade, J.,** Properties of immobilized PEG films and the interaction with proteins, in *Poly (ethylene Glycol) Chemistry: Biotechnical and Biomedical Applications*, Harris, J. M., Ed., Plenum Press, New York, 1992, 221.
9. **Blume, G. and Cevc, G.,** Liposomes for the sustained drug release in vivo, *Biochim. Biophys. Acta*, 1029, 91, 1990.
10. **Lasic, D.D.,** *Liposomes from Physics to Applications*, Elsevier, Amsterdam, 1993.
11. **Needham, D., McIntosh, T., and Lasic, D.,** Repulsive interactions and mechanical stability of polymer-grafted lipid membranes, *Biochim. Biophys. Acta*, 1108, 40, 1992.
12. **Needham, D., Hristova, K., McIntosh, T., Dewhirst, M., Wu, N., and Lasic, D.,** Polymer-grafted liposomes: physical basis for the "stealth" property, *J. Liposome Res.*, 2, 411, 1992.
13. **Kuhl, T., Leckband, D., Lasic, D., and Israelachvili, J.,** Modulation of interaction forces between bilayers exposing short-chained ethylene oxide headgroups, *Biophys. J.*, 66, 1479, 1994.
14. **Gaines, G.,** *Insoluble Monolayers at Liquid-Gas Interfaces*, John Wiley & Sons, New York, 1969.
15. **Israelachvili, J.,** Thin film studies using multiple beam interferometry, *J. Colloid Interface Sci.*, 44, 259, 1973.
16. **Israelachvili, J. and Adams, G.,** Measurement of forces between two mica surfaces in aqueous electrolyte solutions in the range 0 = 100 nm, *J. Chem. Soc. Faraday Trans. 1*, 74, 975, 1978.
17. **Alexander, S.,** Adsorption of chain molecules with a polar head, a scaling description, *J. Phys. (Paris)*, 38, 983, 1977.
18. **de Gennes, P.,** Polymer solutions near an interface. I. Adsorption and depletion layers, *Macromolecules*, 14, 1637, 1981.
19. **de Gennes, P.,** Polymers at an interface; a simplified view, *Adv. Colloid Interface Sci.*, 27, 189, 1987.
20. **Derjaguin, B.,** Friction and adhesion. IV. The theory of adhesion of small particles, Kolloid Z., 69, 155, 1934.
21. **Milner, S., Witten, T., and Cates, M.,** Theory of the grafted polymer brush, *Macromolecules*, 21, 2610, 1988.

22. **Milner, S.,** Compressing polymer "brushes": a quantitative comparison of theory and experiment, *Europhys. Lett.,* 7, 695, 1988.
23. **de Gennes, P. G.,** *Scaling Concepts in Polymer Physics,* Cornell University Press, New York, 1979.
24. **Dolan, A. and Edwards, F.,** Theory of the stabilization of colloids by adsorbed polymer, *Proc. R. Soc. London Ser. A,* 337, 509, 1974.
25. **Israelachvili, J.,** *Intermolecular and Surface Forces,* Academic Press, London, 1991.
26. **de Gennes, P.,** Interactions between polymers and surfactants, *J. Phys. Chem.,* 94, 8407, 1990.
27. **Kuzmenka, D. and Granick, S.,** The collapse of poly(ethylene oxide) monolayers, *Macromolecules,* 21, 779, 1988.
28. **Marra, J. and Israelachvili, J.,** Direct measurements of forces between phosphatidylcholine and phosphatidylethanolamine bilayers in aqueous electrolyte solutions, *Biochemistry,* 24, 4608, 1985.
29. **Grabbe, A.,** Double layer interactions between silylated silica surfaces, *Langmuir,* 9, 797, 1993. The program was kindly supplied by Alexis Grabbe following the earlier work of D. Chan and R. Horn.
30. **Arnold, K., Herrmann, A., Pratsch, L., and Gawrisch, K.,** The dielectric properties of aqueous solutions of poly(ethylene glycol) and their influence on membrane structure, *Biochim. Biophys. Acta,* 815, 515, 1985.
31. **Chen, S. and Shen, E.,** Interparticle correlations in concentrated charged colloidal solutions — theory and experiment, in *Micellar Solutions and Microemulsions,* Chen, S. and Rajagopalan, R., Ed., Springer-Verlag, New York, 1990, 3.
32. **Leckband, D., Helm, C., and Israelachvili, J.,** Role of calcium in the adhesion and fusion of bilayers, *Biochemistry,* 32, 1127, 1993.
33. **Israelachvili, J., Tandon, R., and White, L.,** Measurement of forces between two mica surfaces in aqueous poly(ethylene oxide) solutions, *J. Colloid Interface Sci.,* 74, 113, 1980.
34. **Klein, J. and Luckham, P.,** Forces between two adsorbed poly(ethylene oxide) layers in a good aqueous solvent in the range 0–150 nm, *Macromolecules,* 17, 1041, 1984.
35. **Claesson, P., Cho, D., Golander, C.-G., Kiss, E., and Parker, J.,** Functionalized mica surfaces obtained by a cold plasma process, *Prog. Colloid Polym. Sci.,* 82, 330, 1990.
36. **Murat, M. and Grest, G.,** Structure of a grafted polymer brush: a molecular dynamics simulation, *Macromolecules,* 22, 4054, 1989.
37. **Pakula, T. and Zhulina, E.,** Computer simulation of polymers in thin layers. II. Structure of polymer melt layers consisting of end-to-wall grafted chains, *J. Chem. Phys.,* 95, 4691, 1991.
38. **Vilanove, R. and Rondelez, F.,** Scaling description of two-dimensional chain conformations in polymer monolayers, *Phys. Rev. Lett.,* 45, 1502, 1980.
39. **Kiss, E., Golander, C.-G., and Ericksson, J.,** Surface grafting of poly ethylene oxide optimized by means of ESCA, *Prog. Colloid Polym. Sci.,* 74, 113, 1987.
40. **Mori, Y., Nagaoka, S., Takiuchi, H., Kikuchi, T., Noguchi, N., Tanzawa, H., and Noishiki, Y.,** Poly tetranfluoroethylene grafts coated with ULTI carbon, *Trans. Am. Soc. Artif. Internal Organs,* 28, 459, 1982.
41. **Nilsson Ekdahl, K., Nilsson, B., Golander, C.-G., Lassen, B., Elwing, H., and Nilsson, U. R.,** *J. Biomed. Mater. Res.,* submitted.
42. **Hristova, K. and Needham, D.,** The influence of polymer-grafted lipids on the physical properties of lipid bilayers; a theoretical study, *J. Colloid Interface Sci.,* submitted.
43. **Blume, G. and Cevc, G.,** Molecular mechanism of the lipid vesicle longevity in vivo, *Biochim. Biophys. Acta,* 1146, 157, 1993.
44. **Needham, D. and Hristova, K.,** personal communication, 1994.
45. **Lasic, D.D., Woodle, M.C., Martin, F.J., and Valentincic, T.,** Phase behavior of "Stealth lipid" — lecithin mixtures, *Periodicum Biologorum,* 93, 287, 1991.

Chapter 9

POLYETHYLENE GLYCOL-LIPID CONJUGATES

Samuel Zalipsky

TABLE OF CONTENTS

I. INTRODUCTION

Great interest in lipid conjugates with polyethylene glycol (PEG)* was generated recently as a result of the discovery that incorporation of PEG-lipids into liposomes yields preparations with superior therapeutic performance as compared to conventional liposomes.[1] Such liposomes remain in blood circulation for extended periods of time ($t_{1/2} \geq 48$ h in humans) and distribute through an organism relatively evenly with most of the dose remaining in the blood compartment and only 10 to 15% of the dose ending up in liver. This constitutes a significant improvement over conventional liposomes; typically over 90% of conventional liposomes end up in liver within a few hours from injection (for a comprehensive review of the subject see Reference 2). Despite a great deal of activity generated in the field of PEG-grafted liposomes hardly any attention was paid to the chemistry of the conjugation processes and to the chemical properties of various PEG-lipids and their relationship to the properties of the corresponding liposomes. This chapter will attempt to fill the gap by reviewing these chemical aspects of the field.

---

* Abbreviations used: BA, bromoacetate; Boc, *tert*-butoxycarbonyl; DCC, dicyclohexylcarbodiimide; DSC, disuccinimidyl carbonate; Hz, hydrazide; MWCO, molecular weight cutoff; NMR, nuclear magnetic resonance; PC, phosphatidylcholine; HSPC, hydrogenated soybean PC; PDP, pyridyldithiopropionate; PEG, polyethylene glycol; mPEG, methoxyPEG; IC-PEG, imidazolylcarbonyloxy-PEG; SC-PEG, succinimidyl carbonate-PEG; SS-PEG, succinimidyl succinate-PEG; PE, phosphatidylethanolamine; DPPE, dipalmitoylPE; DSPE, distearoylPE; DOPE, dioeoylPE; PG, phosphatidylglycerol; RES, reticuloendothelial system; TEA, triethylamine; TLC, thin layer chromatography.

0-8493-8383-8/95/$0.00+$.50

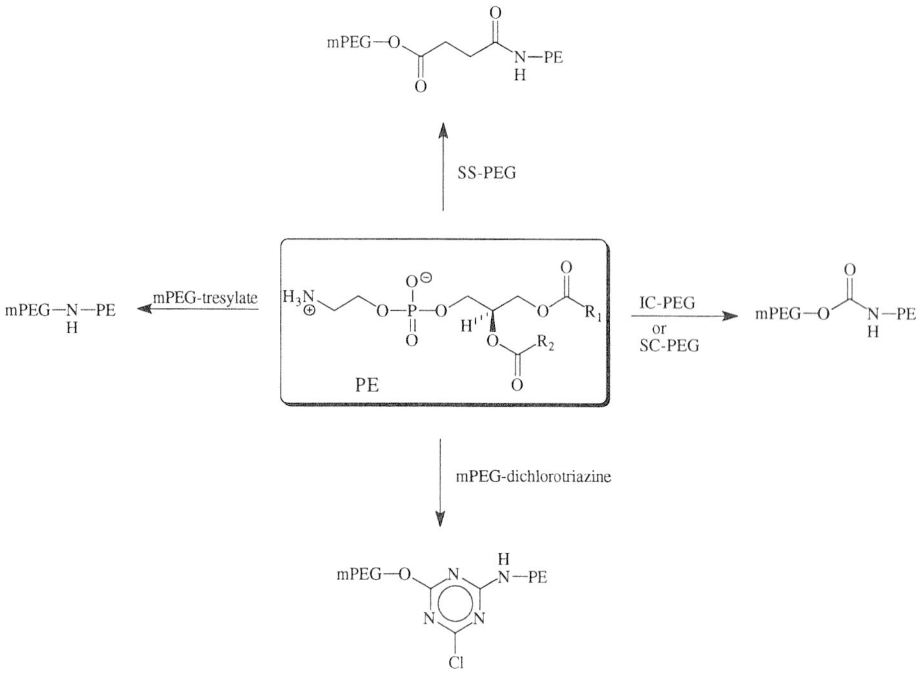

FIGURE 1.   Commonly used methods for attachment of mPEG to PE.

II. SYNTHESIS OF PEG-LIPIDS

PEG-based reagents originally introduced for modification of amino groups of proteins (for a review of the subject see Reference 3) were used for attachment of methoxyPEG (mPEG) to natural lipid derivatives, phosphotidylethanolamines (PEs). Figure 1 summarizes the synthetic pathways leading to various PE conjugates. Klibanov et al.[4] used succinimidyl succinate of mPEG-5000 (SS-PEG) for preparation of mPEG-succinamide-dioleoyl-PE. Urethane-linked mPEG-PE conjugate was obtained by reacting mPEG-oxycarbonylimidazole (IC-PEG) with distearoyl-PE.[5,6] The coupling reaction was carried out in refluxing tetrachloroethylene (6 h) or benzene (20 h). Recently it was demonstrated that the same product is formed cleanly upon warming up of DSPE suspension in chloroform with mPEG-succinimidyl carbonate (SC-PEG) in the presence of triethylamine, within a few minutes.[7] Dichlorotriazine and tresylate derivatives of mPEG-5000 introduced originally for protein modification by Abuchowski et al.[8] and Nilsson and Mosbach,[9] respectively, were also reacted with PE producing the corresponding lipid conjugates.[10,11] With the data available up to date there is no reason to assume that linkage between PE and mPEG influences the ability of the conjugates to form liposomes. Although there were some differences observed in the properties of such liposomes those could be attributed to subtle differences in formulations.[2] On the other hand, the presence of slightly reactive chloride in dichlorotriazine-derived conjugate as well as general concern about toxicity of the cyanuryl halide derivatives could be viewed as disadvantages.

The succinate interlinked mPEG-DSPE can hardly be considered the conjugate of choice due to the inherent susceptibility of the linker to hydrolysis.[12] Recently it has been demonstrated that succinate residue acting as a linker between primary amine- and hydroxyl-containing components of a conjugate, connecting two entities via amide and ester linkages, respectively, tends to cyclize with concomitant cleavage of the hydroxyl component and formation of succinimide derivative of the amine component as the main product.[13] Using model compounds it was shown that this base-catalyzed reaction takes place at physiological pH, and it is approximately 10 times faster than hydrolysis of a simple succinate ester.

Since the lipid part of mPEG-DSPE contains several biodegradable linkages: aliphatic esters and phosphodiester, the linkage between the polymer and the lipid components does not have to be biodegradable. In order to maximize the stability of a conjugate and its preparations it might be preferable to use simple, compact, nonbiodegradable linkage between PEG and DSPE. In this respect secondary amine- and urethane-containing PEG-PE conjugates appear to be more appealing. Although no direct comparison of the two conjugates has been published so far, there is no reason to assume that the small structural differences between the two conjugates (linkage and net charge) translate into differences in performance of the liposomes derived from them.

For some specific applications it might be desirable to design PEG-lipids containing a labile linkage between the conjugate components; so that the "PEG-coat" of liposomes containing such conjugates could be stripped in response to a specific set of conditions existing in a particular biological environment or in response to an artificially triggered event. Such conjugates might be useful in improving selectivity and/or reducing toxicity of some liposomal drugs. For example, acidic conditions are thought to exist in some tumor tissues. Hence loss of PEG from acid-sensitive PEG-liposomes could have enough destabilizing effect to induce release of the drug content of the vesicles. Hydrazone linkages appeared to have an ideal array of properties: they are known to be generally stable above neutral pH and undergo acid-catalyzed hydrolysis at lower pH.[14] Hydrazone-linked conjugate was prepared by condensation of periodate-oxidized DSPG with mPEG-hydrazide (unpublished results, Figure 2). Currently the properties of the hydrazone-linked PEG-lipid and its liposomes are being studied.

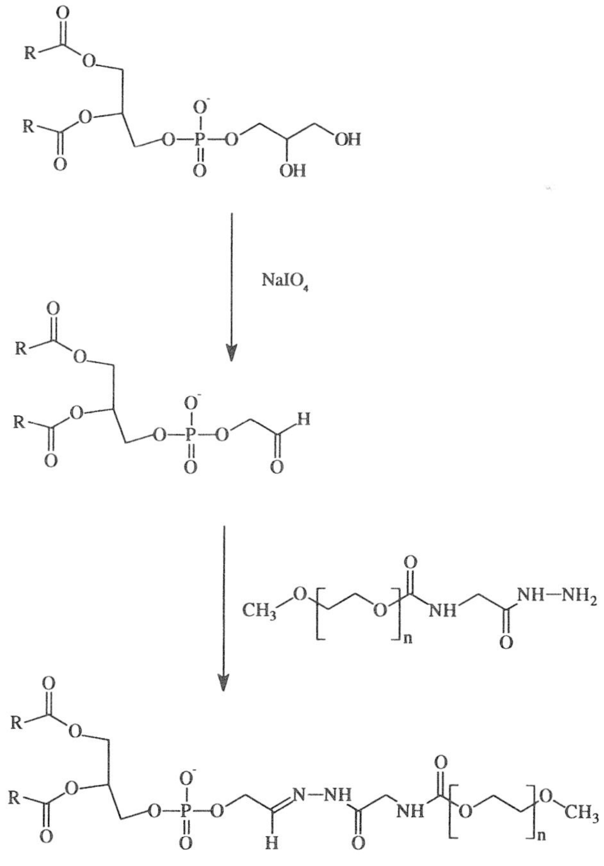

FIGURE 2. Formation of mPEG-hydrazone conjugate with oxidized PG.

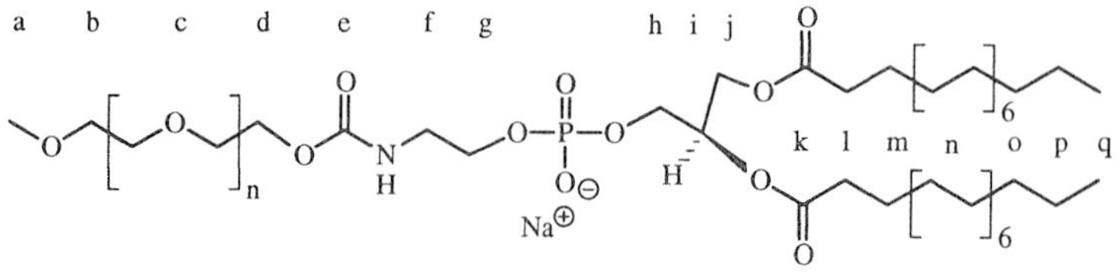

PEG			Glycero-phospho-ethanolamine			Distearoyl		
	H-	^{13}C-		H-	^{13}C-		H-	^{13}C-
a	s, 3.37	59.0	f	br m, 3.35	42.4	k	NA	173.0 & 173.3
b	s, 3.64	72.0	g	m, 3.93	63.4	l	2 x t, 2.28	34.2 & 34.4
c	s, 3.64	70.7	h	dd, 4.17	63.5	m	br m, 1.58	25.0
d	br t, 4.2	64.3	i	m, 5.20	69.7	n	s, 1.26	29.7
e	NA	156.7	j	dd, 4.39	62.9	o	s, 1.26	31.9
						p	s, 1.26	22.7
						q	t, 0.88	14.0

FIGURE 3. H- and ^{13}C-NMR chemical shifts of mPEG-DSPE sodium salt in CDCl$_3$ relative to tetramethylsilane (δ, ppm). Chemical shifts are listed in the table to match the letter assignments on the structure. The assignments from **k** to **q** refer to both of the stearate moieties. Both residues **k** and **l** are represented by two very closely positioned peaks in the ^{13}C-NMR spectrum. This is due to slight differences between the two carbonyls and the corresponding α-carbons. The hydrogens of the α-carbons (**l**) appear as two overlapping triplets on the H-NMR spectrum. The ^{31}P-NMR spectrum of the mPEG-DSPE sodium salt showed one singlet at $\delta = 1.89$ ppm, relative to 85% phosphoric acid, used as an external standard. Abbreviations: br, broad; dd, doublet of doublets; m, multiplet; NA, not applicable; s, singlet; t, triplet.

III. PURIFICATION AND CHARACTERIZATION OF PEG-LIPIDS

The PEG-lipid conjugates can be purified from the conjugation reaction mixtures by silica-gel chromatography. Since such conjugates exist in water almost exclusively as micelles they are readily separated from unreacted PEG-reagents by aqueous size-exclusion chromatography. Taking advantage of their very low critical micelle concentration, PEG-PEs can be retained in dialysis bags having very large pores (≥300,000 MWCO), which allows free PEG to diffuse through the membrane. This method was used for purification of various PEG-PE derivatives.[7,15] Despite the fact that PEG-PEs can be conveniently identified by TLC and NMR,[7] heretofore, inadequate attention was paid to the chemical characterization of such conjugates. To demonstrate the value of NMR-based characterization of PEG-PE, Figure 3 summarizes chemical shifts of urethane-linked mPEG-DSPE, the conjugate of choice in our laboratory. It is clear that both H- and ^{13}C-NMR offer an abundance of structural information on the conjugate.

IV. GRAFTING PEG ONTO PREFORMED VESICLES

Instead of using PEG-lipid conjugate to form liposomes Senior et al. modified preformed DSPC:DPPE:cholesterol vesicles with tresylate ester of mPEG-5000.[16] This reaction leads to formation of secondary amine-linked mPEG-DPPE conjugates (shown in Figure 1) from some of the PE on the exterior of the liposomal bilayer. This approach avoids the presence of mPEG residues inside the liposomes, yet suffers from the inability to bring the reaction to completion, thus leaving unmodified PE as a liposomal component. The asymmetrical lipid bilayer formed as a result of mPEG grafting might cause vesicle destabilization.

Incorporation of maleimide-containing lipids into liposomal vesicles followed by a very efficient reaction with thiol-PEG provides for an alternative way to modify the liposomal exterior with PEG. This approach was used for preparation of immunoliposomes by reacting the maleimide-vesicles with thiol-containing Fab' fragments, followed by treatment with mPEG-SH to consume the remaining maleimide groups.[17,18] The possible drawbacks of this approach are twofold: maleimide residues are only partially stable to the conditions of liposome formation, sizing (and in some cases, drug loading;[19] and the maleimide residues facing the interior of the liposome are not able to react with thiol residues at all, and therefore, they are likely to undergo some side reactions (hydrolysis, reactions with nucleophilic residues of other lipids or of the drug molecules).

V. STRUCTURE-ACTIVITY RELATIONSHIP

Several urethane-linked mPEG-PEs were tested with various lipid compositions for their ability to form liposomes and for variation in properties of those liposomes.[6] It was reported that the prolonged plasma presence of PEG-PE-containing liposomes is independent of degree of saturation of the parent PE or other lipids coincorporated into the same liposomes. However, better results were obtained with 18-carbon fatty acid-containing PEs than with shorter chain derivatives. For example, mPEG-DSPE was superior to mPEG-HSPE conjugates, which contain significant amounts of palmitoyl residues. Conjugates of DSPE derivatives from mPEG of molecular weight 1900 and 5000 incorporated into lecithin/cholesterol liposomes of mean diameter 100 nm, showed comparable blood lifetimes and biodistributions in rats. Maruyama et al. studied molecular weight dependence on performance of SS-PEG-derived mPEG-PE conjugates in large unilamellar liposomes (190 nm mean diameter).[20] In DSPC/cholesterol formulations mPEG-DSPEs of molecular weight 1000 and 2000, while showing the lowest accumulation in liver, extended the circulation time of liposomes more than conjugates derived from mPEG-5000 and -12000. On the other hand, the ability of mPEG-DOPE to extend blood residence time of egg-PC/cholesterol formulations was increasing with the increase of molecular weight of mPEG.[21] The best overall results, in terms of achieving highest blood-to-RES ratio combined with the longest blood circulation time, were achieved with mPEG-DSPE derived from mPEG-2000.[5,6,21]

VI. FUNCTIONALIZED PEG-LIPIDS FOR LIGAND COUPLING

Perhaps one of the most interesting applications of liposomes is in ligand-mediated targeting. Since all the attempts to achieve such targeting of conventional liposomes failed due to their quick liver uptake, long circulating PEG-liposomes should have better chances for successful implementation. Attachment of a targeting moiety, e.g., an antibody, to this type of liposomes with good preservation of activity proved to be a formidable challenge.[19] Since liposomes are normally prepared by hydration of an appropriate lipid mixture, the initial attempts to achieve such conjugates involved coincorporation of functionalized lipid component (maleimide- or carboxyl-PEs) with mPEG-PE into the formulation. After the preparation of the vesicles they were subjected to an appropriate immunoglobulin-conjugation procedure. It was learned that the presence of mPEG chains on the liposomal surface interfered with both the antibody attachment reaction and eventual interaction of the immunoliposome with its target antigen.[15,22] Both interferences grew more significant with increased length of the polymer chains. In light of these findings attachment of immunoglobulins to the external terminal of PEG appeared as a more attractive alternative approach. This in turn prompted development of new PE conjugates which differ from the original mPEG derivatives in that the inert terminal methoxy group is replaced with a reactive functionality suitable for conjugation reactions.

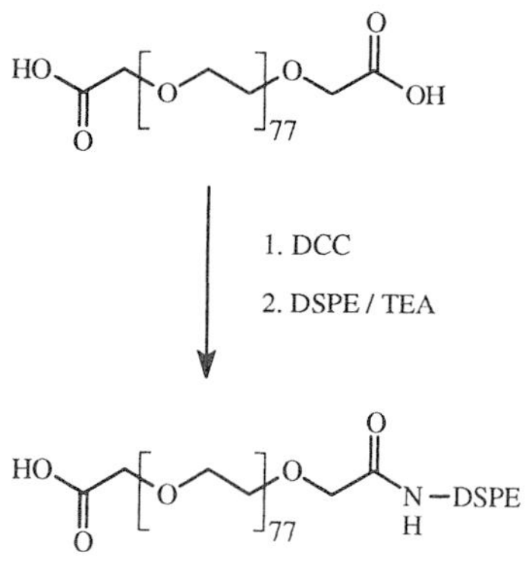

FIGURE 4.   Preparation of carboxylate-DSPE-PEG according to Blume et al. [23,24]

Blume and Cevc[23] reported preparation of DSPE-PEG-carboxylate by reacting DSPE with carbodiimide-activated *bis*-carboxymethyl-PEG-3000 (Figure 4). The product was purified by silica gel chromatography and after incorporation into liposomes was used for water-soluble carbodiimide-mediated conjugation of plasminogen. It was demonstrated that plasminogen linked to the PEG chains of liposomes was effective as a homing device without significantly sacrificing plasma circulation longevity characteristic of PEG-vesicles.[24] In contrast mPEG-coated liposomes, containing plasminogen linked directly to the bilayer surface via DSPE-glutarate, showed persistence in the bloodstream but lost the capacity to bind to their target, fibrin.

An efficient synthesis of hydrazide-PEG-DSPE was recently described utilizing a new heterobifunctional PEG carrying a reactive succinimidyl carbonate group on one terminal and a protected hydrazide residue at the opposite end (Figure 5).[7] The synthesis utilized the ω-hydroxyacid derivative of PEG, which is readily available in two chemical and one ion-exchange chromatographic steps.[25] Clean carbodiimide-mediated coupling with *tert*-butyl carbazate converts the carboxyl end of the polymer into a protected hydrazide without affecting the hydroxyl group. Disuccinimidyl carbonate (DSC) treatment in the presence of pyridine results in complete conversion of the terminal hydroxyl into a succinimidyl carbonate group. This amino-reactive group allows facile attachment of the polymer to DSPE. The resulting conjugate was then subjected to acidolysis to expose the hydrazide group. This functionalized PEG-lipid (Hz-PEG-DSPE) was used to replace mPEG-DSPE in various liposomal preparations without noticeable effects on their pharmacokinetics and biodistribution.

The usefulness of the Hz-PEG-lipid was demonstrated by attachment of periodate-oxidized antibodies to the PEG-tethered hydrazide groups on liposomal surfaces.[19,26] This mode of conjugation takes advantage of the specific reactivity of oligocaccharide residues of immuno-globulin, thus leaving untouched the antigen binding sites on the protein. The results of these studies indicated that it is possible to preserve the advantages of mPEG-grafted liposomes pertaining to their persistence *in vivo* and biodistribution, after substituting the terminal methoxy groups on the polymer chains with hydrazide groups and even after attachment of multiple immunoglobulin molecules. The versatility of the hydrazide group, its use as a starting material for introduction of other functional groups, and its ability to participate in numerous types of conjugation reactions, which is well documented in the literature,[27,28] offers a host of additional possibilities for use of Hz-PEG-liposomes.

Several other functionalized PEG-DSPEs were recently prepared in our laboratory utilizing α-amino-ω-hydroxy-PEG as a starting material (Figure 6). The amino group of this heterobifunctional PEG was first protected with a *tert*-butoxycarbonyl (Boc) group. In the following steps, very similar to the ones used in the synthesis of Hz-PEG-DSPE, the hydroxy

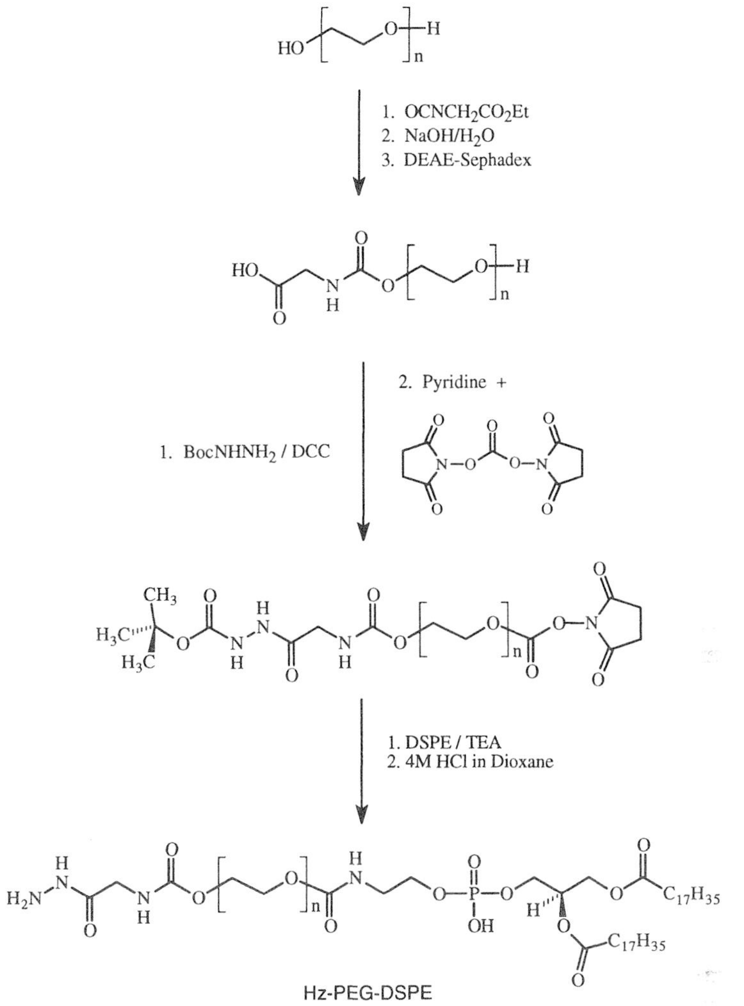

FIGURE 5.   Preparation of hydrazide-PEG-DSPE.[7]

end of the polymer was converted into succinimidyl carbonate and attached to the amino group of DSPE, followed by acidolytic deprotection of the amino group. The H_2N-PEG-DSPE by itself is an interesting conjugate, and studies are currently underway to explore its potential.[29] However, it is most useful as a starting material for preparation of other functionalized PEG-DSPE derivatives, utilizing known heterobifunctional reagents. Coupling of succinimido 3-(2-pyridyldithio)propionate (SPDP) to H_2N-PEG-DSPE yielded PDP-PEG-DSPE. The usefulness of the PDP-group is twofold: it can be reduced (e.g., by mild thiolysis) to expose a reactive thiol group; it can also react with thiol-containing ligands, resulting in their attachment via disulfide linkage. A discussion on the use of PDP-PEG-DSPE for coupling of maleimide-containing immunoglobulins is included in Chapter 20. Bromoacetamide-terminated PEG-DSPE (BA-PEG-DSPE) was prepared by reacting H_2N-PEG-DSPE with nitrophenyl bromoacetate (Figure 6). Reactivity of the bromoacetate group towards thiol-containing molecules is similar to the often-used maleimide, yet it offers a somewhat higher degree of selectivity and yields a more compact thioether linkage.[30] BA-PEG-DSPE is proving to be very useful for linking thiol-containing ligands (for example, Fab′ fragments of immunoglobulins) to PEG-liposomes.

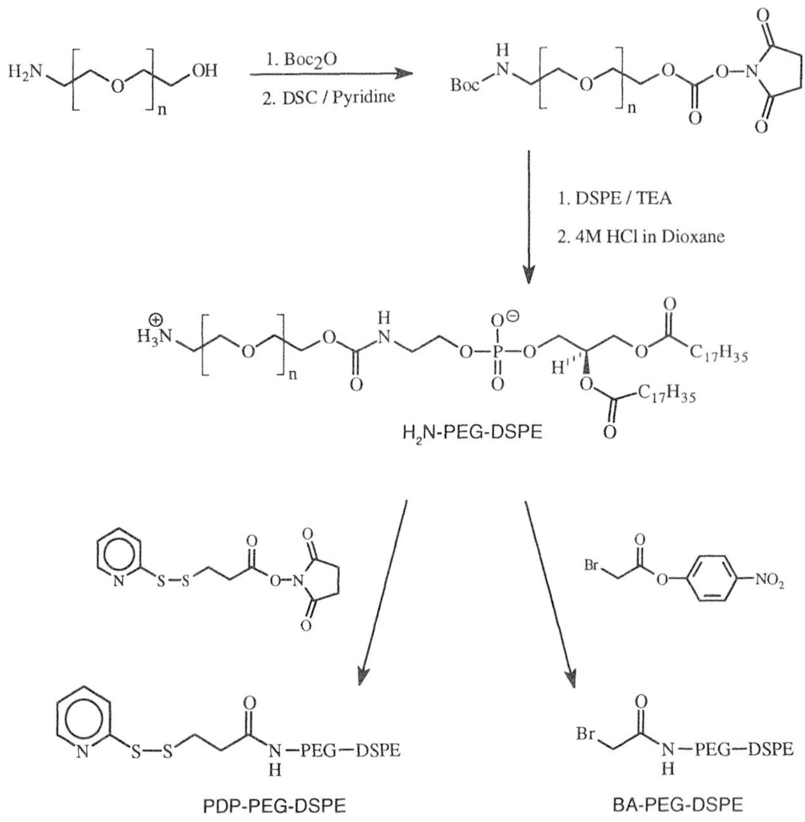

FIGURE 6. Preparation of various functionalized PEG-DSPE conjugates.

VII. FUTURE PERSPECTIVES

It appears that the discovery of long circulating PEG-grafted liposomes solved a number of problems associated with liposomal drug delivery. It also opened up a set of new opportunities as well as new challenges. One can envision a variety of useful ligands: polypeptides, oligosaccharides, fusogenic residues, etc., presented on the exterior of PEG-liposomes for a number of applications. The concept of ligand-mediated targeting was viewed as the ultimate goal of liposomal drug delivery for many years, yet it was severely hampered by very short blood lifetimes of conventional liposomes, their preferential accumulation in liver, and dose-dependent pharmacokinetics. The discovery of long circulating mPEG-grafted vesicles facilitated renewed interest in liposomal targeting (see also Chapter 20).

Development of a broad arsenal of functionalized PEG-lipids, each tailor-made for a specific conjugation strategy, is crucial for preparation of ligand-bearing, long circulating liposomes. Antibodies and their fragments are already under development as targeting moieties of PEG-liposomes in a number of laboratories. Since successful formulation of immunoliposomes requires optimization of a multitude of variables it is likely to remain a formidable challenge for years to come.

The methodologies for PEG-lipid synthesis are likely to undergo further refinement. Useful conjugates could be derived not only from PEs but also from other phospholipids. Rational design of such conjugates containing linkages of a defined stability/lability to specific physiological conditions is another promising area of research. It should lead to creation of "smart" liposomes that would circulate under normal serum conditions and then self-destruct in response to a change in environment (e.g., pH, reducing conditions, etc.) caused by reaching a specific target site or in response to an externally triggered event. These approaches can lead to design of highly specific formulations, which could be extremely useful for delivery of cytotoxic drugs.

REFERENCES

1. **Papahadjopoulos, D., Allen, T. M., Gabizon, A., Mayhew, E., Matthay, K., Huang, S. K., Lee, K.-D., Woodle, M. C., Lasic, D. D., Redemann, C., and Martin, F. J.,** Sterically stabilized liposomes: improvements in pharmacokinetics and antitumor therapeutic efficacy, *Proc. Natl. Acad. Sci. U.S.A.,* 88, 11460, 1991.
2. **Woodle, M. C. and Lasic, D. D.,** Sterically stabilized liposomes, *Biochim. Biophys. Acta,* 1113, 171, 1992.
3. **Zalipsky, S. and Lee, C.,** Use of functionalized polyethylene glycols for modification of polypeptides, in *Poly(ethylene glycol) Chemistry: Biotechnical and Biomedical Applications,* Harris, J. M., Ed., Plenum Press, New York, 1992, 347.
4. **Klibanov, A. L., Maruyama, K., Torchilin, V. P., and Huang, L.,** Amphipathic polyethyleneglycols effectively prolong the circulation time of liposomes, *FEBS Lett.,* 268, 235, 1990.
5. **Allen, T. M., Hansen, C., Martin, F., Redemann, C., and Yau-Young, A.,** Liposomes containing synthetic derivatives of poly(ethylene glycol) show prolonged circulation half-lives in vivo, *Biochim. Biophys. Acta,* 1066, 29, 1991.
6. **Woodle, M. C., Matthay, K. K., Newman, M. S., Hidayat, J. E., Collins, L. R., Redemann, C., Martin, F. J., and Papahadjopoulos, D.,** Versatility in lipid compositions showing prolonged circulation with sterically stabilized liposomes, *Biochim. Biophys. Acta,* 1105, 193, 1992.
7. **Zalipsky, S.,** Synthesis of an end-group functionalized polyethylene glycol-lipid conjugate for preparation of polymer-grafted liposomes, *Bioconjugate Chem.,* 4, 296, 1993.
8. **Abuchowski, A., Van Es, T., Palczuk, N. C., and Davis, F. F.,** Alteration of immunological properties of bovine serum albumin by covalent attachment of polyethylene glycol, *J. Biol. Chem.,* 252, 3578, 1977.
9. **Nilsson, K. and Mosbach, K.,** Immobilization of ligands with organic sulfonyl chlorides, *Methods Enzymol.,* 104, 56, 1984.
10. **Blume, G. and Cevc, G.,** Liposomes for the sustained drug release in vivo, *Biochim. Biophys. Acta,* 1029, 91, 1990.
11. **Tilcock, C., Ahkong, Q. F., and Fisher, D.,** Polymer-derivatized technetium ^{99m}Tc-labeled liposomal blood pool agents for nuclear medicine applications, *Biochim. Biophys. Acta,* 1148, 77, 1993.
12. **Zalipsky, S., Seltzer, R., and Menon-Rudolph, S.,** Evaluation of a new reagent for covalent attachment of polyethylene glycol to proteins, *Biotechnol. Appl. Biochem.,* 15, 100, 1992.
13. **Tadayoni, B. M., Friden, P. M., Walus, L. R., and Musso, G. F.,** Synthesis, in vivo kinetics, and in vivo studies on protein conjugates of AZT: evaluation as a transport system to increase brain delivery, *Bioconjugate Chem.,* 4, 139, 1993.
14. **Meuller, B. M., Wrasidlo, W. A., and Reisfeld, R. A.,** Antibody conjugates with morpholinodoxorubicin and acid-cleavable linkers, *Bioconjugate Chem.,* 1, 325, 1990.
15. **Klibanov, A. L., Maruyama, K., Beckerleg, A. M., Torchilin, V. P., and Huang, L.,** Activity of amphipathic poly(ethylene glycol) 5000 to prolong the circulation time of liposomes depends on the liposome size and is unfavorable for immunoliposome binding to target, *Biochim. Biophys. Acta,* 1062, 142, 1991.
16. **Senior, J., Delgado, C., Fisher, D., Tilcock, C., and Gregoriadis, G.,** Influence of surface hydrophilicity of liposomes on their interaction with plasma protein and clearance from the circulation: studies with poly(ethylene glycol)-coated vesicles, *Biochim. Biophys. Acta,* 1062, 77, 1991.
17. **Herron, J. N., Gentry, C. A., Davies, S. S., and Lin, J.-N.,** Antibodies as targeting moieties: affinity measurements, conjugation chemistry, and applications in immunoliposomes, *J. Controlled Release,* 28, 155, 1994.
18. **Tagawa, T., Hosokawa, S., and Nagaike, K.,** Drug-containing protein-bonded liposome, U.S. Patent 5,264,221, 1993.
19. **Allen, T. M., Agrawal, A. K., Ahmad, I., Hansen, C. B., and Zalipsky, S.,** Antibody-mediated targeting of long-circulating (Stealth) liposomes, *J. Liposome Res.,* 4, 1, 1994.
20. **Maruyama, K., Yuda, T., Okamoto, A., Ishikura, C., Kojima, S., and Iwatsuru, M.,** Effect of molecular weight in amphipathic polyethyleneglycol on prolonging the circulation time of large unilamellar liposomes, *Chem. Pharm. Bull.,* 39, 1620, 1991.
21. **Maruyama, K., Yuda, T., Okamoto, A., Kojima, S., Suginaka, A., and Iwatsuru, M.,** Prolonged circulation time in vivo of large unilamellar liposomes composed of distearoyl phosphatidylcholine and cholesterol containing amphipathic poly(ethylene glycol), *Biochim. Biophys. Acta,* 1128, 44, 1992.
22. **Mori, A., Klibanov, A. L., Torchilin, V. P., and Huang, L.,** Influence of the steric barrier activity of amphipathic poly(ethylene glycol) and ganglioside GM1 on the circulation time of liposomes and on the target binding of immunoliposomes in vivo, *FEBS Lett.,* 284, 263, 1991.

23. **Blume, G. and Cevc, G.,** Molecular mechanism of the lipid vesicle longevity in vivo, *Biochim. Biophys. Acta,* 1146, 157, 1993.

24. **Blume, G., Cevc, G., Crommelin, M. D. J. A., Bakker-Woudenberg, I. A. J. M., Kluft, C., and Storm, G.,** Specific targeting with poly(ethylene glycol)-modified liposomes: coupling of homing devices to the ends of the polymeric chains combines effective target binding with long circulation times, *Biochim. Biophys. Acta,* 1149, 180, 1993.

25. **Zalipsky, S. and Barany, G.,** Facile synthesis of α-hydroxy-ω-carboxymethylpolyethylene oxide, *J. Bioact. Compatible Polym.,* 5, 227, 1990.

26. **Zalipsky, S., Newman, M., Punatambekar, B., and Woodle, M. C.,** Model ligands linked to polymer chains on liposomal surfaces: application of a new functionalized polyethylene glycol-lipid conjugate, *Polym. Mater. Sci. Eng.,* 67, 519, 1993.

27. **Inman, J. K.,** Covalent linkage of functional groups, ligands and proteins to polyacrylamide beads, *Methods Enzymol.,* 34, 30, 1974.

28. **Wilchek, M. and Bayer, E. A.,** Labeling glycoconjugates with hydrazide reagents, *Methods Enzymol.,* 138, 429, 1987.

29. **Zalipsky, S., Brandeis, E., Newman, M.S., and Woodle, M.C.,** Long circulating, cationic liposomes containing amino-PEG-phosphatidylethanolamine, *FEBS Lett.,* in press, 1994.

30. **Bernatowicz, M. S. and Matsueda, G. R.,** Preparation of peptide-protein immunogens using *N*-succinimidyl bromoacetate as a heterobifunctional crosslinking reagent, *Anal. Biochem.,* 155, 95, 1986.

Chapter 10

BIOLOGICAL PROPERTIES OF STERICALLY STABILIZED LIPOSOMES

Martin C. Woodle, Mary S. Newman, and Peter K. Working

TABLE OF CONTENTS

I. INTRODUCTION: LIPOSOMES IN DRUG DELIVERY

The rapid uptake of traditional liposomes composed of natural phospholipids, those now frequently referred to as "conventional", has adversely limited many of their applications for delivery of therapeutic agents, particularly in cases utilizing parenteral routes, as described in several recent reviews and books.[1-4] Uptake by the mononuclear phagocyte system (MPS) cells usually results in the irreversible sequestering of the encapsulated drug in the MPS, where it can be degraded, not only destroying the beneficial effects of the drug, but also posing the potential problem of inducing MPS toxicity and/or the acute impairment of the functions of these cells.[5] This represents a severe limitation of conventional liposomes, or any other colloidal system with similar rapid uptake from blood, for therapeutic applications as drug delivery vehicles. Furthermore, the endothelial barrier of the blood vessels prevents adequate active targeting of liposomes *in vivo*, for example through antibodies attached to the liposome surface, despite very encouraging *in vitro* data.[6] As a result, early development of liposome therapeutics relied either on applications not limited by MPS uptake or on approaches circumventing it, such as use of small, rigid, cholesterol-rich liposomes that exhibited increased stability in plasma[7-9] or by utilizing (pre)saturation of the MPS with empty liposomes or some other type of particle to limit MPS uptake of the liposomal drug.[10,11]

The important need to overcome the limitation of rapid clearance of liposomes in order to expand their usefulness for drug delivery led to efforts by many laboratories to reduce MPS-mediated liposome recognition and uptake from blood. A wide range of approaches were considered, but primarily coating liposomes with virtually every natural biological material, including proteins, polysaccharides, and glycolipids, which were often chosen from components of red blood cells, given their ability to remain in circulation (see review, Reference 12). Little success was reported until recent reports of studies incorporating relatively small amounts of specific purified glycolipids, primarily ganglioside G_{M1} and, subsequently, hydrogenated phosphatidylinositol (HPI) with the best previously identified alternative, small rigid unilamellar vesicles.[13,14] Efforts subsequently began to identify alternative materials that would convey similar physical properties (albeit virtually undefined at the time), but which would be more amenable to the requirements of pharmaceutical products. Eventually, further success was achieved with a phospholipid conjugated with the synthetic hydrophilic polymer, polyethyleneglycol-phosphatidylethanolamine (PEG-PE).[15-17] Liposomes containing PEG-PE, which are referred to as sterically stabilized liposomes (SSL) due to the hypothesis that a steric barrier is formed by the PEG coating, have been found to have the best capabilities to reduce MPS uptake.

The use of a relatively inexpensive and easily prepared lipid, PEG-PE, given its abilities to avoid MPS uptake, has transformed the pharmaceutical development of liposomes and restored much of the original promise of liposomes, as well as opened up entirely new applications ranging from drug delivery systems with dramatically changed pharmacokinetics and biodistribution to providing a more realistic model for the study of biological membranes, red blood cells, cell-cell, protein-cell, and receptor interactions. The current hypothesis is that the long circulation results from reduced adherence of plasma "opsonin" components, based on extension of the known property of polymers to stabilize colloids through what is often called steric stabilization.[12,18,19] Many varied physical measurements appear to substantiate this proposed mechanism.[20,21] Although this aspect needs further investigation and confirmation,[22] such liposomes have become known as sterically stabilized liposomes or SSL, including extensions of this terminology, e.g., SSIL for sterically stabilized immunoliposomes.[19,23]

Fortunately, the lack of a complete understanding of the mechanism of action has not impeded the evaluation of these novel liposomes for their usefulness in a number of different therapeutic applications. SSL containing anticancer drugs have demonstrated improved efficacy in preclinical animal models[24] and these findings have been reflected in encouraging initial results from ongoing clinical studies.[25] Other efficacy studies have evaluated the use of SSL in prolonged delivery of a bioactive peptide hormone[26] and antibiotics.[27]

The aim of this chapter is to review the general understanding of the *in vivo* behavior of SSL, contrasting their properties to those of "conventional" liposomes. Several recent reviews have appeared describing and interpreting the results obtained from many laboratories over the last few years. Here, the goal is to describe the current state of the understanding of the biological fate of SSL and, in particular, surface-modified liposomes made by incorporation of PEG-PE. The changes that occur in pharmacokinetics and biodistribution with inclusion of PEG-PE will be discussed.

II. BIOLOGICAL PROPERTIES OF LIPOSOMES *IN VIVO:* MPS UPTAKE AND ITS CONSEQUENCES

The current understanding of liposome interaction with cells is based primarily on studies *in vitro*, which are reviewed extensively elsewhere.[28-30] The generally accepted aspects that relate to the fate of liposomes *in vivo* and, thus, form the basis for differences observed with SSL will be described here. The interactions fall into four categories: (1) exchange of materials, primarily lipids and proteins, with cell membranes; (2) adsorption or binding of liposomes to cells; (3) cell internalization of liposomes by endocytosis or phagocytosis once bound to the cell; and (4) fusion of bound liposome with the cell membrane. In all of these interactions, a strong dependence exists on lipid composition, type of cell, presence of specific receptors, and many other parameters.

Lipid transfer occurs by two separate processes: either by associated transfer proteins or, in some instances, with no other apparent interactions, i.e., via molecular solubility in the aqueous phase or upon events during particle-particle collisions. It was recognized early on that lipid exchange, including the exchange of cholesterol, is apparently not dependent on enzymatic activity.[30,31] A number of reports of liposome fusion measured by the transfer of lipid markers were later shown to be the result of lipid exchange,[32,33] demonstrating that it can occur in the absence of lipid transfer proteins. The exact mechanisms of lipid exchange are unresolved.

Liposomes adsorb avidly to the surface of a variety of cell types, as revealed by electron microscopic evidence. Such adsorption may or may not lead to other interactions. In some cases, bound liposomes remain bound for extended periods with no further change.[34] Despite extensive studies, the mechanisms of liposome uptake leading to their localization in endosomal and lysosomal compartments are still poorly understood. Perhaps best characterized is phagocytosis by freshly isolated monocytes, macrophages, and Kupffer cells. The studies have shown that binding of the liposomes occurs as a result of specific opsonization,[35] followed by uptake. The liposomes and their contents are degraded in the endosomes or subsequently by lysosomal digestion. Targeted liposomes containing antibodies (Ab) covalently attached to the surface can be endocytosed after binding of the Ab with its cellular receptor, but only if the receptor falls within a class exhibiting internalization. Otherwise, as mentioned above, the liposomes may remain bound to the cell surface until they are degraded by one of several possible mechanisms, including macrophage recognition and uptake, degradation by lipases, physical disintegration, or they are simply washed away.

One early model of drug delivery with liposomes envisioned fusion of liposomes with cells and delivery of their contents directly to the cytoplasm. Membrane fusion is an essential cellular process, for example in exocytosis and endocytosis, and it is highly regulated and controlled. As described above, initial reports of liposome-cell fusion were subsequently attributed to lipid exchange, and it now appears that liposome fusion with cells occurs very rarely. Apparently, this process is largely controlled by membrane proteins, either those of a cell or of a virus. Enhanced fusion by reconstitution of viral surface proteins in liposomes (known as virosomes)[36] supports this vision. In general, rigid liposomes show even less fusion with cells, and only a few reports of their endocytosis exist. They also show no measurable lipid or protein exchange, despite retaining the ability to adsorb onto cells.[28] Consequently, it is unrealistic to expect that simple lipid bilayers will fuse with cells without the action of a fusogen *in vivo*, even though some *in vitro* studies may show good results. This appears to be especially true for the rigid liposomes favored for drug delivery.

On the other hand, endocytosis and phagocytosis can be expected to play a significant role in liposome interactions with cells, both *in vitro* and *in vivo*. Macrophages, one of the most important components of the complex immune defense system, play a major role in clearance of foreign particulate matter, including liposomes and other colloidal particles, from blood circulation. These cells originate as premonocytes in bone marrow and, after circulating in blood as monocytes, attach in various organs and tissues forming a resident population: in connective tissue (histiocytes), liver (Kupffer cells), lung (alveolar macrophages), free and fixed macrophages in spleen and lymph nodes, macrophages in bone marrow, peritoneal macrophages in serous cavity, osteoclasts in bone tissue, and microglia in nerve tissue. Depending on the route of administration and *in vivo* distribution, liposome uptake by all of these macrophage populations can occur. Following IV administration, liposomes primarily come in contact with macrophages in liver, spleen, and bone marrow, where most are removed from the circulation.

On the molecular level, in a first approximation, liposome clearance is thought to consist of two steps: opsonization by blood proteins, followed by macrophage uptake of these marked liposomes. Opsonins are immune and nonimmune serum factors which bind to foreign particles and promote phagocytosis. Immune opsonins are immunoglobulins and complement proteins. Although each opsonin has its own interaction and confers different fates, in general liver sequestration appears to be complement-mediated, whereas the spleen removes foreign particulate matter via antibody Fc receptors.[35] Kupffer cells, however, also have immunoglobulin antigens, but it is not clear whether this is important for immune clearance.

Other circulatory factors have the potential to act as nonimmune opsonins. The most studied is plasma fibronectin, a long-chain glycoprotein. Foreign substances in the blood circulation may cause thrombus formation on their surfaces. Microscopically, this is either a deposition of fibrin or aggregation of platelets, and these coatings also act as opsonins in enhancing clearance of the particle.

This uptake of liposomes by the MPS has been referred to as passive targeting and for some drugs is a limited fulfillment of Ehrlich's magic bullet concept of selective targeting. Basically, parasitic diseases of the MPS, vaccination, or drug-mediated activation of macrophages into a tumoricidal, bactericidal, or virucidal state are examples of such targeting, with achievements of up to 1000-fold improvements in therapy. Some drugs can survive macrophage uptake and be released back into the blood, with consequent therapeutic advantages from this sustained release.[37] Most therapeutic agents, though, are not intended for these tissues nor can survive macrophage uptake.

III. BIOLOGICAL PROPERTIES OF CONVENTIONAL LIPOSOMES

A. STABILITY OF LIPOSOMES IN BIOLOGICAL FLUIDS

Upon parenteral administration, liposomes are diluted extensively into body fluids such as blood, lymphatic, or extracellular fluid (the exact amount depends on the volume of the original sample, the animal size, and route of administration). Most current liposome-based therapeutic applications require IV injections, although blood (or plasma) may be the most detrimental *in vivo* environment for liposomes. The plasma opsonin proteins, the lipoproteins, and the phospholipases are by far the most important factor in interactions with and destabilization of liposomes. The presence of various ions and the extent of dilution also have to be considered.[38]

The interactions of liposomes with various proteins may involve simple charge-charge interactions and/or hydrophobic interactions that lead to protein adhesion or adsorption on the liposome surface or penetration of proteins into the bilayer.[39,40] In the absence of MPS cells, these two types of interactions normally do not have much influence on liposome integrity. The latter interaction, however, can result in the loss of encapsulated or bound molecules, liposome disintegration, or both, as does lipid exchange between liposomes and other particles, especially high density lipoproteins (HDL). Although most, if not all, lipoproteins may be involved in interactions with liposomes, the major interaction occurs with HDL, which is probably the most important for liposome disintegration *in vitro*.[39] *In vivo* behavior may be very different because the opsonized proteins, which may not necessarily induce major changes *in vitro*, can result in enhanced cellular uptake by the MPS *in vivo*. The transfer of phospholipid molecules to HDL is greatly enhanced by plasma phospholipid transfer proteins. After incubation with liposomes, liposomal lipid is found associated with the HDL. This exchange also depends on the vesicle-to-HDL ratio, showing a saturation at about 0.25 mg of phosphatidylcholine (PC) per milligram of protein.

Several studies have shown that liposome stability in plasma depends upon the relative concentrations, liposome size and lamellarity, lipid composition, and the incubation temperature.[39-41] The kinetics of liposome leakage depend on the type of liposome label used, but often show single exponential release profiles.

From many studies it is clear that the liposome lipid composition has the dominant role in conventional liposome stability in plasma (see review, Reference 1). Early experiments have shown that liposomes composed of cholesterol in mixtures with lipids with long and saturated hydrocarbon chains that maximize van der Waals attraction with steroids exhibit the most refractory lipid composition, i.e., are the most stable in blood. Such effects have been found *in vitro* by rigidification through polymerization of the lipids.[42]

B. PHARMACOKINETICS AND TISSUE DISTRIBUTION

As described above, MPS uptake of liposomes results in their rapid removal from blood and accumulation in those tissues, primarily the liver and spleen, as has been reviewed extensively elsewhere.[1,2,43] Generally, liposome clearance from blood shows several kinetic components. Pharmacokinetic (PK) studies of clearance of liposomes from blood have been described most often using a two-compartment model, considered indicative of elimination by cells located either intra- or extravascularly.[44] An alternative view is that this represents clearance by two distinct cell populations, a high-affinity population that can be saturated and a low-affinity population that is not saturated.[8] Clearance with a single rate, consistent with a one-compartment model, occurs in some cases, primarily liposomes with a large particle size.

Some of the first observations with conventional liposomes revealed a strong dependence on lipid dose. Blood circulation time increases considerably at high lipid doses saturating, with a maximum lifetime of up to about 20 h, depending on the lipid composition.[1,8,12] Subsequently, studies showed that liver and spleen uptake can be saturated temporarily, prolonging the apparent blood circulation at high lipid doses. The saturation concentration threshold is species-dependent. In mice, for example, it occurs at around 120 mg lipid/kg (1 to 2 μmol/25 g) body weight.[1,8] In a not too unrelated comparison, intravenous fat nutritional emulsions, which form small cholesterol-rich liposomes in the plasma, are infused at up to 100 g/person/d. With these materials, it was found that the lipid exhibits monophasic kinetics with a half-life of 2 d.[12] While limited success was achieved utilizing MPS saturation techniques with conventional liposomes, the potential for prolonged impairment of MPS function or the induction of other adverse effects has limited this approach.[5]

Besides lipid dose, a number of other liposome properties have been found to influence their clearance. One of the most important is lipid composition, in which the phase transition temperature of the phospholipid component and cholesterol content have a considerable influence.[1] This effect results primarily from their influence on bilayer fluidity. Highly saturated phospholipids (with a high-temperature phase transition) in combination with cholesterol, giving so-called rigid bilayers, exhibit slower clearance from blood compared to liposomes composed with unsaturated acyl forms of the same lipid (with a low-temperature phase transition).[1] The mechanism of action of high-temperature phase transition phospholipids in prolonging circulation is not clear, especially in the presence of high levels of cholesterol, which eliminates the phase transition. Nevertheless, an increase in bilayer rigidity and lateral packing occurs when cholesterol is added, as evidenced by decreased release of entrapped species,[45] which could increase stability towards opsonization.

In addition, an apparently special lipid composition giving reduced clearance was identified: distearoyl phosphatidylcholine (DSPC) mixed with cholesterol. In combination with DSPC, cholesterol-rich formulations show prolonged circulation compared with dipalmitoyl PC (DPPC) under similar conditions, despite very little difference in these two lipids; DSPC differs only by two additional methylene units in both acyl chains. This increases the phase transition temperature, but from an already high value. Differences in their interaction with cholesterol have been documented[46] that may explain the advantage of DSPC. In any event, the prolonged circulation exhibited by either of these formulations is severely limited by particle size, lipid composition, lipid dose, and surface charge, which must be neutral, so their therapeutic applications are also limited. In general, the best results were obtained with small unilamellar vesicles (SUV) composed entirely of DSPC and cholesterol, typically with a molar ratio in the range of 2:1 to 1:1, respectively.[1,8,47] For example, DSPC:Chol at 2:1 is reported to show a blood half-life of up to 20 h, but only at a saturating lipid dose. Below saturation, the half-life is less than 10 h. In some cases with cholesterol-rich formulations, a significant advantage of DSPC over DPPC has not always been observed.[47]

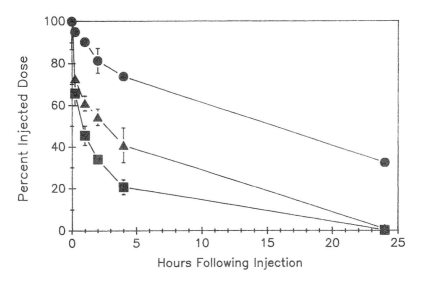

FIGURE 1.   Effect of size on blood clearance of DSPC:Chol at saturating lipid dose in mice: triangles, SUV; circles, 0.2 µm pore extruded REV; squares, 0.4 µm pore extruded REV. Adapted from Reference 1.

Particle size is also an important parameter. Generally, increasing the size of liposomes increases their uptake. Even lipid compositions exhibiting relatively long circulation when prepared as SUVs (diameter <100 nm) show rapid clearance when the diameter is increased to 200 nm.[1] For example, the clearance of different particle sizes of DSPC:Chol at saturating lipid doses are shown in Figure 1.

Such liposomes are not easy to work with, and therefore it is not surprising that even when tested at a comparable size, the circulation half-life measured by several laboratories differs.[1,8,47] It is not clear if such discrepancies are due to sample aggregation, MPS saturation, which are both common problems with this formulation, or from other mechanisms. One aspect often not considered is the strong dependence of the rate of clearance on the lipid dose.[12,48,49] In any case, such SUV are unattractive for therapeutic applications because of several limitations, including low aqueous entrapped volume, particle instability due to the lack of charge stabilization,[50,51] and a strong dependence of circulation time on both particle size and length of the saturated acyl chains.[1] Consequently, preparation of DSPC:Chol liposomes is difficult and release of drugs from them is neither likely nor controllable. For these reasons, alternative approaches have been considered.

Surface properties have also been shown to play a major role in liposome pharmacokinetics and has been investigated extensively, with somewhat confusing results. A negative charge can be attained in liposomes by the incorporation of a variety of lipids, e.g., PA, PS, PG, gangliosides, DCP, and cardiolipin. Studies indicate a general increase in clearance of the negatively charged liposomes. Studies with positively charged liposomes have been more limited, as fewer positively charged lipids are available. The most frequently used method, incorporation of stearyl amine (SA), was found to result in liposomes that were intrinsically toxic. The effect on circulation (or biocompatibility) of other positively charged lipids has not been evaluated extensively. One report, though, with a derivative referred to as BisHOP showed that positive charge decreased clearance.[52] However, evidence of adsorption of blood components to an initially positively charged liposome, leading to a negative charge, may indicate that there is no advantage to a positive charge.[53]

IV. BIOLOGICAL PROPERTIES OF SURFACE-MODIFIED LIPOSOMES

A. METHODS FOR SURFACE MODIFICATION

Of the many approaches based on surface modification designed to decrease biological interactions and thereby increase the blood circulation, covalent attachment of 1000- to 5000-

molecular-weight PEG at about a 5 mol% density appears to give the best results.[12] The results are somewhat analogous to those obtained by adsorption of PEG-based amphipathic polymer surfactants onto latex-based colloidal particles[54] and subsequently applied to lipid-stabilized emulsions and liposomes.[12,55,56] These studies have indicated that these polymers do adsorb and in some cases give rise to coatings on the liposomes that are similar to those obtained on latex. However, decreased retention of aqueous entrapped components may limit this approach to surface modification for drug delivery.[48]

In contrast to coating the surface, methods for doping or co-mixing of amphipathic molecules with special polar heads at low molar ratios in the lipid bilayer have been used to alter the surface properties of liposomes. In fact, the initial success leading to the current state of activity was achieved by incorporation of a few specific natural glycolipids, primarily GM1 and HPI.[13,14,57,58] More recent efforts have identified other similar lipids which can prolong circulation.[59-61] In fact, long circulating formulations containing these surface sugars were evaluated further, showing an ability to provide substantial improvements in therapeutic applications.[62-65]

Despite the success obtained with incorporation of specific glycolipids with small headgroups, perhaps with specific biological interactions responsible for the effect, many continuing efforts for use with drug delivery are focused upon SSL containing PEG-PE. In most studies, small liposomes, those with mean diameters ≤ 100 nm, have been investigated, although incorporation of the polymer also transforms formulations prepared with larger particle sizes, providing significant effects on circulation time and tissue distribution for liposomes up to 200 or even 300 nm.[38,66,67] The initial results obtained in mice and rats have now been confirmed by studies with encapsulated doxorubicin in several clinical studies. Typically, a circulation half-life of 12–20 h in rats or mice increases to 40–60 h in humans.[68,69] Such differences can be attributed to different pulse rates and blood flow patterns, including blood-to-MPS ratios.[19]

Preparation of sterically stabilized liposomes by incorporation of single-chain detergents into a bilayer has also been reported.[49,70] For instance, it was shown that 4 mol% of Tween 80 (a nonionic surfactant with approximately 20 total ethylene glycol groups in four branches) stabilizes liposomes in serum, as well as against size growth upon addition of Ca^{2+}. However, addition of PEG-SA did not increase the circulation time of liposomes.[50] A PEG coating may also be formed in nonionic vesicles, known as niosomes. They are normally composed of roughly equimolar mixtures of cholesterol and nonionic alkyl polyoxyethylene glycol detergents (C_nE_m, typically n = 12 to 16 and m = 3 to 8). In some cases, dialkyl detergents are also used. Although these particles may have a higher surface density of ethylene oxide groups, they do not appear to show prolonged circulation times. This is probably due to shorter PEG segments, which cannot form an effective barrier to prevent interactions with plasma components. Other polymer-derivatized lipid bilayer forms with novel properties are also possible,[71,72] but have not been examined for their *in vivo* properties. In contrast, polymerized bilayers show little advantage *in vivo*.

Evaluations of other polymers besides PEG are just beginning. Attempts to use dextran for a similar increase in blood circulation have failed.[22] Efforts with a more hydrophilic carboxylic acid end group derivative of PEG-PE have also been interpreted as less effective,[73] contrary to expectations that the effect is based upon achieving a hydrophilic coating.[49,74,75] In fact, the partitioning of the PEG-coated liposomes into the PEG phase can be interpreted as indicative of less hydrophilic properties. A number of other more or less hydrophilic polymers might be considered. Success was achieved with polyvinyl alcohol (PVA), polylactic acid (PLA), and polyglycolic acid (PLG)[76] and a similar effect of polyvinylpyrrolidone or polyacrylamide has been suggested,[4,22] followed by initial successful results.[76a] Interestingly, while reports of these and yet other polymers showing positive effects are now finally beginning to appear, including polyglycerols,[76b] none of these appears to provide the same extent of circulation as that provided by PEG-PE. Now, however, two reports have shown effects equal to that provided by PEG-PE: an amino terminated PEG-PE[76c] and an entirely new class of water soluble polymer-lipid derivatives, polyaxozlines.[76d] Thus, now the effect may be more firmly described as a physical phenomena rather than a specific chemical effect of PEG.

Studies indicate that surface-modified liposomes using synthetic lipids with polymeric headgroups amenable to large-scale pharmaceutical-grade production can control the pharmacokinetic properties and thus may permit liposomes to fulfill their original promise as drug delivery vehicles. In addition, these new formulations present an opportunity to gain a fundamental understanding of the interactions of lipid bilayers leading to their functions as biological membranes. For these reasons, and the rapidly growing status of the field, a description of the state of understanding of their biological fate is of interest.

B. PHARMACOKINETICS AND TISSUE DISTRIBUTION

The clearance and tissue distribution of sterically stabilized liposomes following IV administration has been examined in a number of laboratories.[12,49,66,67,74,77-81] These reports are based on different liposome labeling methods, and the reported results are somewhat different. The labels used include aqueous entrapped labels (^{125}I-inulin, ^{67}Ga-DF, etc.) and lipid labels (^{111}In-DTPA-SA). With the continued development of therapeutic agents entrapped in these novel liposomes, additional *in vivo* distribution information is becoming available. A difficulty in comparing all the reports is inconsistency in the choice of time points, especially since only one or a few time points have been examined in many cases. In addition, many investigators combine the uptake by liver and spleen into one category, which is usually considered representative of the MPS, whereas others have reported these two tissues separately. Consideration must also be given to differences in the method used for reporting of results; most are given as percent of injected label per tissue, some as per gram of tissue, and in other cases the percent of label remaining *in vivo* is reported. In the last case, the tissue levels are adjusted according to the extent of label loss, which usually changes over time. While this corrects for label leaked from the liposomes and then excreted, it makes comparisons difficult between different time points and different formulations, since leakage varies with these conditions. It also generally increases the apparent tissue levels, and the results can be misleading if the amount of recovered label is not also reported.

In several reports, blood circulation kinetics have been described.[12,49,66,67,74,77-81] It is interesting that similar interpretations and conclusions are drawn in virtually every investigation reported with PEG-PE as to its general ability to prolong circulation and reduce MPS uptake, despite significant discrepancies between some of the results, as reviewed elsewhere.[12] Nonetheless, the generally accepted conclusion drawn from all the reports is well founded: inclusion of PEG-PE in liposomes results in a dose-independent reduction in hepatosplenic uptake (i.e., MPS uptake) and prolonged circulation significantly beyond that obtained by non-PEG-containing DSPE:Chol controls. Typical recent results of blood circulation from our laboratory are shown in Figure 2. This finding is largely based on studies with a small particle size (≤100 nm). While fewer investigations have evaluated a larger particle size (often about 200 nm), similar findings are reported, but with an even greater distinction from DSPC:Chol liposomes.[66] Other significant differences also exist between smaller and larger liposomes, primarily in tissue distribution,[12,66,67] as discussed further below. The blood levels are compared with tissue levels after 24 h in Table 1.

Studies of blood clearance utilizing entrapped drugs as the label have shown similar enhancement of circulation time, but the increase in time generally has been reduced relative to that reported using radiolabels.[12] In contrast with radiolabels, the clearance of drugs also has been shown to have multiple kinetic components.[24,26] The initial rate is faster than that reported with entrapped label, but is followed by a second rate slower than that of entrapped label, resulting in prolonged blood levels for up to 3 d. One possibility is that drug is initially released more rapidly from the circulating liposomes, producing a faster rate of clearance, followed by slow release indicating that clearance of the liposome, and not the drug, domi-

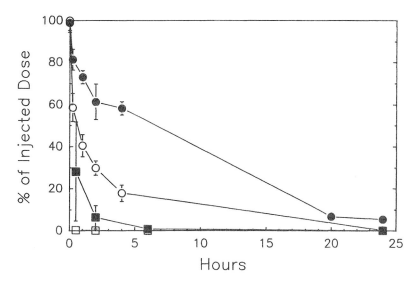

FIGURE 2. Blood clearance of radiolabeled liposomes in rats. Closed circles are PEG-PE/PC/cholesterol, closed triangles are PC/cholesterol, and closed squares are PG/PC/cholesterol.

TABLE 1
Effect of Lipid Composition on Blood and Tissue Distribution of Liposomes in Rats

Lipid Composition	Blood	Liver	Spleen	Heart	Kidney	Lung
PEG-PE/PC/Chol	32.3 ± 0.4	8.4 ± 1.5	2.7 ± 2.3	0.7 ± 0.1	0.9 ± 0.3	0.3 ± 0.3
PG/PC/Chol	0.2 ± 0.2	13.8 ± 1.7	1.3 ± 0.5	1.1[a]	3.7[a]	Not done
PC/Chol	0.8 ± 0.6	28.4a	3.3[a]	0.4[a]	0.7[a]	Not done

Note: After intravenous injection (n = 3 to 5 rats per formulation). Percent of injected dose remaining at 24 h post-dose.

[a]   n = 2 rats.

nates at later times. Despite these differences it is clear that the presence of drugs in the liposomes does not substantially affect the prolonged circulation.

Tissue distribution, in particular uptake by the MPS, is also greatly altered as compared with conventional liposomes. Uptake by tissues other than the MPS still has not been reported on in any detail to date. Results obtained in one study examining the effect of PEG molecular weight indicate that there is not a good correlation between blood circulation and hepatosplenic uptake. In one case, with liposomes containing 5 mol% of [750]PEG-PE, rapid clearance was observed, but without a concomitant increase in the hepatosplenic uptake.[82] This poses the question as to what other tissues are responsible for these liposomes being removed from the blood, assuming that this phenomenon results from an active process and is not simply a filtration or extravasation into one or more tissues.

The common site of uptake of liposomes is the liver and spleen, and their combined uptake is usually reported. The results from different laboratories with [2000]PEG-PE-containing liposomes indicate fairly uniformly a reduction in hepatosplenic uptake, even though the absolute values of the different studies vary. Interestingly, these two tissues differ in their uptake with respect to particle size; the liver uptake is largely size-independent, whereas uptake by the spleen is not. Larger particles show enhanced uptake by the spleen despite the presence of PEG-PE.[67,68] Similarly, a comparison of uptake by liver and spleen as a function of particle

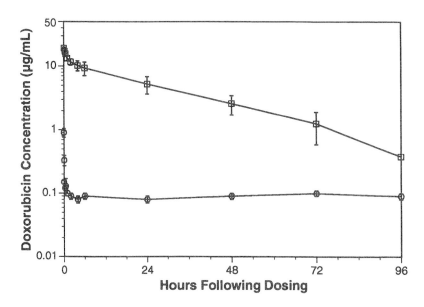

FIGURE 3. Plasma concentration of doxorubicin (μg/ml) in rabbits that received a single intravenous dose of 1.5 mg/kg DOX-SL™ (open square) or Adriamycin (open circle). Data shown are means ± SDs.

size with liposomes containing GM1 showed a strong dependence of splenic uptake on size, but not that by the liver.[67,68] Up to 50% of the injected dose was taken up by the spleen at the larger particle sizes. As found for conventional liposomes, the mechanisms of uptake by these two tissues are dissimilar. A careful evaluation of the differences in uptake by these two tissues as a function of lipid composition could be helpful in gaining a better understanding of the differences in these two important MPS tissues and its consequences for liposome delivery of drugs.

In general, uptake by other tissues, which has been limited to studies with ≤100 nm mean diameter liposomes, is either unaffected or increases. Although large changes may be observed when the uptake is expressed as percent change from that seen with conventional liposomes, this can be misleading since the levels with conventional liposomes tend to be low. The actual amount of lipid that distributes into each tissue should be evaluated for consideration of drug delivery and toxicity. The muscle, adipose, and lung tissues are unaffected, whereas the gut, bone marrow, skin, carcass, kidney, and implanted tumors generally show increased uptake.[63,80] Increases in the kidney can be discounted as due to free label, as this is its primary route of clearance. One set of tissues which appears to show significant enhanced uptake are the skin and carcass, which might give rise to new therapeutic opportunities considered unlikely previously. For example, delivery of many types of therapeutic agents to the skin tissue is of interest. In the case of studies with implanted tumors, a significant uptake was shown that may be indicative of increased extravasation in areas of leaky vasculature as a result of prolonged circulation.[63,80]

C. SURFACE-MODIFIED LIPOSOMES AS THERAPEUTIC DRUGS

DOX-SL™ is doxorubicin HCl encapsulated in SSL containing [2000]PEG-PE, developed as an oncologic drug by Liposome Technology, Inc. Preclinical studies conducted with this drug have validated and extended many of the observations made earlier with SSLs containing radiolabels, including Ga-DF.

The plasma pharmacokinetics of DOX-SL™ was evaluated in rats, rabbits, and dogs. In all three species, plasma data were best fit with a biexponential curve with a brief first phase (half-

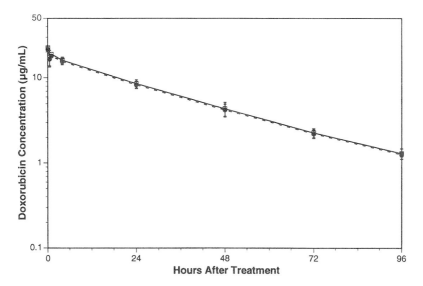

FIGURE 4. Plasma concentration of total (solid line) and liposome-encapsulated (dashed line) doxorubicin in rats that received a single intravenous dose of DOX-SL™. Data shown are means ± SDs.

life = 1 to 3 h) and a prolonged second phase (half-life = 20 to 30 h) that accounted for the majority of the area-under-the-curve (AUC). Plasma concentration and AUC varied dose-dependently, but plasma kinetics, including clearance, plasma half-lives, mean residence time, and volume of distribution, were independent of dose. This is in distinct contrast to studies with conventional liposomes, which show a striking dose-dependence in their kinetics that is related to saturation of the MPS. Plasma levels of doxorubicin are significantly higher (up to 2000-fold) after equivalent doses of DOX-SL™ and nonliposomal doxorubicin (Adriamycin) (Figure 3). Despite the higher plasma level of doxorubicin, the actual amount of free doxorubicin is very low, with the majority of the doxorubicin in the plasma remaining within the liposomes owing to their low leakage rate (Figure 4). Clearance was decreased over 400-fold in rabbits that received DOX-SL™ compared to Adriamycin animals, the volume of distribution was significantly smaller and mean residence time was significantly increased because of the long circulating liposome.

The tissue distribution of doxorubicin was also altered after DOX-SL™ treatment compared to Adriamycin, with generally lower peak tissue concentrations that occurred later after dosing.[83] Adriamycin is a cardiotoxic drug in animals and humans, and its cardiotoxic action is apparently dependent upon peak plasma and tissue levels achieved after treatment.[84] After DOX-SL™ administration, heart levels of doxorubicin are approximately 25% lower than after the same dose of Adriamycin, suggesting that DOX-SL™ may be less cardiotoxic. Indeed, multiple-dose studies in rats demonstrated that DOX-SL™ was markedly less toxic than Adriamycin in the heart.[85] Tissue and tumor AUCS, however, were higher in the animals that received DOX-SL™, suggesting that the liposomal drug may be more efficacious than the same dose of Adriamycin. This supposition is supported by observations in several murine tumor models, including P388 lymphocytic leukemia, C26 colon carcinoma, and MC2 mammary tumor models,[86,87] and in human xenograft tumor models, including ovarian, prostatic, and non-small cell lung tumors.[88-90]

Overall, these studies with DOX-SL™ suggest that the promise of SSL may be borne out in actual application of therapeutic drugs. The combination of small size, low rate of leakage, and long circulation time may both increase the efficacy and decrease the toxicity relative to conventional nonliposomal drugs.

V. CONCLUSIONS

A number of independent reports have demonstrated that surface modification of liposomes with relatively low-molecular-weight PEG, typically 2000 to 5000, results in reduced MPS uptake, primarily measured as hepatosplenic uptake, and prolonged blood circulation. The results from many studies reported from many laboratories establish that these surface-modified liposomes show significant advantages, especially for therapeutic applications: prolonged circulation and reduced MPS uptake virtually independent of (1) lipid dose, (2) other lipid components, and (3) particle size.

Several proposed mechanisms by which these effects are realized are largely based on a hypothesis of steric stabilization of the liposome inhibiting the protein adsorption that normally gives rise to the cellular interactions responsible for the rapid uptake of the liposomes. Refinement of our understanding of the exact nature of the steric stabilization is ongoing, and many new developments can be expected. Ultimately, though, it seems that the high affinity of water for the PEG can create a coating of bound water, leading to the achievement of the goal envisioned long before the advent of PEG-PE: as Bangham has said repeatedly "if you want to be invisible, look like water".[91]

Regardless of the exact nature of the steric stabilization mechanism, the resulting prolonged blood circulation and altered tissue distribution is based on the presence of the polymer that appears to override the other details of the liposomes, at least in terms of uptake and circulation. Other functionalities of the lipid composition, though, such as control of release, loading, and even binding properties of attached antibodies, appear to remain intact.

The many studies and results reviewed here comprise a major advance in the understanding of liposomes and their biological interactions. However, a great deal of knowledge is still lacking. The novel lipids developed are just beginning to be investigated, and they promise to permit a substantial gain in the fundamental understanding of bilayers and membranes formed from them. Their use in therapeutic applications is also just beginning. The results obtained to date support the notion that there are different ways to stabilize liposomes, and we believe that by appropriate selection or optimization of liposome properties in combination with appropriate targeting ligands, selective tissue localization can be achieved.

REFERENCES

1. **Senior, J.H.**, Fate and behaviour of liposomes in vivo: a review of controlling factors, *Crit. Rev. Ther. Drug Carr. Syst.*, 3, 123, 1987.
2. **Gregoriadis, G., Ed.**, *Liposomes as Drug Carriers. Recent Trends and Progress.* Wiley, Winchester, 1988.
3. **Gregoriadis, G., Ed.**, *Liposome Technology*, Vols. 1–3, 2nd ed., CRC Press, Boca Raton, FL, 1993.
4. **Lasic, D.D.**, *Liposomes: From Physics to Applications*, Elsevier, New York, 1993.
5. **Storm, G., Oussoren, C., Peeters, P.A.M., and Barenholz, Y.**, Tolerability of liposomes in vivo, in *Liposome Technology*, Vol. 3, Gregoriadis, G., Ed., CRC Press, Boca Raton, 1993, 345.
6. **Storm, G., Nassander, U.K., Vingerhoeds, M.H., Steerenberg, P.A., and Crommelin, D.J.A.**, Antibody-targeted liposomes to deliver doxorubicin to ovarian cancer cells, *J. Liposome Res.*, 4, 641, 1994.
7. **Kirby, C., Clark, J., and Gregoriadis, G.**, Effect of cholesterol content of small unilamellar liposomes on their stability in vivo and in vitro, *Biochem. J.*, 186, 591, 1980.
8. **Hwang, K.J.**, in *Liposomes from Biophysics to Therapeutics*, Ostro, M., Ed., Marcel Dekker, New York, 1987, 109.
9. **Beaumier, P.L., Hwang, K.J., and Slattery, J.T.**, Effect of liposome dose on the elimination of small unilamellar sphingomyelin-cholesterol vesicles from the circulation, *Res. Commun. Chem. Pathol. Pharmacol.*, 39, 227, 1983.
10. **Mauk, R.M. and Gamble, R.C.**, Stability of lipid vesicles in tissue of the mouse. A gamma-ray perturbed angular correlation study, *Proc. Natl. Acad. Sci. U.S.A.*, 74, 4991, 1979.
11. **Abra, R.M. and Hunt, C.A.**, Liposome disposition in vivo. III. Dose and vesicle-size effects, *Biochim. Biophys. Acta*, 660, 493, 1981.
12. **Woodle, M.C. and Lasic, D.D.**, Sterically stabilized liposomes, *Biochim. Biophys. Acta*, 1113, 171, 1992.

13. **Allen, T.M. and Chonn, A.,** Large unilamellar liposomes with low uptake by the reticuloendothelial system, *FEBS Lett.,* 223, 42, 1987.

14. **Gabizon, A. and Papahadjopoulos, D.,** Liposome formulations with prolonged circulation time in blood and enhanced uptake in tumors, *Proc. Natl. Acad. Sci. U.S.A.,* 85, 6949, 1988.

15. **Woodle, M.C.,** Surface-modified liposomes: assessment and characterization for increased stability and prolonged blood circulation, *Chem. Phys. Lipids,* 64, 249, 1993.

16. **Huang, L., Ed.,** Forum: covalently attached polymers and glycans to alter the biodistribution of liposomes, *J. Liposome Res.,* 2, 289, 1992.

17. **Gabizon, A.A., Ed.,** Forum: liposomes as carriers of anthracyclines, *J. Liposome Res.,* 4, 445, 1994.

18. **Napper, D.H.,** *Polymeric Stabilization of Colloidal Dispersions,* Academic Press, New York, 1983.

19. **Lasic, D.D., Martin, F.J., Gabizon, A., Huang, S.K., and Papahadjopoulos, D.,** Sterically stabilized liposomes: a hypothesis on the molecular origin of the extended circulation times, *Biochim. Biophys. Acta,* 1070, 187, 1991.

20. **Needham, D., McIntosh, T.J., and Lasic, D.D.,** Repulsive interactions and mechanical stability of polymer-grafted lipid membranes, *Biochim. Biophys. Acta,* 1108, 40, 1992.

21. **Woodle, M.C., Collins, L.R., Sponsler, E., Kossovsky, N., Papahadjopoulos, D., and Martin, F.J.,** Sterically stabilized liposomes. Reduction in electrophoretic mobility but not electrostatic surface potential, *Biophys. J.,* 61, 902, 1992.

22. **Torchillin, V.P. and Papisov, M.I.,** Hypothesis: why do polyethylene glycol-coated liposomes circulate so long?, *J. Liposome Res.,* 4, 725, 1994.

23. **Emanuel, N., Bolotin, E., Kedar, E., and Barenholz, Y.,** Tumor targeting of anti-cancer drugs by sterically stabilized immunoliposomes, in Liposomes in Drug Delivery: The Nineties and Beyond. Proceedings, 1993, p. 72.

24. **Mayhew, E., Lasic, D.D., Babbar, S., and Martin, F.J.,** Pharmacokinetics and antitumor activity of epirubicin encapsulated in long circulating liposomes, *Int. J. Cancer,* 51, 302, 1992.

25. **Gabizon, A.A., Barenholz, Y., and Bialer, M.,** Prolongation of the circulation time of doxorubicin encapsulated in liposomes containing a polyethylene glycol-derivatized phospholipid: pharmacokinetic studies in rodents and dogs, *Pharm. Res.,* 10, 703, 1993.

26. **Woodle, M.C., Storm, G., Newman, M.S., Jekot, J., Collins, L.R., Martin, F.J., and Szoka, F.C.,** Prolonged systemic delivery of peptide drugs by long-circulating liposomes; illustration with vasopressin in the Brattleboro rat, *Pharm. Res.,* 9, 260, 1992.

27. **Bakker-Woudenberg, I.A.J.M., Lokerse, A.F., Ten Cate, M.T., Mouton, J.W., Woodle, M.C., and Storm, G.,** Liposomes with prolonged blood circulation and selective localization in *Klebsiella pneumoniae* infected lungs, *J. Infect. Dis.,* 168, 164, 1993.

28. **Margolis, L.,** Cell interactions with solid and fluid liposomes, in *Liposomes as Drug Carriers,* Gregoriadis, G., Ed., Wiley, Winchester, 1988, 75.

29. **Heath, T.D.,** Interactions of liposomes with cells, *Meth. Enzymol.,* 149, 135, 1987.

30. **Pagano, R.E., Schroit, A.J., and Struck, D.K.,** in *Liposomes from Physical Structure to Therapeutic Applications,* Knight, C.G., Ed., Elsevier, Holland, 1981, 324.

31. **Bruckdorfer, K.R., Demel, R.A., DeGier, J., and van Deenen, L.L.M.,** The effect of partial replacements of membrane cholesterol by other steroids on the osmotic fragility and glycerol permeability of erythrocytes, *Biochim. Biophys. Acta,* 183, 334, 1969.

32. **Straubinger, R.M., Hong, K., Friend, D.S., and Papahadjopoulos, D.,** Endocytosis of liposomes and intracellular fate of encapsulated molecules: encounter with low pH compartment after internalization in coated vesicles, *Cell,* 32, 1067, 1979.

33. **Sandra, A. and Pagano, R.E.,** Liposome-cell interactions. Studies of lipid transfer using isotopically asymmetric vesicles, *J. Biol. Chem.,* 254, 2244, 1979.

34. **Poste, G.,** The interaction of lipid vesicles (liposomes) with cultured cells and their use as carriers for drugs and macromolecules, in *Liposomes in Biological Systems,* Gregoriadis, G. and Allison, A.C., Eds., Wiley, New York, 1980, 101.

35. **Moghimi, S.M. and Patel, H.M.,** Differential properties of organ-specific serum opsonins for liver and spleen macrophages, *Biochim. Biophys. Acta,* 984, 379, 1989.

36. **Martin, F.J. and MacDonald, R.C.,** Lipid vesicle-cell interactions. III. Introduction of a new antigenic determinant into erythrocyte membranes, *J. Cell Biol.,* 70, 494, 1976.

37. **Storm, G., Nassander, U.K., Roerdink, F.H., Steerenberg, P.A, Jong, W.H., and Crommelin, D.J.A.,** (1989), pp. 105–116.

38. **Barenholz, Y.,** Liposomal formulations as therapeutic devices: What is still missing?, Liposomes in Drug Delivery: The Nineties and Beyond. Proceedings, 1993, p. 49.

39. **Scherphof, G., Damen, J., and Wilschut, J.,** Interactions of liposomes with plasma proteins, in *Liposome Technology,* Vol. 3, Gregoriadis, G., Ed., CRC Press, Boca Raton, FL, 1984, 205.

40. **Scherphof, G. and Morselt, H.,** On the size-dependent disintegration of small unilamellar phosphatidyl vesicles in rat plasma. Evidence of complete loss of vesicle structure, *Biochem J.,* 221, 423, 1984.

41. **Jones, M.V. and Nicholas, A.R.,** The effect of blood serum on the size and stability of phospholipid liposomes, *Biochim. Biophys. Acta,* 1065, 145, 1991.

42. **Krause, H.J., Juliano, R.L., and Regen, S.,** In vivo behavior of polymerized lipid vesicles, *J. Pharm. Sci.,* 76, 1, 1987.

43. **Poznansky, M. and Juliano, R.L.,** Biological approaches to the controlled delivery of drugs: a critical review, *Pharm. Rev.,* 36, 277–336.

44. **Hwang, K.J. and Beaumier, P.L.,** Disposition of liposomes in vivo, in *Liposomes as Drug Carriers. Recent Trends and Progress,* Gregoriadis, G., Ed., Wiley, Winchester, 1988, 19.

45. **Kirby, C. and Gregoriadis, G.,** Plasma-induced release of solutes from small unilamellar liposomes is associated with pore formation in the bilayers, *Biochem. J.,* 199, 251, 1981.

46. **Needham, D. and Nunn, R.S.,** Elastic deformation and failure of liposome bilayer membranes containing cholesterol, *Biophys. J.,* 58, 997, 1990.

47. **Ogihara-Umeda, I. and Kojima, S.,** Increased delivery of gallium-67 to tumors using serum-stable liposomes, *J. Nucl. Med.,* 29, 516, 1988.

48. **Woodle, M.C., Matthay, K.K., Newman, M.S., Hidayat, J.E., Collins, L.R., Redemann, C., Martin, F.J., and Papahadjopoulos, D.,** Versatility in lipid compositions showing prolonged circulation with sterically stabilized liposomes, *Biochim. Biophys. Acta,* 1105, 192, 1992.

49. **Allen, T.M., Hansen, C., Martin, F., Redemann, C., and Yau-Young, A.,** Liposomes containing synthetic lipid derivatives of poly(ethylene glycol) show prolonged circulation half-lives in vivo, *Biochim. Biophys. Acta,* 1066, 29, 1991.

50. **Gamon, B.L., Virden, J.W., and Berg, J.C.,** The aggregation kinetics of an electrostatically stabilized dipalmitoyl phosphatidylcholine vesicle system, *J. Colloid Interface Sci.,* 132, 125, 1989.

51. **Crommelin, D.J.A.,** Influence of lipid composition and ionic strength on the physical stability of liposomes, *Pharm. Sci.,* 73, 1559, 1984.

52. **Tan, L. and Gregoriadis, G.,** Effect of positive charge of liposomes on their clearance from the blood and its relation to vesicle lipid composition, *Biochem. Soc. Trans.* 17, 690, 1989.

53. **Black, C.D. and Gregoriadis, G.,** Interactions of liposomes with blood plasma proteins, *Biochem. Soc. Trans.,* 4, 253, 1976.

54. **Illum, L., Davis, S.S., Muller, R.H., Mak, E., and West, P.,** The organ distribution and circulation time of intravenously injected colloidal carriers sterically stabilized with a block copolymer — polyoxamine 908, *Life Sci.,* 40, 367, 1987.

55. **Moghimi, S.M., Porter, C.J.H., Illum, L., and Davis, S.S.,** The effect of poloxamer-407 on liposome stability and targeting to bone marrow: comparison with polystyrene microspheres, *Int. J. Pharm.,* 68, 121, 1991.

56. **Woodle, M.C., Newman, M.S., and Martin, F.J.,** Liposome leakage and blood circulation: comparison of adsorbed block copolymers with covalent attachment of PEG, *Int. J. Pharm.,* 88, 327, 1992.

57. **Burkhanov, S.A. and Torchillin, V.P.,** The effect of liposome lipid composition on liposome clearance from the circulation and accumulation in mouse liver, *Bull. Eksp. Biol. Med.,* 103, 299, 1987.

58. **Burkhanov, S.A., Kosykh, V.A., Repin, V.S., Saatov, T.S., and Torchillin, V.P.,** Interaction of liposomes with different lipid content with hepatocytes in vitro, *Int. J. Pharm.,* 46, 31, 1988.

59. **Namba, Y., Sakakibara, T., Masada, M., Ito, F., and Oku, N.,** Glucuronate-modified liposomes with prolonged circulation time, *Chem. Pharm. Bull.,* 38, 1663, 1990.

60. **Oku, N., Namba Y., and Okada, S.,** Tumor accumulation of novel RES-avoiding liposomes, *Biochim. Biophys. Acta,* 1126, 255, 1992.

61. **Park, Y.S., Maruyama, K., and Huang, L.,** Some negatively charged phospholipid derivatives prolong liposome circulation in vivo, *Biochim. Biophys. Acta,* 1108, 257, 1992.

62. **Gabizon, A., Shiota, R., and Papahadjopoulos, D.,** *J. Natl. Cancer Inst.,* 81, 1484, 1989.

63. **Gabizon, A., Price, D.C., Huberty, J., Bresalier, R.S., and Papahadjopoulos, D.,** *Cancer Res.,* 50, 6371, 1990.

64. **Bakker-Woudenberg, I.A.J.M., Lokerkerse, A.F., and Storm, G.,** Enhanced localization of liposomes with prolonged blood circulation time in infected lung tissue, *Biochim. Biophys. Acta,* 1138, 318, 1992.

65. **Oku, N., Namba, Y., Takeda, A., and Okada, S.,** Tumor imaging with technetium-99m-DTPA encapsulated in RES-avoiding liposomes, *Nucl. Med. Biol.,* 20, 407, 1993.

66. **Maruyama, K., Yuda, T., Okamoto, A., Kojima, S., Suginaka, A., and Iwatsuru, M.,** Prolonged circulation time in vivo of large unilamellar liposomes composed of distearoyl phosphatidylcholine and cholesterol containing amphipathic poly(ethylene glycol), *Biochim. Biophys. Acta,* 1128, 44, 1992.

67. **Litzinger, D.C., Buiting, A.M.J., van Rooijen, N., and Huang, L.,** Effect of liposome size on the circulation time and intraorgan distribution of amphipathic poly(ethylene glycol)-containing liposomes, *Biochim. Biophys. Acta,* 1190, 99, 1994.

68. **Liu, D., Mori, A., and Huang, L.,** Large liposomes containing ganglioside GM1 accumulate effectively in spleen, *Biochim. Biophys. Acta,* 1066, 159, 1991.

69. **Juliano, R.L. and Daoud, S.S.,** Liposomes as a delivery system for membrane-active antitumor drugs, *J. Controlled Release,* 11, 225, 1990.

70. **Kronberg, B., Dahlman, A., Carltors, J., Karlsson, J. J., and Artursson, P.,** Preparation and evaluation of sterically stabilized liposomes: clinical stability, serum stability, macrophage uptake and toxicity, *J. Pharm. Sci.,* 79, 667, 1990.

71. **Assadullahi, T., Hider, R.C., and McAuley, A.J.,** Liposome formation from synthetic polyhydroxyl lipids, *Biochim. Biophys. Acta,* 1083, 271, 1991.

72. **Janas, T., Kotowski, J., and Tien, H.T.,** Polymer-modified bilayer lipid-membranes — the polypyrrole lecithin system, *Bioelectron. Bioenerg.,* 19, 405, 1988.

73. **Blume, G. and Cevc, G.,** Molecular mechanism of the lipid vesicle longevity in vivo, *Biochim. Biophys. Acta,* 1146, 157, 1993.

74. **Senior, J., Delgado, C., Fisher, D., Tilcock, C., and Gregoriadis, G.,** Influence of surface hydrophobicity of liposomes on their interaction with plasma protein and clearance from the circulation: studies with poly(ethylene glycol)-coated vesicles, *Biochim. Biophys. Acta,* 1062, 77, 1991.

75. **Allen, T.M., Hansen, C., Martin, F., Redemann, C., and Yau-Young, A.,** Liposomes containing synthetic lipid derivatives of poly(ethylene glycol) show prolonged circulation half-lives in vivo, *Biochim. Biophys. Acta,* 1066, 29, 1991.

76. **Woodle, M.C., Redemann, C., Newman, M., and Martin, F.,** unpublished results.

76a. **Torchillin, V.P., Shtilman, M.I., Trubetskoy, V.S., Whiteman, K., and Milstein., A.M.,** Amphiphilic vinyl polymers effectively prolong liposome circulation time *in vivo, Biochim. Biophys. Acta,* 1195, 181, 1994.

76b. **Maruyama, K., Okuizumi, S., Ishida, O., Yamauchi, H., Kikuchi, H., and Iwatsuru, M.,** Phosphatidyl polyglycerols prolong liposome circulation *in vivo, Int. J. Pharm.,* 111, 103, 1994.

76c. **Zalipsky, S., Brandeis, E., Newman, M.S., and Woodle, M.C.,** Long Circulating, Cationic Liposomes Containing Amino-PEG-Phosphatidylethanolamine, *FEBS Letters,* 353, 71, 1994.

76d. **Woodle, M.C., Engbers, C.M., and Zalipsky, S.,** New Amphipatic Polymer-Lipid Conjugates Forming Long-Circulating Reticuloendothelial System-Evading Liposomes, *Bioconjuate Chem.,* 5, 493, 1994.

77. **Woodle, M.C., Newman, M., Collins, L., Redemann, C., and Martin, F.,** Improved long circulating (Stealth®) liposomes using synthetic lipids, *Proc. Int. Symp. Controlled Release Bioact. Mater.,* 17, 77, 1990.

78. **Klibanov, A.L., Maruyama, K., Torchillin, V.P., and Huang, L.,** Amphipathic polyethylene glycols effectively prolong the circulation time of liposomes, *FEBS Lett.,* 268, 235, 1990.

79. **Blume, G. and Cevc, G.,** Liposomes for the sustained drug release in vivo, *Biochim. Biophys. Acta,* 1029, 91, 1990.

80. **Papahadjopoulos, D., Allen, T., Gabizon, A., Mayhew, E., Matthay, K., Huang, S.K., Lee, K.-D., Woodle, M.C., Lasic, D.D., Redemann, C., and Martin, F.J.,** Sterically stabilized liposomes: significant improvements in blood clearance, tissue disposition and therapeutic index of encapsulated drugs against implanted tumors, *Proc. Natl. Acad. Sci. U.S.A.,* 88, 11460, 1991.

81. **Mori, A., Klibanov, A.L., Torchilin, V.P., and Huang, L.,** Influence of the steric barrier of amphipathic poly(ethylene glycol) and ganglioside GM1 on the circulation time of liposomes and on the target binding of immunoliposomes in vivo, *FEBS Lett.,* 284, 1991.

82. **Newman, M.S. and Woodle, M.C.,** manuscript in preparation.

83. **Newman, M.S., Dickrell, L., and Working, P.K.,** Comparative pharmacokinetics and tissue distribution of free and liposome-encapsulated doxorubicin after a single intravenous dose in rats, *Toxicologist,* 14, 217, 1994.

84. **Speth, P.A.J., van Hoesel, Q.G.C.M., and Haanen, C.,** Clinical pharmacokinetics of doxorubicin, *Clin. Pharmacokinet.,* 15, 15, 1988.

85. **Working, P.K., Newman, M.S., Carter, J.L., Kerl, R.E., and Kiorpes, A.L.,** Comparative target organ toxicity of free and liposome-encapsulated doxorubicin in rats, *Toxicologist,* 14, 217, 1994.

86. **Vaage, J., Mayhew, E., Lasic, D., and Martin, F.,** Therapy of primary and metastatic mouse mammary tumors with doxorubicin encapsulated in long circulating liposomes, *Int. J. Cancer,* 51, 942, 1992.

87. **Vaage, J., Barbera-Guillem, E., Abra, R., Huang, A., and Working, P.K.,** Tissue distributions and therapeutic effects of intravenous free and liposomal doxorubicin against human prostatic carcinoma xenografts, *Cancer,* 73, 1478, 1994.

88. **Vaage, J., Donovan, D., Mayhew, E., Uster, P., and Woodle, M.C.,** Therapy of mouse mammary carcinomas with vincristine and doxorubicin encapsulated in sterically stabilized liposomes, *Int. J. Cancer,* 54, 959, 1993.

89. **Vaage, J., Donovan, D., Mayhew, E., Abra, R., and Huang, A.,** Therapy of human ovarian carcinoma xenografts using doxorubicin encapsulated in sterically stabilized liposomes, *Cancer,* 72, 3671, 1993.

90. **Williams, S.S., Alosco, T.R., Mayhew, E., Lasic, D.D., Martin, F.J., and Bankert, R.B.,** Arrest of human lung tumor xenograft growth in severe combined immunodeficient mice using doxorubicin encapsulated in sterically stabilized liposomes, *Cancer Res.,* 53, 3964, 1993.

91. **Bangham, A.D.,** Liposomes, in nuce, *Biol. Cell,* 47, 1, 1983.

Chapter 11

MECHANISM OF STEALTH® LIPOSOME ACCUMULATION IN SOME PATHOLOGICAL TISSUES

Shi Kun Huang, Frank J. Martin, Daniel S. Friend, and
Demetrios Papahadjopoulos

TABLE OF CONTENTS

I. INTRODUCTION

To better understand the mechanism of uptake of sterically stabilized (Stealth®) liposomes by various tissues, and to the improve therapeutic efficacy of encapsulated agents, it would be advantageous to know the pathway and the final localization of liposomes in tissues following IV injection. Radioactive agents, such as ^{67}Ga or ^{111}In, were very commonly employed as liposome markers in pharmacokinetics and biodistribution studies.[1,2] The resultant levels of radioactivity measured in relevant tissues proved that Stealth liposomes have prolonged circulation time in blood and reduced uptake by the reticuloendothelial system (RES),[3,4] and increased accumulation in solid tumors.[5-7] Instead of radioactivity measurements, people also use images of radiolabeled liposomes to monitor tissue distribution by gamma scanning.[8] Despite extensive studies on tissue distribution, it is still not known where liposomes are localized at the cellular level in most tissues following IV injection. Are there any tissues or pathologic conditions where the liposomes can cross blood vessel endothelium and basal lamina? If liposomes can extravasate and accumulate in pathologic regions, such as tumors, are they internalized by tumor cells?

Fluorescent molecules (pyranine and carboxy fluorescein)[9,10] have been encapsulated in liposomes and used to monitor their interaction with cultured cells *in vitro*. *In vivo*, fluorescent markers, encapsulated in Stealth liposomes, demonstrated extravasation of liposomes from blood vessels to the interstitium space of a tumor region, using a transparent access chamber for mouse and rat dorsal skin fold.[11,12] We used rhodamine-labeled phospholipid and applied the same dorsal window preparation to study liposome tissue distribution in an acute inflammation model that will be described in the following chapter. We also used rhodamine-dextran as a water-soluble liposome-encapsulated marker to study the distribution and extravasation of Stealth liposomes within the liver and tumor mass in frozen sections by fluorescence microscopy.[5,13] However, although we were able to prove liposome extravasation from blood vessel to tumor region, it was difficult to obtain satisfactory resolution of tissue morphology to identify various types of cells and to see whether liposomes were internalized by cells.

A variety of cytochemical or histological markers such as proteins (ferritin)[14] and enzymes (horseradish peroxidase),[15] have been encapsulated in liposomes and used *in vitro*. However, it is difficult to apply similar markers *in vivo* because of interference of the natural occurrence of these markers. Recently, we have developed a liposome preparation with encapsulated colloidal gold particles and applied silver enhancement of the colloidal gold to monitor the localization of liposomes.[13,16] Using electron microscopy, colloidal gold particles can provide an unambiguous image due to their high electron density. Using optical microscopy, silver particles indicate localization of the colloidal gold-liposomes. The amplified silver signal is much easier to detect, because the tissue sections used for light microscopy are about 30 times thicker and 30 times larger than those used for electron microscopy. Combined with hematoxylin and Eosin staining, this system can be used to clearly identify tissue morphology, cell structure, as well as show the interaction of liposomes with cells. Applying this new method, we have investigated liposome localization under some pathological conditions discussed below.

II. LOCALIZATION OF STEALTH LIPOSOMES IN SOLID TUMORS

A silver-enhanced staining procedure of liposome-entrapped colloidal gold was developed for light microscopic visualization of liposomes. Liposome size is very important for their biodistribution, and is directly related to the rate of their clearance from the blood circulation. Large liposomes (~300 nm) are cleared from the blood approximately three times faster than small (~100 nm) liposomes.[16] High-pressure extrusion is widely used for sizing down of liposomes. However, the conventional encapsulation methods, either by using commercial colloidal gold encapsulated directly into liposomes, or gold particles formed inside of liposomes before sizing, often produces a release of gold particles from liposomes during the high pressure of extrusion. For our new methodology, colloidal gold particles were formed after liposome sizing by extrusion. In this way the majority of the liposomes (80%) within 80 to 90 nm in diameter contained one to three gold particles.

Preparation of colloidal gold-liposomes was modified from previous methodology.[17] A solution of citric acid (120 mM) and K_2CO_3 (30 mM) was freshly prepared and mixed with gold tetrachloride HAuCl$_4$ (12.72 mM) in a ratio of 1:1, pH 3.4. Liposomes composed of PC/Chol/PEG-DSPE* (mole ratio 10:5:0.8) were prepared by reverse-phase evaporation[18] with gold chloride/citrate included in the aqueous phase.[13] The liposomes underwent three cycles of freezing and thawing, and then were extruded under pressure[19] through Nuclepore membranes (Pleasanton, CA), twice through pore-size 0.1 μm and five times through 0.05 μm. Immediately after final extrusion, the pH of the liposome suspension was raised to 6.0 by adding NaOH and the suspension was then incubated at 55°C for 30 min. After gold particles had formed, unencapsulated free gold and excess citrate were removed by passing the liposome suspension through a column (1 × 15 cm) of Sephacryl S-500 (Pharmacia, Piscataway, NJ). The average size of the liposomes was 80 to 100 nm in diameter, determined by electron microscopy. Most liposomes contained more than one colloidal gold particle

Light microscopy was used to visualize liposome distribution following silver enhancement of colloidal gold. The tissue specimens were embedded in water-soluble JB-4 resin, from Polysciences, Inc., Warrington, PA. All procedures involving tissue handling were performed

---

* PC/Chol/PEG-DSPE, phosphatidylcholine/cholesterol/polyethylene glycol-distearoyl phosphatidylethanolamine.

at 4°C. Sections were cut from embedded specimens with a Sorvall JB-4 microtome at a thickness of 2.5 µm. Reagents A (enhancer) and B (initiator) for silver enhancement were purchased from Amersham (Arlington Heights, IL). The sample area on the slide was covered with mixture (A and B) for 15 min at 22°C. Then the thin sections were stained with hematoxylin for 1 min and Eosin Y (1%) for 15 min.

We used silver-enhanced liposome-entrapped colloidal gold to study localization of Stealth liposomes in C-26 murine colon carcinoma.[16] Female BALB/c mice, 4 to 5 weeks old (West Seneca Laboratory, NY), were inoculated with a single cell suspension of C-26 mouse colon carcinoma injected directly into subcutaneous tissue. Liposome localization experiments were performed on mice 2 weeks after tumor cell seeding when tumor sizes ranged from 5 to 10 mm in diameter. Liposomes were injected into mice via the tail vein and 24 h later tissues were collected. Sections were processed for silver enhancement of the gold particles and examined by light microscopy. In the tumor tissue sections, blood vessels penetrating into the tumor mass are clearly visible (Plate 1A*). The tumor exhibited numerous vessels of various diameters. Dense, silver-enhanced colloidal gold was often seen surrounding blood vessels (Plate 1B). Silver-enhanced colloidal gold particles were found predominantly in the extracellular space among tumor cells (Plate 1B). Silver was rarely found in tumor cell cytoplasm. In addition, silver-enhanced gold particles could be observed to cross the blood vessel endothelium (Plate 1C), extensively penetrating into the extravascular, interstitial space among tumor cells, as well as scattered in the proximity to nonendothelial-bound streams of erythrocytes, possibly in a region of angiogenesis (Plate 1D).

We obtained a similar liposome distribution pattern following IV injection in LS174T human colon carcinoma implanted in nude mice (provided by Dr. R.S. Bresalier, UCSF). The tumor was implanted subcutaneously in nude mice and was processed by the same method as C-26 colon carcinoma. The dense concentration of silver particles was detected predominantly in the tumor extracellular space surrounding the blood vessels (Plate 2A). The silver-enhanced gold particles infiltrated into the extracellular space of tumor cells, and scattered around the entire tumor region (Plate 2B).

Transgenic mice bearing the HIV tat gene develop dermal lesions resembling a common malignant tumor in AIDS, Kaposi's sarcoma (KS). These mice were provided by Dr. Gilbert Jay, Laboratory of Virology, Jerome H. Holland Laboratory, Rockville, MD (Plate 3A). We also investigated the localization of liposomes in this Kaposi's sarcoma-like dermal lesion of transgenic mice.[20] In contrast to the distribution found in mouse and human colon carcinomas described above, in this case liposomes were found to be internalized by KS tumor cells. By light microscopy, the KS-like lesions showed areas of hypercellularity with fibroblastic or spindle-shaped cells in the dermis (Plate 3B). In this region, the abnormal thickness of the papillary dermis, composed primarily of spindle-shaped cells with elongated cytoplasmic processes and oval nuclei, was apparent. Silver-enhanced colloidal gold particles were seen predominantly localized within the lesion in the region close to the epidermis. The dense concentration of silver particles was observed to surround abnormal blood vessels and to be scattered around nonendothelial-bound streams of erythrocytes (Plate 3B). Some particles could be seen within the cytoplasm of some spindle-shaped cells with elongated nuclei (Plate 3C). The early lesions on the skin of an 8-month-old transgenic mouse were predominantly composed of collagen.[20] There were not as many spindle cells along the dermal-epidermal junction, but colloidal gold particles could still be seen to have crossed the blood vessel endothelium, extensively penetrating into the extravascular interstitial space between a few spindle-shaped cells. The silver deposits around the endothelium of blood vessels within the lesions are much denser and occur more frequently than in adjacent normal skin (Plate 3D).

* All plates follow page 124.

III. LOCALIZATION OF STEALTH LIPOSOMES IN PSORIATIC LESIONS

Mice, strain FSN, with single-gene immunologic mutations were obtained from Dr. Leonard D. Shultz, the Jackson Laboratory, Bar Harbor, ME. These animals developed dermal lesions resembling human psoriasis.[21] Some of the mice showed visible nude desquamating patches on the skin, as shown in Plate 4A.

To evaluate the therapeutic potential of drug-carrying liposomes, we have studied the localization of Stealth liposomes labeled with colloidal gold particles. Colloidal gold-containing liposomes (0.25 ml, 2 μmol phospholipid) were injected into mice via the tail vein. The mice were sacrificed 24 h after liposome injection. Tissues were collected following perfusion with heparinized phosphate-buffered saline (PBS) and fixative (1.5% glutaraldehyde, 0.1 M Na cacodylate, 1% sucrose, pH 7.4).

The mice used for examining the localization had visible desquamating patches on the skin. Plate 4A shows a typical mouse which had one 1×1 cm nude patch on the back of the neck. Erythematous and large parakeratotic crusts could often be observed on the psoriasis lesions.

Histologic examination of the psoriatic lesions shows thinning of the epidermal cell layer overlying the tips of dermal papilla with focal paraketosis (Plate 4B). Increased number of mast cells line the superficial dermis and epidermis. Silver-enhanced gold particles are located predominantly at the dermal and epidermal junction evenly distributed in the tip of the papilla. In some early and developed lesions, inflammatory foci were highly proliferated with macrophages, polymorphonuclear leukocytes, and mast cells. The silver-enhanced colloidal gold was scattered around these inflammatory cells (Plate 4C). In addition, silver-enhanced gold particles could be observed to surround hair follicles as well.

The high distribution of silver-enhanced gold particles was found exclusively in the areas around the psoriatic lesions (Plate 4D). This might be due to the leakage of the superficial postcapillary venules in the inflammatory areas, which gave the opportunity for liposomes to extravasate through the gaps between endothelial cells of the blood vessel. It is likely that the prolonged circulation time of the sterically stabilized liposomes favors their accumulation. Due to the property of high accumulation of Stealth liposomes in psoriatic lesions, they appear highly attractive as carriers of drugs for this disease.

IV. EXTRAVASATION OF STEALTH LIPOSOMES IN ACUTE INFLAMMATION

It is known that opening of inter-endothelial junctions or gaps can apparently be induced by most inflammatory mediators and some nucleotides and pro-inflammatory factors (such as histamine, bradykinin, and substance P).[22] Intravital microscopy is a versatile tool for the study of inflammation, bridging *in vitro* observations to their *in vivo* relevance. In the present study, we have examined the extravasation of fluorescence-labeled liposomes from the postcapillary venules to the interstitial space following topical application of bradykinin, as a model of acute inflammation, using a transparent access chamber of the rat dorsal skin fold. This work was performed in collaboration with Drs. N.Z. Wu and M.W. Dewhirst (Duke University Medical Center, Durham, NC).

Fluorescently labeled (rhodamine-PE) liposomes composed of PC/Chol/PEG-DSPE/Rho-PE in molar ratio of 10:5:0.8:0.1 were prepared by reverse-phase evaporation[18] and extruded under pressure.[19] The liposomes used in these experiments were approximately 80 to 95 nm in diameter. Rhodamine labeled liposomes (1 mol% rhodamine, dose: 10 μmol PL/ml; 0.1 ml) were injected IV into a rat bearing a skin flap window. Light intensity of emission fluorescence was detected from the interstitial and vascular spaces in the region of the skin flap window. Excitation and emission wavelengths used were 546 and 590 nm, respectively. A

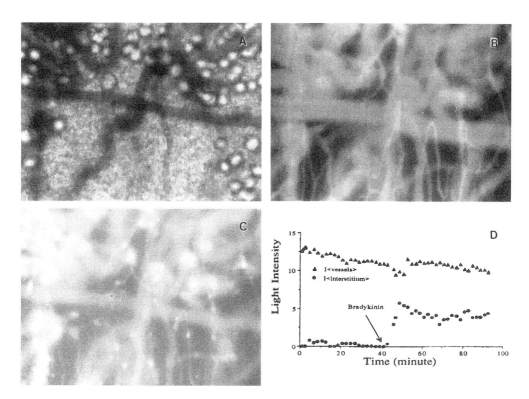

FIGURE 1. (A) Applying a transparent access window chamber, subcutaneous vasculature is observed through the window chamber by transmitted light microscopy. (B) Fluorescence micrograph shows the region in (A), 40 min after rhodamine-PE-labeled liposomes injected IV. (C) The same region 20 min after topical application of bradykinin (1 μM). Many leakage sites at postcapillary venules are shown by extravasation of rhodamine-PE-labeled liposomes. (D) Light intensity as a function of time before and after bradykinin treatment (at 40 min after liposome injection).

dorsal flap window-chamber preparation is used on Fischer-344 rats. Bradykinin solution (1 μM) was applied directly into the window chamber of the rat skin flap model following removal of one side of glass, and light emission was monitored continuously by fluorescence videomicroscopy.

For visualization, an appropriate region within blood vessels was selected in the window chamber preparation. Figure 1A shows a micrograph of vasculature from the dorsal flap window by transillumination. Rhodamine-labeled liposomes were injected intravenously. After liposome injection, blood vessels were exclusively identified by epi-illumination (Figure 1B, in the same view of Figure 1A). Obviously, no fluorescence intensity could be detected in the interstitial space. The edges of blood vessels were sharp and clear. Soon after topical treatment with bradykinin was initiated, patches with high fluorescence intensity could be observed in some interstitial space close by the blood vessels, indicating that fluorescence-labeled Stealth liposomes had begun to extravasate. Figure 1C is a fluorescence image of Rhodamine-labeled liposomes in the same region of Figure 1A and 1B obtained 30 minutes after bradykinin treatment. Several sites of leakage from vessels are shown (bright areas) by extravasation of rhodamine-labeled liposomes.

Changes in fluorescence intensity were monitored by videocamera and quantified by digital image processing. In Figure 1D, a plot shows the fluorescence intensity that was measured in vascular (solid triangles) and interstitial regions (solid circle) before and after bradykinin application. Sharp increases in fluorescence attributable to liposome-associated label were observed in the interstitial region only after application of bradykinin, as indicated. The fluorescence intensity in interstitial spaces increased more than 10-fold after bradykinin stimulation. Visual assessment of the regions confirmed the accumulation of liposomes in the interstitial region, by the appearance of bright spots therein following bradykinin treatment. Such visually apparent fluorescence were not observed prior to the application of bradykinin to the region.

V. DISCUSSION AND SUMMARY

In morphological, physiological, and pharmacological studies of microvascular permeability, leakage of particles or macromolecules was observed at postcapillary venules by electron microscopy and intravital fluorescent microscopy soon after the local application of histamine, bradykinin, or several other inflammatory mediators used to mimic the inflammatory pathological condition.[22,23] It is known that this leakage occurs through widened inter-endothelial junctions or gaps that are believed to result from endothelial cell contraction.[24] Occasionally such endothelial gaps are sufficiently wide to allow erythrocytes to extravasate.[25]

Inflammatory cells, such as mast cells, which were seen proliferating in the boundary between the dermis and epidermis of psoriatic lesions, as well as in some tumor regions, are known to be capable of synthesizing histamine and other vasoactive amines. Defects (gaps) in the tight junctions between venular endothelial cells could form in response to such vasoactive mediators, permitting liposomes to flow into the interstitial space.

Besides the leakage of the endothelium induced by inflammatory mediators, in general, tumor vessels are inherently leaky, due to the wide inter-endothelial junctions, large number of fenestrae and transendothelial channels formed by vesicles, and discontinuous or absent basement membrane.[26,27] An earlier study documented the localization of colloidal gold-containing Stealth liposomes in the interstitial space between tumor cells by a process of extravasation from tumor vessels.[16] Stealth liposomes can avoid rapid uptake by the RES and can remain in the blood circulation for a relatively long period of time. Thus, they may encounter more opportunities to extravasate through discontinuous capillaries, as well as to escape from the gaps between adjacent endothelial cells and openings at the vessel termini during tumor angiogenesis.[16]

In contrast, the blood vessels in most normal tissues are nonfenestrated capillaries. These blood vessels are composed of a single layer of endothelial cells with tight junctions. The endothelial barrier may prevent liposomes to traverse intact vessels, except by other mechanisms, such as transcytosis. This process was observed in postcapillary venules of skin in normal and trasgenic mice (Reference 20 and unpublished data), but is quite likely to be a slow process.

In sites of tumors and inflammation, either gaps induced by vasodilators (secreted by inflammatory cells) or inherently leaky endothelia are the two most likely major pathways for enhanced liposome accumulation in the perivascular space. Irrespective of the detailed mechanism for their extravasation, the observed accumulation of Stealth liposomes makes them highly attractive as drug carriers against neoplasms and site of inflammation. In addition they are also able to reduce systemic toxicity caused by some drugs, thus significantly increasing their therapeutic indices.

REFERENCES

1. **Hwang, K.J., Luke, K.K., and Beaumier, P.L.,** Hepatic uptake and degradation of unilamellar sphingomyelin/ cholesterol liposomes — kinetic study, *Proc. Natl. Acad. Sci. U.S.A.,* 77, 4030, 1980.
2. **Gabizon, A. and Papahadjopoulos, D.,** Liposome formulations with prolonged circulation time in blood and enhanced uptake by tumors, *Proc. Natl. Acad. Sci. U.S.A.,* 85, 6949, 1988.
3. **Allen, T.M., Hansen, C., Martin, F., Redemann, C., and Yau-Young, A.,** Liposomes containing synthetic lipid derivatives of polyethylene glycol show prolonged circulation half-lives in vivo, *Biochim. Biophys. Acta,* 1066, 29, 1991.

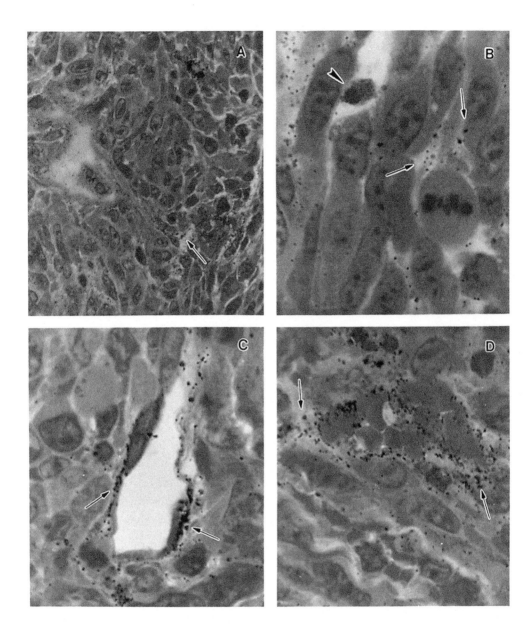

Plate 11.1. Histological preparation of C-26 colon carcinoma showing liposome-encapsulated colloidal gold with silver enhancement. Tissues were collected 24 h following colloidal gold liposome injection (IV). **(A)** Silver-enhanced gold particles indicated that Stealth liposomes infiltrate in the tumor region surrounding blood vessels. **(B)** The silver-enhanced colloidal gold particles are seen to be predominantly scattered in the extracellular space between tumor cells. A mast cell displays to metachromatic staining (arrowhead). **(C)** A penetrating blood vessel shows dense accumulation of silver-enhanced colloidal gold markers on the parenchymal side of the endothelial cells. **(D)** In C-26, silver-enhanced colloidal gold was scattered around nonendothelial-bound streams of erythrocytes, possibly in a region of angiogenesis. H & E. (A) x600; (B), (C), and (D) x1500. (Adapted from Huang, S.K. et al., *Cancer Res.*, 52, 5135, 1992.)

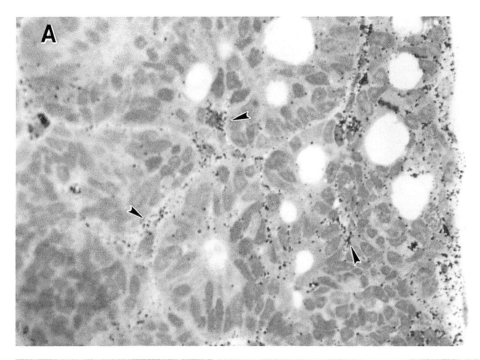

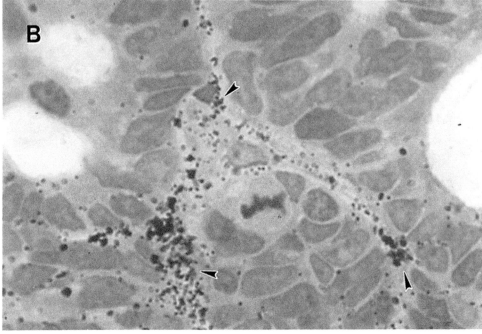

Plate 11.2. A well-differentiated LS174T human colon carcinoma. Tumor was collected 24 h following colloidal gold liposome injection intravenously. (A) The dense concentration of silver-enhanced gold particles indicate that Stealth liposomes infiltrate in the entire tumor region surrounding penetrating blood vessels. (B) The silver-enhanced colloidal gold particles are seen to be predominantly scattered in the extracellular space between tumor cells. H & E. (A) x600, (B) x1500.

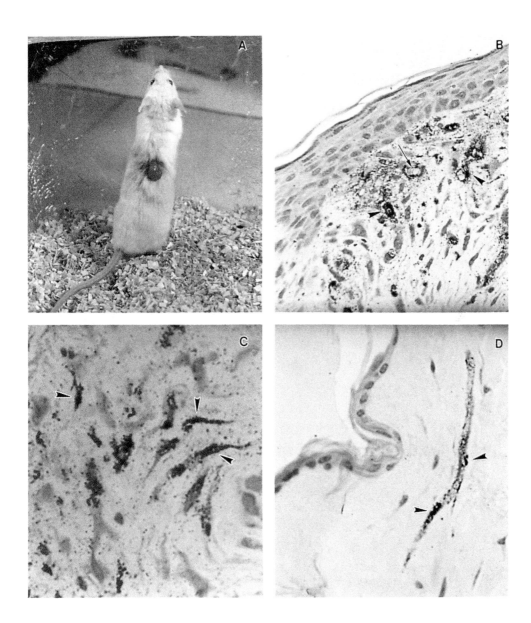

Plate 11.3. (A) Transgenic mice bearing the HIV tat gene develop dermal lesions resembling Kaposi's sarcoma. The DNA fragment was micro-injected into fertilized eggs. A 16-month-old F2 mouse had a localized 5 mm spherical erythematous tumor on its back. (B) The KS section shows silver-enhanced gold particles predominantly in the lesion region. (C) Silver-enhanced gold particles heavily labeling some of the spindle cells (arrows) in dermal lesion. (D) Normal skin in the same mouse, adjacent to the tumor. H & E. (B) and (D) x250; (C) x1000. (Adapted from Huang, S.K. et al., *Am. J. Pathol.*, 143, 10, 1993.)

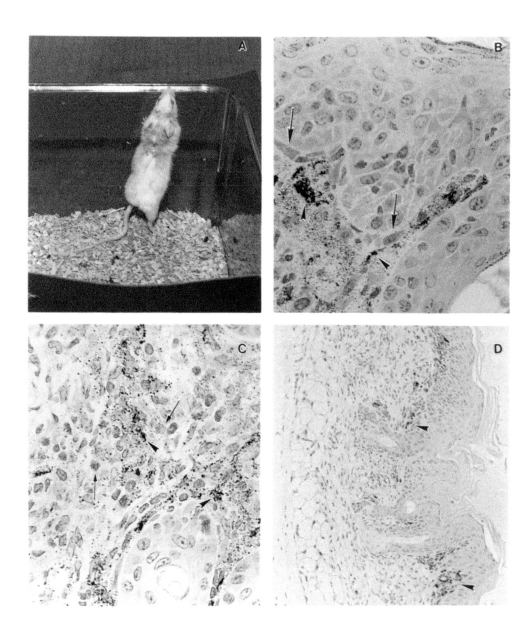

Plate 11.4. (A) A FSN mouse with gene mutation resembling psoriasis. A nude patch with erythematous and large parakeratotic crusts can be observed on the lesion in the back of the mouse neck. (B) Silver-enhanced gold particles indicate that Stealth liposomes are located predominantly in the superficial dermis. (C) The silver markers can be observed with scattered macrophages, mast cells, and polymorphonuclear leukocytes in the inflammatory area, and around the hair follicles as well. (D) The low magnification of the micrograph shows silver-enhanced gold particles exclusively around the inflammatory lesions between epidermis and dermis. Very few silver-enhanced gold particles were observed in the region of connective tissues with many blood vessels. H & E. (B) x1000, (C) x600, (D) x100.

Plate 12.1. Photomontage of an implanted tumor. The tumor mass is in the center of the micrograph and it is ~3 mm in diameter. It is surrounded by normal, well-branched microvasculature.

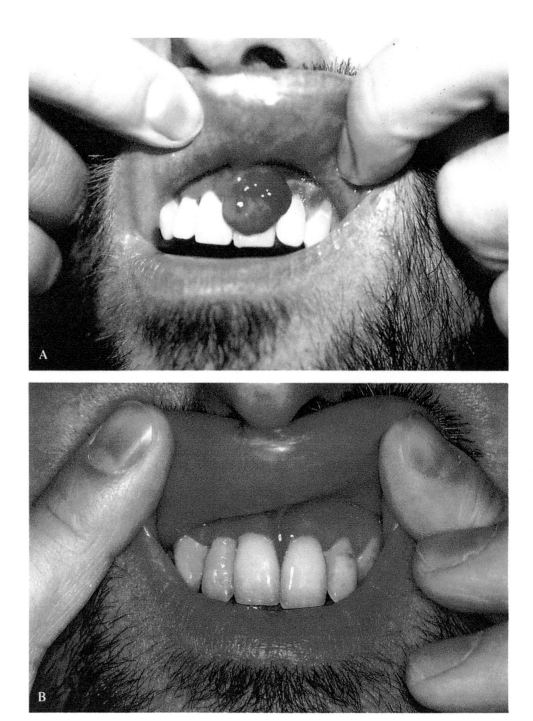

Plate 23.1. Oral KS lesion located at the frontal gingiva. (A) Prior to therapy. (B) After 6 cycles of "Stealth" liposomal doxorubicin.

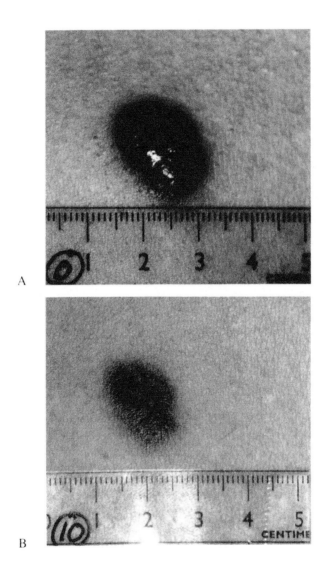

Plate 23.2. Representative cutaneous lesion before (A) and after (B) therapy with "Stealth" liposomal doxorubicin.

4. **Lasic, D.D., Martin, F.J., Gabizon, A., Huang, S.K., and Papahadjopoulos, D.,** Sterically stabilized liposomes — hypothesis on the molecular origins of the extended circulation times, *Biochim. Biophys. Acta,* 1070, 187, 1991.

5. **Papahadjopoulos, D., Allen, T., Gabizon, A., Mayhew, E., Matthay, K., Huang, S.K., Lee, K.-D., Woddle, M.C., Lasic, D.D., Redemann, C., and Martin, F.J.,** Sterically stabilized liposomes — improvements in pharmacokinetics and anti-tumor efficacy, *Proc. Natl. Acad. Sci. U.S.A.,* 88, 11460, 1991.

6. **Vaage, J., Mayhew, E., Lasic, D., and Martin, F.,** Therapy of primary and metastatic mouse mammary carcinomas with doxorubicin encapsulated in long circulating liposomes, *Int. J. Cancer,* 51, 942, 1992.

7. **Huang, S.K., Mayhew, E., Lasic, D., Martin, F.J., and Papahadjopoulos, D.,** Pharmacokinetics and therapeutics of sterically stabilized liposomes in mice bearing C-26 colon carcinoma, *Cancer Res.,* 52, 6774, 1992.

8. **Gabizon, A., Amselem, S., Goren, D., Cohen, R., Druckmann, S., Former, I., Chisin, R., Peretz, T., Sulkes, A., and Barenholz, Y.,** Preclinical and clinical experience with a doxorubicin-liposome preparation, *J. Liposome Res.,* 1(4), 491, 1990.

9. **Straubinger, R.M., Hong, K., and Papahadjopoulos, D.,** Endocytosis and intracellular fate of liposomes using pyranine as a probe, *Biochemistry,* 29, 4929, 1990.

10. **Weinstein, J.N., Yoshikami, S., Henkart, P., Blumenthal, R., and Hagins, W.A.,** *Science,* 196, 489, 1977.

11. **Wu, N., Da, D., Rudoll, T.L., Needham, D., Whorton, A.R., and Dewhirst, M.W.,** Increased microvascular permeability contributes to preferential accumulation of Stealth liposomes in tumor tissue, *Cancer Res.,* 53, 2912, 1993.

12. **Yuan, F., Leunig, M., Huang, S.K., Berk, D.A., Papahadjopoulos, D., and Jain, R.K.,** Microvascular permeability and interstitial penetration of sterically stabilized (Stealth) liposomes in a human tumor xenograft, *Cancer Res.,* 54, 3352, 1994.

13. **Huang, S.K., Hong, K., Lee, K.-D., Papahadjopoulos, D., and Friend, D.S.,** Light microscopic localization of silver enhanced liposome entrapped colloidal gold in mouse tissues, *Biochim. Biophys. Acta,* 1069, 117, 1991.

14. **Magee, W.E., Goff, C.W., Schoknecht, J., Smith, M.D., and Cherian, K. J.,** *Cell Biol.,* 63, 492, 1974.

15. **Wu, P., Tin, G.W., and Baldeschwieler, J.D.,** *Proc. Natl. Acad. Sci. U.S.A.,* 78, 2033, 1981.

16. **Huang, S.K., Lee, K.-D., Hong, K., Friend, D., and Papahadjopoulos, D.,** Microscopic localization of liposomes with prolonged circulation time in tumor bearing mice, *Cancer Res.,* 52, 5135, 1992.

17. **Hong, K., Friend, D.S., Glabe, C.G., and Papahadjopoulos, D.,** Liposomes containing colloidal gold are a useful probe of liposome/cell interactions, *Biochim. Biophys. Acta,* 732, 320, 1983.

18. **Szoka, F., Jr. and Papahadjopoulos, D.,** Procedure for preparation of liposomes with large internal aqueous space and high capture by reverse-phase evaporation, *Proc. Natl. Acad. Sci. U.S.A.,* 75(9), 145, 1978.

19. **Olson, F., Mayhew, E., Maslow, D., Rustum, Y., and Szoka, F.,** Characterization, toxicity and therapeutic efficacy of adriamycin encapsulated in liposomes, *Eur. J. Cancer,* 18, 167, 1982.

20. **Huang, S.K., Martin, F.J., Jay, G., Vogel, D., Papahadjopoulos, D., and Friend, D.,** Sterically stabilized liposomes as a drug delivery system against Kaposi's sarcoma, *Am. J. Pathol.,* 143, 10, 1993.

21. **Sundberg, J.P., Beamer, W.G., Shultz, L.D., and Dunstan, R.W.,** Inherited mouse mutations as models of human adnexal, cornification, and papulosquamous dermatoses, *J. Invest. Dermatol.,* 95(5), 615, 1990.

22. **Svesjo, E. and George, J.G.,** Evidence for endothelial cell-mediated regulation of macromolecular permeability by postcapillary venules, *Fed. Proc.,* 45, 89, 1986.

23. **Majno, G., Gilmore, V., and Leventhal, M.,** On the mechanism of vascular leakage caused by histamine-type mediators, *Circ. Res.,* 21, 833, 1976.

24. **Majno, G. and Palade, G.E.,** Studies on inflammation. I. The effect of histamine and serotonin on vascular permeability. An electron microscopic study, *J. Biophys. Cytol.,* 11, 571, 1961.

25. **McDonald, D.M.,** Neurogenic inflammation in the rat trachea. I. Changes in venules, leucocytes and epithelial cells, *J. Neurocytol.,* 17, 583, 1988.

26. **Jain, R.K.,** Transport of molecules across tumor vasculature, *Cancer Metastasis Rev.,* 6, 559, 1987.

27. **Dvorak, H.F., Nagy, J.A., Dvorak, J.T., and Dvorak, A.M.,** Identification and characterization of the blood vessels of solid tumors that are leaky to circulating macromolecules, *Am. J. Pathol.,* 133(1), 95, 1988.

Chapter 12

EXTRAVASATION OF STEALTH® LIPOSOMES INTO TUMORS: DIRECT MEASUREMENT OF ACCUMULATION AND VASCULAR PERMEABILITY USING A SKIN FLAP WINDOW CHAMBER

Mark W. Dewhirst and David Needham

TABLE OF CONTENTS

I. INTRODUCTION

This chapter will focus on some of our recent work concerning the *in vivo* deployment of the Stealth® liposome drug delivery system. This system incorporates polyethyleneglycol (PEG)-grafted lipids at ~5 mol% in drug-carrying liposomes. Our work aims to understand the mechanisms whereby such PEG liposome-encapsulated drugs and other agents might be selectively delivered to specific target sites in the body, especially to solid tumors. In the nascent state we have shown that the inherently high permeability of tumor microvasculature leads to the preferential accumulation of both conventional and Stealth liposome constructs within tumor tissues. We are also interested in specific modifications of permeability that might further enhance the specificity of uptake within tumors. For example, our laboratory has a strong interest in hyperthermia as a therapeutic treatment for cancer, both alone,[1,2,3] and coupled to radiation[4,5] and drug/agent therapies.[6,7]

We can directly observe normal and tumor microcirculation by using the dorsal skin flap chamber in the Fischer-344 rat. This observation chamber has been a central tool in obtaining critical information that allows us to evaluate the effects of a variety of physiologic manipulations (e.g., hyperthermia, radiation, vasoactive drugs) on microvascular function and solute transport in implanted tumor models.[8-10] This model gives us a direct evaluation of drug delivery to the tissue extravascular space that is separated from effects of treatment on vascular volume.

The dorsal flap chamber, then, provides a literal "window on the microcirculation". Plate 1* shows a low-magnification montage of the tissue and microvascular blood system of a

___
* All plates follow page 124.

transplanted tumor in the window preparation. In the approximately 2-week period since the tumor was seeded, it has grown to a size of ~3 mm and has established its own blood supply. It is clear that, unlike the normal blood vessels around it, the tumor microvessels are not well organized; morphologically they run a tortuous and convoluted path. In addition to this vessel tortuosity, such tumor vasculature is typified by venous lakes, regurgitant flow, and stasis.[11] At the endothelial cell level the vessel walls are not structurally complete. It has been shown by others that the blood vessels of many solid tumors are often leaky to circulating macromolecules, such as low-molecular-weight dextrans (70,000 and 150,000 g/mol) and even colloidal carbon (sizes up to 50 nm).[12] It is this feature of *increased permeability* (especially to colloidal-sized particles) that, when coupled to an ability to keep the liposomal drug supply circulating for an extended period of time, has provided the means to deliver drugs and agents "passively" into the tumor tissue.

In our liposome experiments, we are using the skin flap technique to monitor the changes in vascular concentration and extravascular accumulation of fluorescently labeled liposomes and the drug (doxorubicin, which is naturally fluorescent) in implanted tumor tissue.[13,14] The technique allows us to gain new, essential information concerning the localization of drugs and drug-carrying liposomes both in the microcirculation and their distributions within the interstitial space of the tumor. Moreover, a theoretical analysis of the distribution of fluorescence intensities associated with the drug and liposomes provides a permeability coefficient that characterizes the passage of liposomes from the blood stream into the tumor.

Methods of localizing drugs and agents by biopsy, sectioning, and electron microscopy have given valuable information with regard to both tumor permeability and drug/agent distribution in cellular and interstitial compartments.[15] The window chamber provides additional information and allows us to observe not only the *in vivo* fate of injected materials from the moment of injection, but also gives us prior knowledge of the tumor to which the drugs are targeted. Thus, it is possible with this technique to evaluate the relatively short-term (90 min) accumulation of a bolus of injected material in the tumor as well as observing the consequential effects of a treatment that is spread over a longer period of time. In this way, dosimetry and efficacy are quantified. Relations can therefore be established between the "performance in service" of a given liposome system and its chemical composition, component structure, and physicochemical properties. Through this integrated engineering approach, these microcarrier systems can thus be rationally engineered and tested, and more effective treatment schedules can be devised accordingly.

II. EXPERIMENTAL METHODS

A. THE DORSAL FLAP WINDOW CHAMBER

In our experiments we use Fischer-344 rats with transparent window chambers transplanted into the dorsal skin flap, as shown schematically in Figure 1. As discussed in more detail elsewhere,[11] a tissue plane of ~200 μm is encased between two cover-slip windows in the microchamber apparatus. The windows provide a barrier to infection and dehydration and allow a transplanted tumor to be positioned and to grow in the skin flap. Tumor transplants are made by placing a 0.1 mm^2 piece of R3230Ac mammary adenocarcinoma into the center of the window chamber. Tumor neovascularization is observed within 7 d, and ~10 d after surgery the tumor has grown to ~3 mm in diameter. Control preparations are simply the window chamber without implanted tumor. In these controls, the tissue has undergone surgery and is therefore in a state of recovery (granulating).

This window chamber can then be viewed by optical microscopy by placing the rat on the heated microscope stage to maintain normal body temperature. The window chamber itself can be heated while the animal is on the microscope stage, thus allowing for direct, on line, visualization of any effects on microcirculatory function or permeability. A full morphological

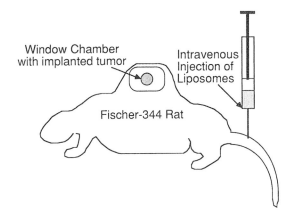

FIGURE 1. Schematic of rat with window chamber preparation as it is used, mounted on the dorsal skin flap (Fischer rat with rat mammary adenocarcinoma R3230Ac). Liposomes are injected into the tail vein, as in the schematic, or into the femoral vein.

and hemodynamic characterization of the tumor or normal tissue can be made.[11,16] In the tumor window, both tumor and normal (granulating) tissue are present and this offers an important *in situ* control. With this microscopic technique then, liposomes[13] and any other microparticle or molecule[14] that is fluorescently labeled can be injected into the bloodstream of the animal, either via the tail vein or femoral vein. These fluorescent species can thus be directly visualized as they flow through the tumor and normal circulation and accumulate in the tumor interstitium.

The details of the experiment are described in a recent publication.[14] Briefly, liposomes were made by the extrusion method and their size was controlled to be ~80 to 90 nm (measured by light scattering and negative stain electron microscopy.[13] The liposomes were made fluorescent by incorporating 2 mol% TRITC-DHPE in the egg PC:cholesterol (65:35) liposome mixture. As shown in Figure 1, a small sample (~0.2 ml) of fluorescently labeled liposomes is injected intravenously into the femoral vein. The femoral artery can also be canulated and used for taking blood samples. While viewing the tumor in transmitted light, a region of tumor vasculature is selected for observation that has "normal" blood flow and a paucity of underlying vessels. As can be seen from the micrograph in Plate 1, the over-riding feature of the tumor circulation is the tortuosity of the vessels and their disorganized arrangement compared to normal, granulating microvasculature around the tumor that shows a straighter, more evenly branched structure.

The selected region of tissue is first viewed and video-recorded in transmitted light. Observation is then switched to epifluorescence and a background image is recorded prior to liposome injection. A 0.2-ml bolus of liposomes is then injected into the femoral vein and the microvasculature is flooded with fluorescence as the liposomes flow around the bloodstream. The baseline fluorescence intensity is set as soon as possible (within 1 min after administration) and the subsequent images are compared to this baseline intensity. The region of tissue is then viewed intermittently (for 5 s every 2 min) for a period of up to 90 min. In these experiments we are forced to end our measurements at ~90 min postinjection because of limitations arising from anesthesia of the animal.

Typical images are shown in Figure 2 for tumor and normal (granulating) microvasculature preparations. In this figure, all intensities have been normalized to a common calibration and so accurately reflect the concentration of fluorescent liposomes in both vasculature and tissue. Generally, tumor microvasculature had larger blood vessels than normal tissues, and were usually devoid of capillaries. For Stealth liposomes in normal vasculature (Figure 2c, f, and i), the fluorescence in the bloodstream is maintained (the Stealth effect) whereas little

Stealth in Tumor   Conventional in Tumor   Stealth in Non-Tumor

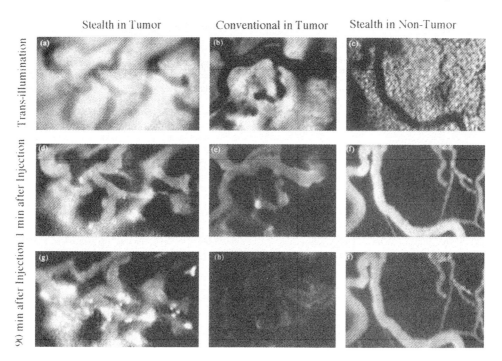

FIGURE 2. Summary of window experiments. Transilluminated and epifluorescent video images of liposomes in microvascular networks from representative experiments. The top row shows transilluminated images of tumor (a and b) and normal (c) microvasculatures. In (d), (e), and (f), observation is by epifluorescence of these three experiments 1 min after the injection of fluorescent liposomes. Stealth liposomes in tumor are shown in (d); conventional liposomes in tumor are shown in (e); and Stealth liposomes in a nontumor preparation are shown in (f). The bottom row (g, h, and i) again shows epifluorescent images of the same experiments at 90 min after injection of the fluorescent liposomes. All the video images were digitized and composed together by image software. The original light intensities in the video images were maintained during this process. (From Wu, N.Z., et al., *Cancer Res.*, 53, 3765, 1993. With permission.)

extravasation of the liposome sample is observed into the normal tissue. However, in some cases a slight diffuse fluorescence indicates some liposome extravasation does occur. The simple visual evaluation provided by examining the videomicrographs, however, does not take into account that granulating tissue microvasculature is often more dense than tumor microvasculature and so intensity values are skewed unless normalized to vascular density (as is done in our analysis described below). It should also be noted that *nongranulating* normal tissue would be expected to be much less permeable to these liposomes.

For the tumor tissue, the behavior is markedly different (Figure 2a, d, and g showing Stealth liposomes; and Figure 2b, e, and h, showing conventional liposomes). In these tumor tissues, a dramatic "hot spot" pattern of fluorescence is observed almost immediately after injection. This indicates that the permeable sites in the endothelial vascular wall are heterogeneously distributed throughout the tumor, especially for this large 100-nm sized particle. As with the nontumor preparation, the fluorescence in the tissue can be a little more diffuse. Diffuse fluorescence in all cases shows that the liposomes (lipid label) are not only extravasating but are also being transported by diffusion and/or hydraulic convection through the interstitium of the tissue. The Stealth effect is clearly seen when a comparison is made between Stealth (Figure 2g) and conventional (Figure 2h) liposomes at 90 min postinjection time. For Stealth liposomes, blood vessels still had strong light intensities, indicating the persistent plasma level of Stealth liposomes. In stark contrast, the light intensity associated with conventional liposomes had clearly deteriorated, suggesting more rapid disappearance from the blood circulation for these "unprotected" liposomes.

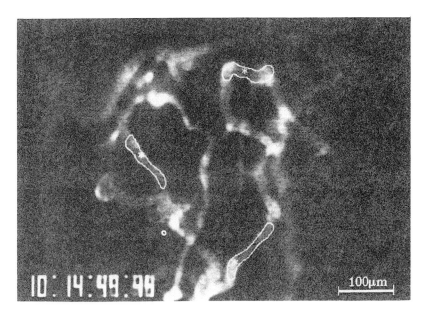

FIGURE 3. Fluorescence video image showing regions of vasculature used in analysis to generate I^*. I_t is the total fluorescence from the whole video image. (From Wu, N.Z., et al., *Microvasc. Res.*, 46, 231, 1993. With permission.)

Analysis of these images then provides a measure of vascular concentration and tissue uptake of the liposomal sample.

B. ANALYSIS OF VIDEO IMAGES

As described in more detail elsewhere,[13,14] we measure two quantities for each image of a recorded sequence: the fluorescence light intensity of the entire image I_t, and the fluorescence light intensity from selected areas within vessels I^*. Assuming that the amount of liposomes is proportional to the amount of fluorescent label within the liposomes, I_t represents the total amount of liposomes within the recorded area at any given time. Similarly, I^* represents the amount of liposomes within the vasculature. So, by using image processing software (JAVA; Jandel Scientific), we obtain I_t and I^* from an image such as that shown in Figure 3. From the values of these two parameters at time equal to zero and time equal to t, we can then calculate the total amount of liposomes remaining in the vasculature $I_v(t)$ and the total accumulation of liposomes in the interstitium of the tissue $I_i(t)$, both as a function of time after injection.

Furthermore, these parameters can also give us the effective permeability of the vasculature to these liposomes. If we assume that liposomes leak out of the bloodstream into the tumor interstitium by diffusion through the endothelial wall, then we can define an apparent permeability coefficient P (apparent implies that this model does not take into account any convection component of the extravasation that arises due to a pressure difference between tissue and vasculature). This coefficient then describes the flux of liposomes crossing the wall and normalizes for the surface area of the vasculature and the (changing) relative concentration gradient between the vasculature and the tissue interstitium.

C. DISCUSSION OF RESULTS

Analysis of the video images shown in Figure 2 by the above method provides plots of liposome concentration in the bloodstream and the interstitium vs. time after injection. Figure 4 compares the behavior of Stealth and conventional liposomes in tumor preparations. As shown in these plots, by the end of the 90-min experiment, the relative vascular concentration of Stealth liposomes maintains a higher value in the vascular system (60%) than conventional

(a)

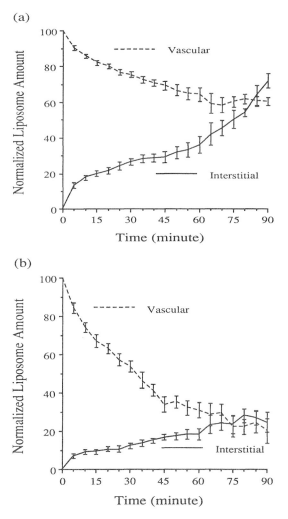

FIGURE 4. Fluorescence intensities vs. time for liposomes in implanted tumor preparations. Shown is the averaged decay in vascular concentration and increase in interstitial accumulation of injected liposomes. (a) Average of nine Stealth liposome experiments (egg PC/cholesterol 2/1, 4 mol% PEG^{2000}-lipid) and (b) seven conventional liposome experiments (egg PC/cholesterol 2/1). (From Wu, N.Z., et al., *Cancer Res.*, 53, 3765, 1993. With permission.)

liposomes (20%). This result agrees with several other earlier studies that measured liposome concentration from blood samples.[17-21] As an internal check we also performed blood sample analysis and found that the video and sample methods gave identical results.[13]

Thus, the half-life for each type of liposome is ~2 h for Stealth and 0.6 h for conventional. These values contrast markedly with the half-lives found from blood sampling that is extended over much longer periods (~24 h). The discrepancy appears to arise because of the difference in length of time over which data are acquired. In the 24-h blood circulation studies it is common to observe a bimodal decay, i.e., a rapid uptake in the early stages and a much slower uptake extending out to 24 h. The time constant for this later stage is ~20 h. In our experiments that extend to only 90 min after injection, our sensitive technique obviously picks up this early, rapid uptake stage. Even in this early stage, the concentration of Stealth liposomes in the tumor interstitium was two times higher after 90 min than for conventional. If we could carry out an extended study (and this will require a new anesthetic protocol) we would expect that at longer times the decay constant for Stealth liposomes will approach 20 h and their accumulation will be even more dramatic.

TABLE 1

**Permeability Coefficient for Stealth and Conventional
Liposomes in Tumor Preparations. Also Compared
Are the Data for Albumin**

Liposome Type	Permeability in Tumor ($\times 10^{-7}$ cm/s)	Permeability in Nontumor ($\times 10^{-7}$ cm/s)
Stealth	3.42 ± 0.78	0.88 ± 0.27
Conventional	1.75 ± 0.38	0.83 ± 0.30
Albumin	7.80 ± 1.20	2.50 ± 0.80

Another check that we made concerned whether the label accurately reflected the location of the liposome, i.e., the validity of the fluorescence measurements relies on the marker only being associated with the liposomes and not with any other blood component. Using column chromatography, we also showed that the fluorescence we measured in the blood was largely (>95%) associated with liposomes and did not involve transfer of the label to other blood components (such as albumin) that might be more permeable through the vascular wall.[13,14] Thus, we are very confident that the fluorescence method gives accurate measures of the whereabouts and concentration of *injected liposomes*.

Returning to Figure 4, the accumulation of liposomes in the tumor interstitium was found to be greater for the Stealth liposomes than for the conventional. At the end point of the experiment, the relative accumulation of Stealth liposomes was almost three times higher than for conventional liposomes. Clearly one reason is that there is a greater supply of Stealth than conventional liposomes to the interstitium from the bloodstream, due to their increased circulation time. But, this is only part of the mechanism. Interestingly, our permeability data show that the Stealth liposomes, with their "protective" PEG coat, are actually more permeable than "unprotected" conventional liposomes. The data for the two types of liposomes are shown in Table 1, and are compared to the behavior of a much smaller molecule, albumin. The permeability coefficients were 3.42×10^{-7} cm/s for Stealth vs. 1.75×10^{-7} cm/s for conventional liposomes. Thus, the PEG polymer seems to exert another effect during the permeation process itself. Morphological studies of tumor microvasculature show depleted endothelial layers and basal lamina. This suggests that the pores between endothelial cells expose surfaces that show a range of hydrophilicity (including charge) and hydrophobicity. The PEG polymer appears to exert its repulsive effect during these kinds of interactions as well, allowing the Stealth liposomes to slip through the intercellular defects more easily than the nonprotected conventional liposomes, producing an intrinsically higher permeability coefficient for extravasation.

Also, both liposomes show a lower permeability coefficient in normal granulating tissue due to their large size (100 nm) and the tighter, more ordered endothelia of the normal vessels, demonstrating the specificity of Stealth delivery to the tumor and not to normal adjacent tissue. For comparison, the same experiment, carried out with a smaller plasma protein, albumin (3×15 nm), showed the highest permeability in tumor.[14]

III. DOXORUBICIN AND DOXIL

Preclinical studies have convincingly shown the ability of anticancer drug treatments based on the Stealth liposome to eliminate implanted mammary carcinomas in mice.[22] The Stealth liposome formulation was significantly more effective than conventional liposome formulations or free drug in treating several aspects of the diseased states associated with solid tumors. For example, for various tumor types, doxorubicin encapsulated in Stealth liposomes (Doxil, Liposome Technology, Inc) was shown to: (1) reduce the incidence of metastases from intramammary implants of tumor, (2) cure mice with recent implants of tumor, and (3) increase the 8-week survival of mice with well-established implants.

These preclinical results raise several questions, such as "is the action of Doxil really due to their accumulation in tumor tissue?"; "how is the drug released and distributed throughout the tissue?"; and "how do the time courses for vascular and tumor tissue compartments for doxil vs. free doxorubicin differ?" To begin answering these kinds of questions and to gain further information about drug delivery to these sites we have carried out some preliminary experiments that compare Doxil to free doxorubicin. We administered a relatively high dose, 20 mg/kg, of doxorubicin to several animals bearing the skin flap preparation with implanted tumor. In free form the doxorubicin is itself fluorescent and so we were able to follow its vascular and interstitial concentrations with the fluorescence assay described above. At 2 min postinjection of free drug, the fluorescence in the tumor tissue rises rapidly and peaks at ~6 to 8 min. The blood vessels, which were initially intensely fluorescent, darken at this time, showing that the vascular concentration was rapidly decreasing over this time period and was actually close to zero at the 8-min mark. After 19 min the vascular compartment was completely devoid of fluorescent drug and the tumor tissue was almost completely washed out! At 30 min postinjection all the drug that entered the tissue had disappeared. (Strictly speaking, all the fluorescence associated with the drug has disappeared.) Thus, as found by others, these new data show that free Dox is rapidly cleared from the circulation (low therapeutic index) and even though its small size results in extravasation into the tumor tissue its concentration in the interstitium also rapidly depletes by mechanisms as yet not known. Total accumulation, however, appears to be small.

This implies one or both of two mechanisms for the reduction in fluorescence: (1) the drug disperses rapidly throughout the tumor tissue (and the rest of the body) and is diluted to lower than efficacious concentrations; (2) the drug is rapidly taken up by the tumor cells and its fluorescence is quenched by binding to intracellular species such as in forming cross-links with DNA. To try and distinguish between these two possibilities we have begun to carry out separate tissue assays for free and bound doxorubicin delivered to tumors as free drug and as Doxil.[23]

Finally, we have also carried out some preliminary experiments in the window chamber preparation using the Doxil product itself. Our first attempts have shown an important effect: the drug is delivered to and remains in the tissue over the 24-h period that we conducted the experiment, as shown in Figure 5. It is known that the fluorescent signal from the drug is somewhat quenched when encapsulated in a gel form inside the liposomes;[24] it has a greater intensity of fluorescence when it is not encapsulated and exists as free drug. With this preparation then, we can again track the vascular and interstitial appearance of liposomes, although this is much more difficult due to the lower fluorescence intensity of the entrapped doxorubicin. Moreover, we can observe the release of free drug from the liposomes into the tumor tissue because of the unquenching effect upon release from liposomes. As Figure 5 shows, after 90 min the characteristic heterogeneous "hot spot" extravasation was again observed and, after 24 h, as the same dose of liposomes continued circulating, we observed a build-up of fluorescence in the tumor. Thus, in contrast to free doxorubicin, doxorubicin delivered by the Doxil formulation is still present in the tumor tissue at much higher concentrations after 24 h. This apparent lack of immediate reduction in fluorescence is interesting and could imply that Doxil doxorubicin is somehow better localized within the interstitium than free doxorubicin. We are still left with not being able to answer the critical question, "how much drug reaches its intended site of action, namely cellular DNA?"

IV. SUMMARY AND CONCLUSIONS

Thus, compared to conventional liposomes there is a greater accumulation of Stealth liposomes in tumors and this is promoted by a sustained concentration of more well-dispersed

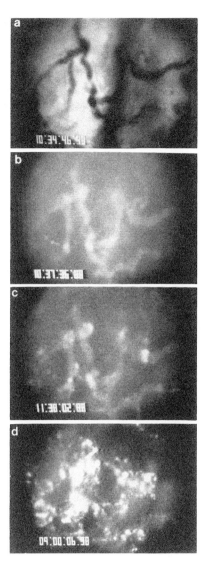

FIGURE 5. Videomicrographs of Doxil in tumor. (a) Transmitted light prior to injection; then fluorescence images postinjection, (b) 1 min; (c) 60 min; (d) 24 h.

liposomes in the blood and an increased permeability coefficient for extravasation. A characteristic hot-spot pattern is observed when fluorescent Stealth liposomes extravasate into tumor tissue, indicating the highly heterogeneous nature of tumor vascular permeability. Such hot-spot patches were also observed for conventional liposomes, although they were not nearly as bright as for Stealth liposomes. This demonstrates that these liposomes do extravasate to some extent; the accumulated amount, however, is compromised by their relatively short half-life. In contrast to tumor vasculature, there were no bright patches in nontumor vasculature, and no discernable increase in the interstitial fluorescence. This result clearly shows that Stealth liposomes are preferentially taken up in tumor tissue vs. nontumor tissue.

Thus, two mechanisms are involved in the preferential accumulation of Stealth liposomes vs. conventional liposomes in tumor tissues: (1) as reported previously the life-time of Stealth liposomes in the circulation is longer than for conventional liposomes; (2) Stealth liposomes show a twofold higher vascular permeability when liposome size, vascular concentration, vessel density, and vessel area/volume ratios are taken into account. Furthermore, the leakiness of tumor vessels promotes accumulation in tumors vs. accumulation in healthy adjacent nontumor tissue.

ACKNOWLEDGMENT

This work was supported in part by a grant from the North Carolina Biotechnology Center.

REFERENCES

1. **Thrall, D.E., Dewhirst, M.W., Page, R.L., Samulski, T.V., McLeod, D.A., and Oleson, J.**, A comparison of temperatures in canine solid tumors during local and whole-body hyperthermia administered alone and simultaneously, *Int. J. Hyperthermia*, 6, 305, 1990.
2. **Dewhirst, M.W., Phillips, T.L., Samulski, T.V., Stauffer, P., Shrivastava, P., Paliwal, B., Pajak, T., Gillim, M., Sapozink, M., Myerson, R., Waterman, F.M., Sapereto, S.A., Corry, P., Cetas, T.C., Leeper, D.B., Fessenden, P., Kapp, D., Oleson, J.R., and Emami, B.**, RTOG quality assurance guidelines for clinical trials using hyperthermia, *Int. J. Radiat. Oncol. Biol. Phys.*, 18, 1249, 1990.
3. **Oleson, J.R., Samulski, T.V., Leopold, K.A., Clegg, S.T., Dewhirst, M.W., Dodge, R.K., and George, S.L.**, Sensitivity of hyperthermia trial outcomes to temperature and time: implications for thermal goals of treatment, *Int. J. Radiat. Oncol. Biol. Phys.*, 25, 289, 1993.
4. **Leopold, K.A., Dewhirst, M., Samulski, T., Harrelson, J., Tucker, J.A., George, S.L., Dodge, R.K., Grant, W., Clegg, S., and Prosnitz, L.R.**, Relationships among tumor temperature, treatment time, and histopathological outcome using preoperative hyperthermia with radiation in soft tissue sarcomas, *Int. J. Radiat. Oncol. Biol. Phys.*, 22, 989, 1992.
5. **Dewhirst, M.W.**, Combining hyperthermia and radiation: how beneficial?, *Oncology*, 5, 110, 1992.
6. **Laskowitz, D.T., Elion, G.B., Dewhirst, M.W., Griffith, O.W., Savina, P.M., Blum, M.R., Prescott, D.M., Bigner, D.D., and Friedman, H.S.**, Hyperthermia-induced enhancement of melphalan activity against a melphalan resistant human rhabdomyosarcoma xenograft, *Radiat. Res.*, 129, 218, 1992.
7. **Laskowitz, D.T., Elion, G.B., Dewhirst, M.W., Griffith, O.W., Casero, R., Scott, P.A., Bigner, D.D., and Friedman, H.S.**, Effects of glutathione or polyamine depletion on in vivo thermosensitization, *Int. J. Hyperthermia*, 8, 199, 1992.
8. **Dewhirst, M.W., Vinuya, R.Z., Ong, E.T., Klitzman, B., Rosner, G., Secomb, T., and Gross, J.F.**, Effects of bradykinin on the hemodynamics of tumor and granulating normal tissue microvasculature, *Radiat. Res.*, 130, 345, 1992.
9. **Dewhirst, M.W., Ong, E.T., Klitzman, B., Secomb, T.W., Vinyua, R.Z., Dodge, R., Brizel, D., and Gross, J.F.**, Perivascular oxygen tensions in a transplantable mammary tumor growing in a dorsal flap window chamber, *Radiat. Res.*, 130, 171, 1992.
10. **Dewhirst, M.W., Ong, E.T., Madwed, D., Klitzman, B., Secomb, T., Brizel, D., Bonaventura, J., Rosner, G., Kavanagh, B., Edwards, J., and Gross, J.**, Effects of the calcium channel blocker flunarizine on the hemodynamics and oxygenation of tumor microvasculature, *Radiat. Res.*, 132, 61, 1992.
11. **Dewhirst, M.W., Tso, C.Y., Oliver, R., Gustafson, C., Secombe, T., and Gross, J.F.**, Morphologic and hemodynamic comparison of tumor and healing normal tissue microvasculature, *Int. J. Radiat. Oncol.*, 17, 91, 1989.
12. **Dvorak, H.F., Nagy, J.A., Dvorak, J.T., and Dvorak, A.M.**, Identification and characterization of the blood vessels of solid tumors that are leaky to circulating macromolecules, *Am. J. Pathol.*, 133, 95, 1988.
13. **Wu, N.Z., Da, D., Rudoll, T.L., Needham, D., and Dewhirst, M.W.**, Increased microvascular permeability contributes to preferential accumulation of Stealth liposomes in tumor tissue, *Cancer Res.*, 53, 3765, 1993.
14. **Wu, N.Z., Klitzman, B., Rosner, G., Needham, D., and Dewhirst, M.W.**, Measurement of protein and liposome extravasation in microvascular networks using fluorescence video microscopy, *Microvasc. Res.*, 46, 231, 1993.
15. **Huang, S.K., Lee, K.-D., Hong, K., Friend, D.S., and Papahadjopoulos, D.**, Microscopic localization of sterically stabilized liposomes in colon carcinoma-bearing mice, *Cancer Res.*, 52, 5135, 1992.
16. **Dewhirst, M.W., Oliver, R., Tso, C.Y., Gustafson, C., Secombe, T., and Gross, J.F.**, Heterogeneity in tumor microvascular response to radiation, *Int. J. Radiat. Oncol. Biol. Phys.*, 18, 559, 1990.
17. **Allen, T.M.**, Stealth liposomes: avoiding reticuloendothelial uptake, UCLA Symp. Molecular and Cellular Biology, 1989.
18. **Blume, G. and Cevc, G.**, Liposomes for the sustained drug release in vivo, *BBA*, 1029, 91, 1990.
19. **Klibanov, A.L., Maruyama, K., Torchilin, V.P., and Huang, L.**, Amphipathic polyethyleneglycols effectively prolong the circulation time of liposome, *FEBS Lett.*, 268, 235, 1990.

20. **Allen, T.M., Hansen, C., Martin, F., Redmann, C., and Y-Young, A.**, Liposomes containing synthetic lipid derivatives of poly(ethylene glycol) show prolonged circulation half-lives in vivo, *Biochim. Biophys. Acta*, 1066, 1991.

21. **Lasic, D.D., Martin, F.J., Gabizon, A., Huang, S.K., and Papahadjopoulos, D.**, Sterically stabilized liposomes: a hypothesis on the molecular origin of the extended circulation times, *Biochim. Biophys. Acta*, 1070, 187, 1991.

22. **Vaage, J., Mayhew, E., Lasic, D., and Martin, F.**, Therapy of primary and metastatic mouse mammary carcinomas with doxorubicin encapsulated in long circulating liposomes, *Int. J. Cancer*, 51, 942, 1992.

23. **Dewhirst, M.W., Da, D., Wu, N.Z., Mays, M., Frazier, D., and Needham, D.**, A comparison of plasma and tumor tissue disposition of doxorubicin: free drug vs. Stealth liposome encapsulated, *Cancer Chemother. Rep.*, submitted.

24. **Lasic, D.D., Frederik, P.M., Stuart, M.C.A., Barenholz, Y., and McIntosh, T.J.**, Gelation of liposome interior: another method for drug encapsulation, *FEBS Lett.*, 312, 255, 1992.

Chapter 13

PHARMACOKINETICS AND ANTITUMOR ACTIVITY OF ANTHRACYCLINES PRECIPITATED IN STERICALLY STABILIZED (STEALTH®) LIPOSOMES

Danilo D. Lasic

TABLE OF CONTENTS

I. INTRODUCTION

Due to their wide spectrum of activity, high potency, and also high toxicity, anthracyclines are probably the most studied family of drugs in the context of liposome encapsulation.[1] Anthracyclines are natural antibiotics which, due to their intercalation into nucleic DNA, kill dividing cells. Selective killing is therefore based on quickly dividing cells. The severe toxic side effects, such as myelosuppression (depletion of blood cells), fever and chills (due to released substances of lysed blood cells), alopecia (hair loss), and vomiting (depletion of lining cells in the digestive tract), are therefore not surprising. Furthermore, the drug action is believed to also involve membrane-related effects and high-affinity binding to cardiolipin makes the drug also cardiotoxic. Most early studies demonstrated the reduced toxicity of liposome-encapsulated doxorubicin (Adriamycin), epirubicin, and daunorubicin, while antitumor efficacy was often not studied in depth.[1-3] At that time many researchers simply believed that reduced toxicity was enough of a benefit to endorse liposome encapsulation. In reality, however, the reduced toxicity was in most cases due to increased liver and spleen uptake of the drug, and this resulted in reduced bioavailability, and in many cases negligible therapeutic efficacy. In other words, drug accumulated in liver could not reach tumors elsewhere. Only few tumors can benefit because of the altered biodistribution and pharmacokinetics. This applies mostly to liver tumors.[4,5] In some cases liposomes can also act as a sustained release system and it was speculated that the release of the drug from liposomes as well as from macrophages after liposome digestion resulted in improved efficacy observed in some tumor models.[6] In general, however, such examples were rather rare. Imaging techniques have shown reduced marker accumulation even in the case of some liver tumors and metastases because in "nondiffuse" tumors drug cannot diffuse from Kuppfer cells in the liver (which engulf most of the liposomes carrying drug) to other liver cells.[7] In other words, despite almost two decades of extensive research there is no formulation close to the market.[8]

I believe that the situation has changed dramatically with the advent of Stealth® liposomes. Preclinical data show remarkable improvements and even complete remissions of tumor mass in models in which free drug or drug encapsulated in conventional liposomes were not effective.[9-16] In contrast to conventional liposomes, these liposomes avoid uptake by the cells of the body's immune system and can circulate for a longer time. The effect is due to the protective coating of polymers which induces steric stabilization.[17] During this prolonged circulation liposomes can extravasate at the sites of leaky vasculature, which often characterizes tumors or their boundaries.[18] It was shown that drug accumulation in tumors is proportional to the blood circulation times and so was the therapeutic efficacy.[19,20] In addition to passive targeting, another mode of action of Stealth liposomes can be as a long circulating microreservoir. This would employ long circulating liposomes with slowly releasing drug in the blood circulation.

Following the pioneering work of Gabizon and Papahadjopoulos, who developed long circulating formulations on the basis of hydrogenated soy phosphatidylinositol as the "stealth component", the first *in vivo* antitumor efficacy studies of an anthracycline (epirubicin) encapsulated in sterically stabilized liposomes were performed by Mayhew, Lasic, and colleagues in 1990.[10] The lipid that provided stealth properties in these formulations was distearoyl phosphatidylethanolamine (DSPE) with covalently attached linear polyethylene glycol chain ($M_w = 1900$ Da, [1900]PEG-DSPE).

II. EXPERIMENTAL

Liposome formulations had the following composition: [1900]PEG-DSPE/lecithin/cholesterol/vitamin E in a molar ratio 15/185/100/2. To determine the optimal fluidity of the membrane several different types of lecithin were used: egg lecithin (iodine value, IV = 65), partially hydrogenated lecithins (IV = 40, 30, 20, 10) and hydrogenated soy lecithin (HSPC, IV = 1). Liposomes were prepared in 125 mM ammonium sulfate, pH = 5.5. After extrusion of large multilamellar liposomes through 0.4-, 0.2-, 0.1-, and 0.05-μm filters (three to ten passes each) liposomes were dialyzed into 5% glucose solution at pH = 7. Epirubicin was loaded by incubating liposomes at 60°C for 1 h with drug solution in glucose. After quick cooling 90 to 95% of the drug was found loaded by gel filtration or dialysis and colorimetric assays. Nonencapsulated drug was removed by ion exchange. Final, sterilized preparation (S-Epi) had the following characteristics: total lipid concentration 20 ± 2 mM, drug concentration 2 ± 0.2 mg/ml, size 90 ± 10 nm, and encapsulation efficiency >95%.

Formulation used in the human tumor xenograph study had the lipid ratio [1900]PEG-DSPE/ HSPC/Chol/antioxidant = 5.3/57/38/1.5. Liposomes were prepared in 250 mM ammonium sulfate which was, after down sizing, replaced for 10% sucrose at pH = 5.5. Instead of epirubicin doxorubicin was used (S-Dox). Unpublished studies[25] found equivalent activity of the two drugs in the colon C26 tumor model.

HSPC-containing formulations were extruded at 55 to 60°C because the gel-liquid crystal phase transition temperature of the fully hydrated suspension, as measured by DSC, is 54.96°C. As expected,[8] dense aqueous suspension of formulation with 38 mol% cholesterol did not show any phase transition in the range from 20 to 70°C.[26]

Conventional liposomes (L-Dox) were prepared from phosphatidyl glycerol/phosphatidylcholine/cholesterol/*dl*-α-tocopherol (1/4/3/0.02). Doxorubicin concentration was 0.8 mg/ml and total lipid concentration was 35 mM. Particle size was around 250 nm.

Stability *in vitro* was tested by dilution-induced leakage and plasma-induced leakage assays. In both cases liposomes were diluted 1:100 with the buffer (or 50% human plasma) incubated for 1 h at 37°C and the encapsulated and released drug were measured by fluorescence after separation on a gel permeation filtration column (BioGel 15M, 25×0.9 cm, eluted with degased saline at flow rate 20 ml/h).

Studies of the growth of C26 colon carcinoma and human lung tumor xenographs as tumor models will be reviewed. C26 colon carcinoma cells were injected subcutaneously into BALB/c female mice (1 million cells in 0.1 ml). Mice were randomized, selected into groups and treatment began 3, 10, or 14 d later. Toxicity of the free and encapsulated drug was measured in healthy animals.[10]

Human lung cancer cells were engrafted in severe combined immunodeficient (scid) mice (CB-17) by subcutaneous injection (2 million cells). Tumor size was measured weekly and tumor volumes were calculated as $a \cdot b^2/2$ with a and b being the longest and shortest dimension of the tumor, respectively. Weekly injections at 2 mg/kg started 1 week after engrafting.[13]

III. RESULTS AND DISCUSSION

Drug release upon dilution- or plasma-induced leakage (DIL and PIL, respectively) has shown that liposomes containing hydogenated lecithin (HSPC) give the most stable encapsulation. Typically DIL show no leakage (at 1 h and 1 d time point) while EPC formulations lost 5 to 10% of drug at those time points. PIL showed about 5% release for HSPC and 30 to 40% release for the EPC-based formulations after 1 h of incubation. Also *in vivo* tumor efficacy was shown to be the highest when this lecithin was a major component in the bilayer (data not shown).

Typical blood circulation half-lives ($t_{1/2}$) in rats and mice of about 20 h were measured. In the case when epirubicin was used as a marker almost 10% of the dose was still circulating after 3 d.[10] Parallel *in vitro* plasma-induced drug release studies indicated releases of 15 to 20% of the drug. This means that the measured circulation time may be an underestimate. I also believe that two-component decays in the percent of the remaining dose = f (time) profiles (which were sometimes observed, especially with the early samples) are due to experimental artifacts. Mostly, contamination of samples with larger liposomes or with colloidal drug and some leaky liposomes may be responsible for the fast clearance component observed with some samples after administration. (Such samples also show larger releases in DIL and PIL.) When only small and stable liposomes remain, they exhibit longer blood circulation times. In the above case (10% of the dose after 3 d) the second release gives a blood circulation half-life of approximately 24 h. After correcting for the drug leaked from the circulating liposomes, one can estimate that $t_{1/2} = 30$ h in rats and mice. In larger animals and in humans the same formulations achieve longer circulation times due to the different volumes of distribution and pulse rates.

A. PHYSICAL STATE OF THE DRUG IN THE LIPOSOMES

The high stability of the drug encapsulation was attributed, in addition to mechanically very cohesive and impermeable bilayers, also to the precipitation of the drug in the liposomes.

Assuming uniform size distribution of 0.08 µm and encapsulation efficiency of 95%, one can calculate that the internal concentration of drug in liposomes is above 120 mM, which is several times above its aqueous solubility.[8] Actual measurements of internal volume by EPR yielded internal concentrations of around 200 mM.[27] This is due to the fact that internal volume is smaller than calculated from light-scattering data. One reason is the nonspherical shape of vesicles, which may be due to the formation of anisotropic gel and/or osmotic imbalances resulting from drug loading. The second reason is the fact that vesicles are smaller than dynamic light scattering results indicate. Small angle neutron scattering experiments,[28] as well as gel chromatography on calibrated Sephacryl S 1000 column, as described by Nozaki et al.,[21,29] and cryoelectron microscopy,[30] show diameters which are 10, 5, and 15% smaller,

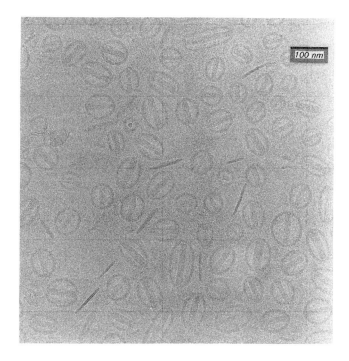

FIGURE 1. Cryoelectron microscopic observation of doxorubicin encapsulated in sterically stabilized liposomes. (Courtesy of P. Frederik, University of Limburg, Maastricht.)

respectively, than the ones observed by light scattering. This may be understood by increased hydrodynamic drag and reduced diffusion constant of sterically stabilized liposomes containing surface attached polymers. High-resolution electron micrographs have shown that drug molecules are precipitated with sulfate ions in fibrilar structures which align and bind themselves into bundles.[31] This gel was visualized by cryoelectron microscopy (Figure 1) and small angle X-ray scattering where the observed periodicity of 2.7 nm was attributed to the thickness of the gel fibers.[22] Test tube studies showed formation of a gel already at much lower drug concentrations. The liquid-solid phase transition increases with increasing concentrations of either reactant. It shows hysteresis that could be clearly observed by naked eye in the thermostated bath. DSC measurements (Perkin Elmer, model DSC-7), however, didn't reveal any phase transition, probably due to small enthalphy of the transition which could well be due to weak H-bonds.[32]

The mechanism of drug encapsulation is shown schematically in Figure 2. Rigorous theoretical explanation of the loading process will be published elsewhere.[23] We should stress the importance of stable drug encapsulation in the case of Stealth liposomes because of their long circulation.

B. THERAPEUTIC EFFICACY

Toxicity studies revealed that encapsulated drug is slightly less toxic. Free epirubicin has $LD_{50} = 11 \pm 1$ mg/kg, whereas liposome-encapsulated epirubicin had $LD_{50} = 13.5 \pm 1$ mg/kg in female BALB/c mice.[10] Maximal tolerated doses of 6 and 9 mg/kg were found, respectively. At this point one should add that the toxicity issue of enhanced accumulation of conventional liposomes in the liver macrophages, which may not exhibit dose-limiting toxicity, has not been addressed critically yet. Apart from possible severe suppression of the immune system, it was mentioned recently that macrophages can be depleted and that this can cause dramatic increase in the development rate of the liver metastases from other tumor cells. Liver-targeted

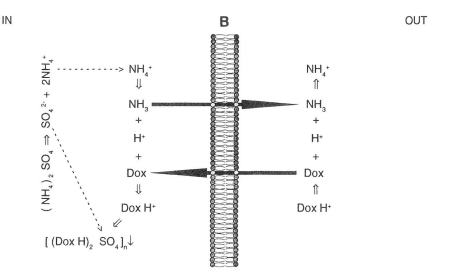

FIGURE 2. Schematic presentation of the mechanism of drug encapsulation in liposomes using ammonium sulfate gradient. Loading is a consequence of exchange of ammonia with uncharged doxorubicin due to the gradient of chemical potential.[23] The driving force for the efflux of ammonia is larger external volume and possibly higher pH outside, while the driving force for doxorubicin accumulation is precipitation with sulfate anion. The four reactions (dissociation reactions) of NH_4^+ and $DoxH^+$ on each side of the membrane are coupled via permeation rates of neutral species, pK values of both reactions, pH at each side, and the ratio of internal and external volume. (Drawing courtesy of Phil LeRoy Carter.)

liposome formulations laden with potent drugs can therefore severely jeopardize the natural defense system.[24]

In contrast to small changes in the safety, more dramatic improvements, however, were observed in therapeutic efficacy studies of drug-laden Stealth liposomes.

Figure 3 shows C26 colon tumor-sized growth in 10 mice as a function of different treatments for two dosing protocols, i.e., drug administration on days 3, 10, and 17 or 10, 17, and 24 after tumor inoculation, as indicated by arrows. Panels show saline controls (A), free epirubicin at 6 mg/kg (B), S-Epi at 6 (C) and 9 mg/kg (D), and a mixture of free drug and placebo liposomes at lipid dose 600 μM/kg (E). Panel F shows the mean tumor size from panels A to E. The percent of surviving animals on an extended time scale is shown in Figure 4. These results clearly show that S-Epi can cause remission of tumors, whereas free drug is practically inactive.

Figure 5 shows the growth of human lung tumor cells after several different treatments. Free drug is compared to Stealth doxorubicin (S-Dox) in panel A, conventional doxorubicin-containing liposomes (L-Dox) in B, while C shows the dose dependence of the treatment with S-Dox. The data show that at equivalent doses S-Dox can arrest the growth while free drug and drug in conventional liposomes do not. Both only decreased the growth from about 1.1 cm³/week in untreated animals to about 0.5 cm³/week. This is in line with previous observations that conventional liposomes are effective mostly in the treatment of experimental liver metastasis.

Treatment with S-Dox also shows a linear dose response (Figure 6). At the dose 2 mg/kg, complete arrest of tumor growth was obtained and 100% of mice survived to week 12. They appeared healthy and active with minimal to small body weight losses (2 to 19%).[13]

Improvements were also achieved in other resistant tumor models, such as mammary carcinoma (see Chapter 14). S-Dox formulation was substantially more effective, not only in curing mice with recent implants from various tumors, but also in reducing the incidence of metastases originating from these intramammary implants.[11]

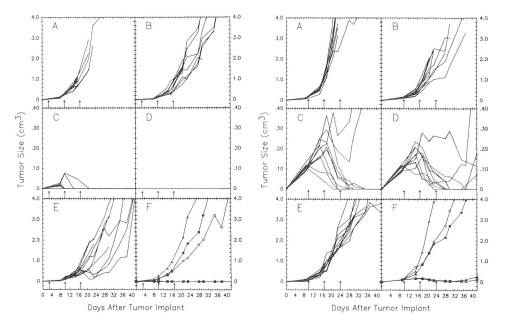

FIGURE 3. Colon C26 tumor growth in BALB/c mice as a function of various treatments: (A) saline control (△); (B) free epirubicin at 6 mg/kg (○); (C) S-Epi at 6 mg/kg (▲); (D) S-Epi at 9 mg/kg (□); (E) placebo liposomes and free drug at 6 mg/kg (●); (E) average tumor sizes from all 10 animals. *Left*: treatment on days 3, 10, 17; *right*: on days 10, 17, 24. (From Mayhew, E. et al., *Int. J. Cancer,* 51, 302, 1992. With permission.)

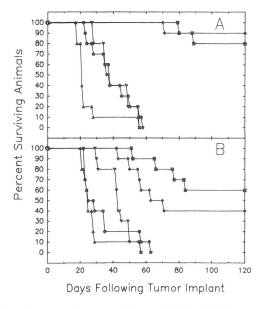

FIGURE 4. Data from Figure 3 presented as percent of survival of mice. (A) Treatment on days 3, 10, and 17; (B) with 1-week delay. Symbols: (▲) saline control; (●) free epirubicin at 6 mg/kg; (▼) mixture of free drug and placebo liposomes; (□) S-Epi at 6 mg/kg; and (♦) S-Epi at 9 mg/kg. (From Mayhew, E. et al., *Int. J. Cancer,* 51, 302, 1992. With permission.)

C. MECHANISM OF ACTION

The results can be explained by the increased concentration of drug in tumors. The prolonged presence of small liposomes in blood enables their extravasation into tumors with a leaky vascular system. Other researchers, including Gabizon, Northfelt, and Dewhirst, have measured the amount of drug in various tumors to be five- to tenfold larger in the case of the

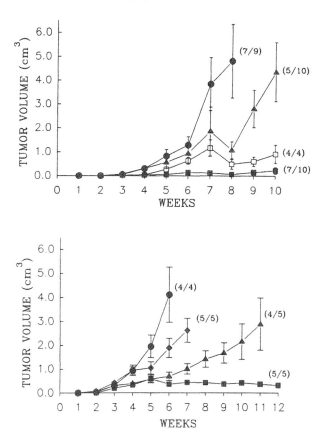

FIGURE 5. Growth of human lung tumor xenographs in scid mice. *Top,* comparison to free drug: (●) control; (▲) free doxorubicin at 3 mg/kg; (□) S-Dox at 3 mg/kg. (▦) shows delayed treatment, commencing at week 4. *Bottom,* Comparison to conventional liposomes: (●) control; (▲) L-Dox at 3 mg/kg; and (▦) S-Dox at the same dose. (From Williams, S.S. et al., *Cancer Res.,* 53, 3964, 1993. With permission.)

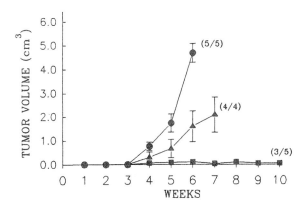

FIGURE 6. Dose dependence of the growth inhibition: (●) control; S-Dox at 0.5 mg/kg (♦); 1 mg/kg (▲); and 2 mg/kg (▦). In parentheses the fraction of mice surviving to final time points. (From Williams, S.S. et al., *Cancer Res.,* 53, 3964, 1993. With permission.)

encapsulated drug (S-Dox) as compared to the free drug. Skin flap window studies have enabled continuous monitoring of the accumulation of doxorubicin in some tumors.[19] In addition to standard mechanisms of action, high drug concentration in tumor and/or its surroundings can cause necrosis of blood vessels and consequent death of tumor cells due to lack of oxygen and nutrients.

A very important factor in microencapsulation systems is also the release of encapsulated molecules. We still do not know exactly what happens to the liposomes after extravasation. We believe that they become trapped in the tissue, where they release the drug. Some liposomes may eventually be digested by macrophages, some chemically degraded and/or mechanically ruptured or disintegrated. In addition to lipid composition and surface properties of liposomes, the physical state of the drug within liposomes, as discussed above, is also very important. Because the drug practically does not leak *in vitro*, we believe that liposomes break and drug molecules are (slowly) dissolved and dispersed. Increase in fluorescence with time clearly indicates that drug molecules are being released from liposomes, in which their fluorescence is quenched.[19] Furthermore, drug release can be accelerated by the extraliposomal ammonium sulfate from ruptured liposomes which can reduce the gradient and causes drug leakage. Regardless of the release mechanism, the therapeutic efficacy of this formulation clearly shows that the drug is bioavailable in an active form.[8]

IV. CONCLUSION

The results of these model studies show that anthracyclines in Stealth liposomes may represent an important breakthrough in cancer chemotherapy. A simple mechanism can explain enhanced therapeutic efficacy of anthracyclines encapsulated in long circulating liposomes. In addition to the preclinical studies described above, and the ones described in Chapter 14, clinical studies of doxorubicin encapsulated in sterically stabilized liposomes have also shown encouraging results. A good response with minimal toxicity was observed in patients with AIDS-related Kaposi's sarcoma, and Phase II studies against other cancers are underway.

Another benefit of long circulating liposomes is the possibility of selectively targeting them to accessible sites, a subject that is described elsewhere in this volume.

Future research and development in the liposome field may therefore concentrate, in addition to very specific formulations for specific uses, on targeted Stealth liposomes with programmable release kinetics or programmable internalization into target cells. It appears that the advent of Stealth liposomes has made these goals feasible again.

ACKNOWLEDGMENTS

I would like to thank Drs. E. Mayhew, F. Martin, and A. Gabizon for helpful discussions and Dr. J. Vaage for the critical reading of the manuscript and several important comments.

REFERENCES

1. **Gabizon, A., Ed.,** Liposomes as carriers for anthracyclines, Forum Issue, *J. Liposome Res.,* 4, 445–687, 1994.
2. **DeDuve, C., Trouet, A., Deprez, D., and Baurain, R.,** Liposomes as lysosomotropic carriers, *Ann. NY Acad. Sci.,* 308, 226, 1978.
3. **Forssen, E.A. and Tokes, Z.A.,** Use of anionic lipids for the reduction of chronic doxorubicin induced cardiotoxicity, *Proc. Natl. Acad. Sci. U.S.A.,* 78, 1873, 1981.
4. **Mayhew, E., Goldrosen, and Vaage, J.,** Effects of liposome entrapped doxorubicin on liver metastases of mouse colon tumors 26 and 38, *J. Natl. Cancer Inst.,* 78, 707, 1987.
5. **Szoka, F.C.,** Liposome drug delivery, in *Membrane Fusion,* Wilshut, J. and Hoekstra, R., Eds., Marcel Dekker, New York, 1990, 845.
6. **Storm, G., Roerdink, F.J., Steerenberg, B., de Jong, W., and Crommelin, D.J.A.,** Influence of lipid composition on the anti-tumor activity exerted by doxorubicin-containing liposomes in a rat solid tumor model, *Cancer Res.,* 47, 3366, 1987.
7. **Gabizon, A., Cherin, R., Amselem, S., Druckmann, S., Cohen, R., Goren, D., Fromer, I., Peretz, T., Sulkes, A., and Barenholz, Y.,** Pharmacokinetics and imaging studies in patients receiving a formulation of liposome associated adriamycin, *Br. J. Cancer,* 64, 1125, 1991.

8. **Lasic, D. D.,** *Liposomes: from Physics to Applications*, Elsevier, Amsterdam, New York, 1993.

9. **Papahadjopoulos, D., Allen, T.A., Gabizon, A., Mayhew, E., Matthay, K., Huang, S.K., Woodle, M.C., Lasic, D.D., Redemann, C., and Martin, F.J.,** Sterically stabilized liposomes: improvements in pharmaco-kinetics and anti-tumor therapeutic efficacy, *Proc. Natl. Acad. Sci. U.S.A.,* 88, 11460, 1991.

10. **Mayhew, E., Lasic, D.D., Babbar, S., and Martin, F.J.,** Pharmacokinetics and antitumor activity of epirubicin encapsulated in long circulating liposomes incorporating a polyethylene glycol-derivatized phos-pholipid, *Int. J. Cancer,* 51, 302, 1992.

11. **Vaage, J., Mayhew, E., Lasic, D.D., and Martin, F.J.,** Therapy of primary and metastatic mouse mammary carcinoma with doxorubicin encapsulated in long circulating liposomes, *Int. J. Cancer,* 51, 9942, 1992.

12. **Huang S.K., Mayhew, E., Gilani, S., Lasic, D.D., Martin, F.J., and Papahadjopoulos, D.,** Pharmacoki-netics and therapeutics of sterically stabilized liposomes in mice bearing C-26 colon carcinoma, *Cancer Res.,* 52, 6774, 1992.

13. **Williams, S. S., Alosco, T.R., Mayhew, E., Lasic, D. D., Martin, F.J., and Bankert, R.B.,** Arrest of human lung tumor xenografts growth in severe combined immunodeficient mice using doxorubicin encapsulated in sterically stabilized liposomes, *Cancer Res.,* 53, 3964, 1993.

14. **Working, P.W., Newman, M., Huang, S.K., Vaage, J., Mayhew, E., and Lasic, D.D.,** Pharmacokinetics, biodistribution, and therapeutic efficacy of doxorubicin encapsulated in Stealth liposomes, *J. Liposome Res.,* 4, 667, 1994.

15. **Ning, S., Macleod, K., Huang, A.H., Abra, R.M., and Hahn, G.M.,** Hyperthermia induces doxorubicin release from long-circulating liposomes and enhances their antitumor efficacy, *Cancer Res.,* submitted.

16. **Vaage, J., Mayhew, E., Abra, R.M., and Huang, A.,** Therapy of human ovarian carcinoma in nude mice using doxorubicin encapsulated in long circulating liposomes, *Cancer,* in press.

17. **Lasic, D.D., Gabizon, A., Huang, S.K., and Papahadjopoulos, D.,** Sterically stabilized liposomes: a hypothesis on the molecular origin of extended blood circulation times, *Biochim. Biophys. Acta,* 1070, 1187, 1991.

18. **Dvorak, H.F., Nagy, J.A., Dvorak, J.T., and Dvorak, A.M.,** Identification and characterization of the blood vessels that are leaky to circulating macromolecules, *Am. J. Pathol.,* 133, 95, 1988.

19. **Needham, D., Hristova, K., McIntosh, T.J., Dewhirst, D., Wu, N., and Lasic, D.D.,** Polymer-grafted liposomes: physical basis for the "stealth" property, *J. Liposome Res.,* 2, 411, 1992.

20. **Wu, N.Z., Da, D., Rudoll, T.L., Needham, D., Whorton, A., and Dewhirst, M.W.,** Increased microvascular permeability contributes to preferential accumulation of Stealth liposomes in tumor tissue, *Cancer Res.,* 53, 3765, 1993.

21. **Nozaki, Y., Lasic, D.D., and Tanford, C.,** Size characterization of phospholipid vesicle preparations, *Science,* 217, 366, 1982.

22. **Lasic, D.D., Frederik, P., Stuart, M.C., Barenholz, Y., and McIntosh, T.J.,** Gelation of liposome interior: a novel method for drug encapsulation, *FEBS Lett.,* 312, 255, 1992.

23. **Ceh, B. and Lasic, D.D.,** Rigorous theory of drug loading into vesicles, submitted.

24. **Scherphof, G.L., van Borssum M., Thomas, K., and Daeman, T.,** Liposomal formulations and anticancer drugs: anti-tumor activity and effects on macrophages, in *Liposomal Drug Delivery: Nineties and Beyond,* Gregoriadis, G. and Florence, A.T., Eds., London, Dec. 1993.

25. **Mayhew, E., Martin, F., and Lasic, D.D.,** unpublished results.

26. **Huang, S. K. and Lasic, D.D.,** unpublished data.

27. **Cafiso, D. and Lasic, D.D.,** unpublished data.

28. **Auvray, L., Auroy, P., and Lasic, D.D.,** in preparation.

29. **Grunwald, T. and Lasic, D.D.,** unpublished data (see Reference 8, p. 160–161).

30. **Frederik, P. and Lasic, D.D.,** unpublished data.

31. **Lasic, D.D., Federick, P., Stuart, M.C., Ceh, B., and Barenholz, Y.,** Transmembrane gradient driven phase transitions within vesicles, submitted.

32. **Huang, S.K. and Lasic, D.D.,** unpublished data.

Chapter 14

TISSUE UPTAKE AND THERAPEUTIC EFFECTS OF STEALTH® DOXORUBICIN

Jan Vaage and Emilio Barbera

TABLE OF CONTENTS

I. TISSUE UPTAKE

A number of studies in animal tumor models have found that the therapeutic effects of several anticancer drugs will be enhanced and prolonged, and toxic side effects reduced, when the drugs are encapsulated in liposomes.[1-4] The improved therapeutic effect is thought to be due to the slow release of drugs from liposomes.[5] The effectiveness of drugs in conventional liposomes is limited, however, by their rapid uptake by the cells of the reticuloendothelial system (RES),[6,7] reducing the amount of drug that may reach the tumor.[7] The uptake of liposomes by the RES is decreased and their circulation time increased, by the covalent attachment of polyethylene glycol to the lipid bilayers (sterically stabilized liposomes, a formulation characteristic named Stealth®).[8,9] The increased accumulation of sterically stabilized liposomes in tumors outside RES organs[10] is supported by observations, using standard and electron microscopy, of the accumulation of sterically stabilized liposomes containing colloidal gold in the extravascular spaces of tumor stroma.[11] The therapeutic efficacy of doxorubicin against several mouse mammary carcinomas was shown to be increased by encapsulating the drug in sterically stabilized liposomes, Doxil*, in comparisons with the drug in conventional liposomes, or free in saline.[12] From the intratumor location, a drug would be slowly released from the liposomes and would maintain the optimum local intracellular and extracellular cytotoxic levels,[13] for prolonged periods.

The tissue uptakes after the IV injection of doxorubicin in saline suspension, and encapsulated in sterically stabilized liposomes, were compared in nude Swiss mice carrying subcutaneous (s.c.) implants of the human prostatic carcinoma PC-3. Confocal laser scanning

---

* Stealth® and Doxil® are registered trademarks of Liposome Technology, Inc., Menlo Park, CA.

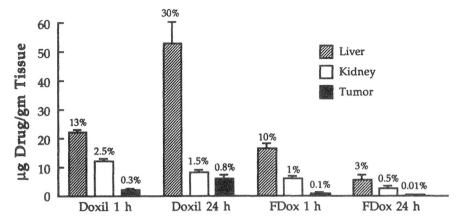

FIGURE 1. Tissue distributions of doxorubicin in Stealth liposomes (Doxil) (adjusted for the auto-quenching factor 2.8), and free doxorubicin (FDox). The percentages of the total quantity of drug injected per mouse (250 µg), present in the liver (1.5 g), in the kidneys (0.25 g each), and in the tumor (0.3 g), are shown at each bar.

microscopy and microfluorimetry were used to determine tissue distribution and to quantitate drug uptake by the tumor implants, and by two major metabolic organs, the liver and the kidneys. Figure 1 shows the relative quantities of free doxorubicin and of Doxil corrected for the auto-quenching factor 2.8. The measurements were made on cryostat sections of tumor, liver, and kidney, removed 1 h and 24 h after the IV injection of 0.9 mg/kg of each drug formulation. The mice, previously untreated, carried 30-d PC-3 implants, 0.3 to 0.4 mm^3.

Then the uptakes of Doxil and free doxorubicin by 30-d PC-3 implants were measured again, with time points at 1, 2, 6, 16, 24, 48, 72, 120, 168, and 216 h after the IV injection of 0.9 mg/kg of each formulation. Figure 2 shows that free doxorubicin was detectable in the tumor up to 24 h, and that Doxil was detectable up to 216 h. The relative values for the areas under the curves for free doxorubicin and Doxil were 36.5 and 919, respectively. This represents a 25-fold increase in the drug at the disease site.

Using confocal laser scanning microscopy, it was found that (1) the elimination of doxorubicin via the liver was reduced when the drug was in the Doxil formulation; (2) the uptake of doxorubicin in the tumor was increased, and the presence of the drug prolonged, when the drug was in the Doxil formulation.

It is likely that the reduced excretion and the increased circulation time of Doxil, which enabled more of the drug to enter the tumor, and the slow drug release from the liposomes inside the tumor, are drug formulation characteristics responsible for the therapeutic efficacy of Doxil. It is also likely that the therapeutic efficacy of doxorubicin encapsulated in sterically stabilized liposomes is less inhibited by the multidrug resistance P-glycoprotein efflux pump than is free doxorubicin.[14,15]

II. THERAPEUTIC EFFECTS

A. HUMAN CARCINOMA HETEROGRAFTS
1. Prostatic Carcinoma

Earlier studies using human tumors implanted into nude mice found that doxorubicin encapsulated in conventional liposomes was no more potent as a therapeutic agent than doxorubicin suspended in saline.[16,17] This was also observed in a mouse mammary tumor model.[12] In that study, empty liposomes were found to be without effect on tumor growth. In the study reported here, the therapeutic effects of doxorubicin in saline and in the Doxil formulation were compared. The drug formulations were injected IV to treat the human prostatic carcinoma PC-3, implanted s.c. into mature female Swiss nude mice. Each mouse

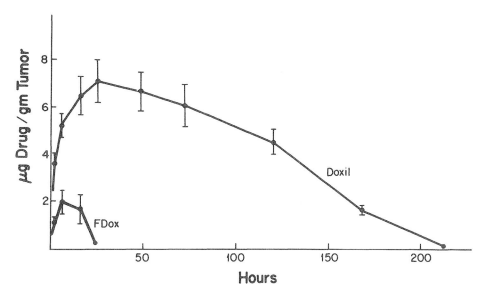

FIGURE 2. Quantitation by microfluorimetry of Doxil (adjusted for the auto-quenching factor 2.8), and free doxorubicin, in 30-d, 0.3-g tumors. Each mouse received 0.9 mg/kg drug IV at 0 h.

was killed by CO_2 asphyxiation when one or both of its s.c. tumors had grown to a size beyond possible regression, or when showing signs of deteriorating health. Mice with smaller tumors and tumor-free mice were killed and necropsied at the termination of each experiment, 64 d after tumor implantation.

The effects of treatments with Doxil and free doxorubicin on the growth of PC-3 implants are presented in Figure 3, and in Table 1. The results presented in Figure 3 show that Doxil inhibited the growth of PC-3 more effectively than free doxorubicin in saline. Comparing the mean tumor volumes on day 29, which was the last time point with enough mice surviving in the groups receiving 9 mg/kg free doxorubicin to permit a statistical comparison with the mice receiving 9 mg/kg Doxil, the t value of comparison was $p = 0.05$.

The effects of treatments on the incidences of measurable tumors are presented in Table 1. The results show that Doxil, at 9 mg/kg/injection, cured a significant number of the tumor implants. Because of the toxicity and early deaths, the test of the long-term therapeutic effect of free doxorubicin was inconclusive. In this experiment, 9 of 10 mice that received 9 mg/kg Doxil survived until the end of the experiment, with a mean survival of 62.5 ± 1.5 d. Of the mice that received 9 mg/kg free doxorubicin, 10 of 10 died or had to be killed before the end of the experiments, with a mean survival of 32.4 ± 3 d (62.5 ± 1.5 vs. 32.4 ± 3 d, $p < 0.0001$).

In the present study, the encapsulation of doxorubicin in sterically stabilized liposomes significantly increased the therapeutic efficacy of the drug against a human prostatic carcinoma. The liposome formulation also reduced the systemic toxicity, in comparisons with free doxorubicin administered in saline suspension. In view of the summary opinion expressed in the National Cancer Institute's Meeting Report on Prostate Cancer,[18] that "prostatic cancer shows little if any responsiveness to chemotherapy", the present observations on the high therapeutic efficacy and the low systemic toxicity of Doxil against heterografts of a human prostatic carcinoma are encouraging.

2. Ovarian Carcinoma

This study compared the therapeutic effects of doxorubicin in saline, and in the Doxil formulation. The drug formulations were injected IV or intraperitoneally (i.p.) to treat the human ovarian carcinoma HEY, implanted s.c. or i.p. into mature female Swiss nude mice.

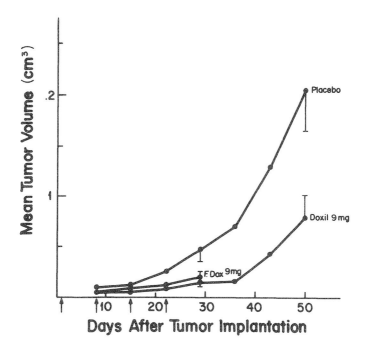

FIGURE 3. The effects of treatments with saline (Placebo), free doxorubicin in saline (F.Dox), and doxorubicin in Stealth liposomes (Doxil), on the growth of s.c. implants of tumor PC-3. The values are the mean volumes of 20 tumors. The arrows indicate the times of the treatment. The SEM are, for the sake of graphic clarity, shown only for the last points of tumor measurement.

TABLE 1
Prostatic Carcinoma PC-3 in Nude Mice: Incidence[a]
of Growth with Treatments on Days 1, 8, 15, 22

				Day		
Treatment[b]	8	15	22	29	36	64[c]
Placebo	15/20	15/20	14/20	14/20	13/20	13/20
F.Dox 9 mg/kg	16/20	17/20	11/16	8/14		
Doxil 9 mg/kg	15/20	16/20	13/20	11/20	10/20	4/18[d]

[a] Incidence of tumors per group of 10 mice. Each mouse carried 2 tumor pieces implanted s.c. in the right and left posterior flanks on day 0.

[b] Placebo = saline; F.Dox = free doxorubicin in saline; Doxil = doxorubicin in Stealth liposomes.

[c] Final incidence. By this time the tumor implants had either regressed, or some had grown to a size beyond possible regression, necessitating killing.

[d] Significantly less than placebo ($p < .05$).

Each mouse was killed by CO_2 asphyxiation when one or both of its s.c. tumors had grown to a size beyond possible regression, and before showing signs of discomfort. The mice with i.p. tumor implants were killed and necropsied when palpation determined progressive tumor growth. Tumor-free mice were killed and necropsied at the termination of the experiment, 70 d after tumor implantation.

Subcutaneous Implants The effects of treatments on the s.c. growth of HEY implants are presented in Figure 4, and the effects of treatments on the incidences of measurable tumors and the final incidences of progressive tumor growth are presented in Table 2. The results in

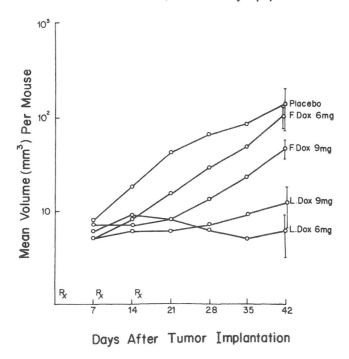

FIGURE 4.   The effects of treatments with saline (Placebo), free doxorubicin in saline (F.Dox), and doxorubicin in Stealth liposomes (L.Dox), on the growth of s.c. implants of tumor HEY. The values are the mean volumes of 16 to 20 tumors per group of 8 to 10 mice.

TABLE 2
Ovarian Carcinoma Hey in Nude Mice: Incidence[a]
of Growth with Treatments on Days 1, 8, 15

	Day					
Treatment[b]	7	14	21	28	35	42[c]
Placebo	3/16	7/16	10/16	11/16	10/16	9/16
F.Dox 6 mg/kg	1/20	2/20	6/20	9/20	10/20	12/20
F.Dox 9 mg/kg	2/20	3/20	3/20	8/20	10/20	13/20
Doxil 6 mg/kg	2/18	4/18	4/18	4/18	4/18	3/18[d,e]
Doxil 9 mg/kg	1/20	1/20	1/20	2/20	5/20	3/20[d,f]

[a]   Incidence of tumors per group of 8–10 mice. Each mouse carried 2 tumor pieces implanted s.c. in the right and left posterior flanks on day 0.

[b]   Placebo = saline; F.Dox = free doxorubicin in saline; Doxil = doxorubicin in Stealth liposomes.

[c]   Final incidence. By this time the tumor implants had either regressed, or some had grown to a size beyond possible regression, necessitating killing.

[d]   Significantly less than placebo ($p < .05$).

[e]   Significantly less than F.Dox ($p < .05$).

[f]   Significantly less than F.Dox ($p < .005$).

Figure 4 show that the higher dose of free doxorubicin (9 mg/kg) delayed only the early growth of the tumor implants (placebo vs. free doxorubicin, F. Dox, day 21, $p < .01$; day 42, $p = 0.2$). Doxil was significantly more effective than free doxorubicin (F. Dox vs. Doxil, day

TABLE 3

Ovarian Carcinoma Hey in Nude Mice.
Incidence[a] of Growth and Survival with
Treatments on Days 1, 8, 15

Treatment[b]	Mean Survival (d)	Final Incidence[c]
Placebo	50 ± 4.5	8/10
F.Dox 9 mg/kg	25 ± 6.2	0/4[d]
Doxil 9 mg/kg	70 ± 0.0[e]	0/10[f]

[a] Incidence of i.p. tumor growth per group of 10 mice. Each
    mouse received an i.p injection of 10^5 cells on day 0.
[b] Placebo = saline; F. Dox = free doxorubicin in saline; Doxil
    = doxorubicin in Stealth liposomes.
[c] Accumulated incidence of i.p. tumor growth at the termina-
    tion of the test with necropsies on day 70.
[d] 6 of 10 mice died before day 26.
[e] Significantly longer than placebo ($p < .001$).
[f] Significantly less than placebo ($p < .005$).

42, $p < .001$). The results in Table 2 show that Doxil, at 6- and 9-mg/kg doses, but not doxorubicin in saline, cured a significant number of the tumor implants. Over the course of treatments, the mice receiving Doxil had an average 5% weight loss, which was recovered over the next 5 weeks. The mice receiving free doxorubicin in saline had an average 11% weight loss over the course of treatments, which increased to 15% by day 42, and was not recovered during the 70-d observation period.

Intraperitoneal Implants The effects of treatments on the i.p. growth of HEY implants are presented in Table 3. Palpable i.p. tumor growth developed in 8 of 10 placebo control mice from 18 to 25 d after implantation. All of the Doxil-treated mice were tumor-free at necropsy, 70 d after tumor implantation. Of the mice treated with free doxorubicin, 6 of 10 died with bloody peritoneal exudate from 7 to 25 d after the first therapeutic i.p. injection. The 4 mice that survived were all tumor-free when the test was terminated on day 70. The mice receiving Doxil experienced an average 3% weight loss, which was recovered after the last injection.

Because three weekly i.v. or i.p. injections of up to 9.0 mg/kg doxorubicin in sterically stabilized liposomes are therapeutic procedures that produced only minor toxic side-effects in this study using a human tumor growing in nude mice, the observed therapeutic advantage of sterically stabilized long circulating liposomes as drug carriers has potential clinical relevance as a new method in cancer drug therapy.

B. MOUSE MAMMARY CARCINOMA ISOGRAFTS
1. Implanted Tumors

Mouse mammary carcinomas have been found in earlier studies to have only moderate susceptibility to drug therapy.[19,20] Tumors seeded in the lungs via IV injection were more susceptible than tumors implanted s.c., and slow-growing tumors were more susceptible than rapidly growing tumors, presumably because small tumors have a higher proportion of cells in the drug-sensitive S phase of the mitotic cycle.[19] A recent study compared the therapeutic effects of doxorubicin in saline, encapsulated in conventional liposomes composed of egg phosphatidylglycerol/egg phosphatidylcholine/cholesterol/*dl*-α-tocopherol, and encapsulated in Doxil liposomes.[12] In that study, the doxorubicin formulations were used to treat the mammary carcinomas MC2 and MC65. The tumors had developed spontaneously in multiparous C3H/He mice, and the second transplant generation of each is stored in liquid N_2. The tumors were used in their third to seventh transplant generations. The immunogenic MC2 is,

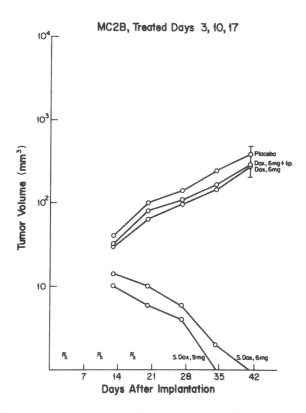

FIGURE 5.   The effect of treatments with doxorubicin in saline (Dox), doxorubicin in saline + empty Stealth liposomes (Dox + lip), doxorubicin in Stealth liposomes (S.Dox), on the growth of implants of tumor MC2B. The values are the mean volumes of 20 tumors.

in addition to the original primary tumor, also established as two variants: MC2A has an *in vivo* doubling time of 4.5 d (from 0.5 to 1.0 cm³ in 4.5 d), MC2B has a doubling time of 15 d. The nonimmunogenic MC65 has a doubling time of 5.5 d.

The tumors were treated under two circumstances: (1) as recently implanted tumors, with the injections started 3 or 10 d after tumor implantation, and the drugs were given in 3 IV injections over 15 d; (2) the slow growth of tumor MC2B made it possible to treat the implants as long-established tumors, with treatments starting an average 38 d after tumor implantation. These mice received four IV drug injections over a period of 22 d. Each mouse was killed by CO_2 asphyxiation when its tumor had regressed completely, or when its tumor had grown to a size beyond possible regression, and before showing signs of discomfort.

Recently Implanted Tumors These experiments compared the therapeutic effects of Doxorubicin with different formulations, doses, and treatment schedules. The results, which are presented in Figures 5 to 7, and in Tables 4 to 6, show the following:

1.   The improved therapeutic effect of doxorubicin in Doxil liposomes compared to doxorubicin in saline or in conventional liposomes, can be seen in growth inhibition (Figures 5 to 7), and in the higher number of cures when the final incidences were recorded (Tables 4 to 6). The therapeutic effect of Doxil was dose-related in each test. The greater growth inhibition at 9 mg/kg/dose can be seen in Figures 5 to 7, and the more frequent and/or earlier cures can be seen in Tables 4 to 6.
2.   The slower-growing MC2B responded better to treatment than the faster-growing variant MC2A (compare Figures 5 and 6, and Tables 4 and 5).

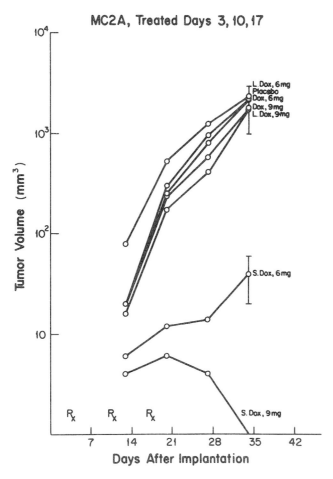

FIGURE 6. The effect of treatments with saline (Placebo), doxorubicin in saline (Dox), doxorubicin in conventional liposomes (L.Dox), doxorubicin in saline + empty Stealth liposomes (Dox + lip), doxorubicin in Stealth liposomes (S.Dox), on the growth of implants of tumor MC2B. The values are the mean volumes of 20 tumors.

3. Treatments with free doxorubicin at 6 mg/kg/dose cured a significant number of mice carrying slow-growing MC2B implants (Table 4) but also fast-growing MC65 implants (Table 6). The Doxil formulation at the same dose was significantly more effective in both tests ($p <.01$).

When a tumor implant eventually grew progressively after therapy that had cured most of the mice carrying implants of MC2A (Table 5), the question presented itself whether a drug-resistant variant had grown out of the original cancer cell population. To study this question, MC2A implants that grew progressively in the mice given placebo and in the mice given 6 mg/kg Doxil were transplanted to two groups of 20 mice each. Half of the mice in each group were given placebo, and half were given 6 mg/kg Doxil, 3, 10, and 17 d after tumor implantation. The results of this test, presented in Figure 8, show that the treatment with Doxil had not resulted in the emergence of a drug-resistant cancer cell population.

Long-Established Tumors Tumor MC2 induces a subacute immune response that is characterized by peritumor monocyte accumulation and collagen deposition. This reaction may take the form of a cellular-fibrous capsule that can interrupt growth in an average 30% of s.c. implants. The dormant state may end in rejection or in renewed growth. Because the slow growth of tumor MC2B and its occasional dormancy has relevance to the long latency of some breast cancers, it was of interest to determine the therapeutic effect of Doxil on

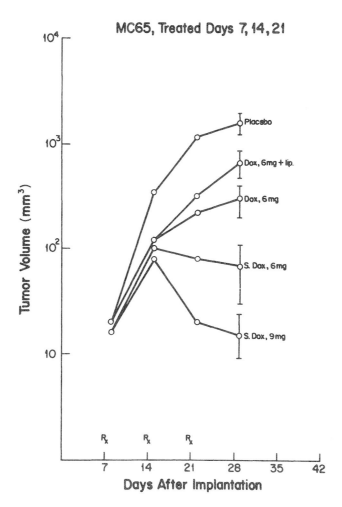

FIGURE 7. The effect of treatments with saline (Placebo), doxorubicin in saline (Dox), doxorubicin in saline + empty Stealth liposomes (Dox + lip), doxorubicin in Stealth liposomes (S.Dox), on the growth of implants of tumor MC65. The values are the mean volumes of 20 tumors.

established, slow-growing MC2B implants. Implants that, after initial growth to no less than 5 mm, then showed at least 3 weeks (mean 4.1 ± 0.2 weeks) of slow or arrested growth, were over a period of 5 months selected from groups of routine implant recipients, and randomly assigned to groups receiving four IV injections of placebo, or 9 mg/kg of doxorubicin in saline, or 9 mg/kg of Doxil, 1, 8, 15, and 22 d after selection. The results presented in Table 7 show that the therapeutic injections of free doxorubicin did not result in a significantly reduced incidence of progressive growth compared to placebo. The Doxil treatments did, however, result in a significant increase in the number of mice alive 8 weeks from the first treatment. Since four weekly injections of 9 mg/kg Doxil is a therapeutic procedure with only minor toxicity, indicated by an average 5% loss of weight, recovered within 3 weeks after the last treatment, the therapeutic benefit against long-established tumors is encouraging.

2. Spontaneous Metastases

Doxorubicin in saline or in the Doxil formulation was used to inhibit the spontaneous development of metastases from mammary carcinomas, implanted into the mammary glands of 8- to 10-week-old female C3H/He mice. The mammary carcinomas MC19 and MC65 developed spontaneously in multiparous C3H/He mice, and the second transplant generation

TABLE 4
Tumor MC2B Incidence[a] with Different Treatments on Days 3, 10, 17

Treatment[b]	Day 13	20	27	34	41	69[c]
Placebo	19/20	19/20	19/20	19/20	17/20	16/20
F.Dox 6 mg/kg	18/20	19/20	15/20	9/20	8/20	8/20[d]
F.Dox + lip	17/20	18/20	17/20	16/20	14/20	14/20
Doxil 6 mg/kg	11/20	7/20	1/20	0/20	0/20	0/20[d,e]
Doxil 9 mg/kg	4/20	0/20	0/20	0/20	0/20	0/20[d,e]

[a] Incidence of tumors per group of 10 mice. Each mouse carried 2 tumor pieces implanted s.c. in the right and left posterior flanks on day 0.

[b] Placebo = saline; F.Dox = free doxorubicin in saline; F.Dox + lip = 6 mg/ kg free doxorubicin + empty Stealth liposomes, Doxil = doxorubicin in Stealth liposomes.

[c] Final incidence. By this time the tumor had either regressed, or had grown to a size beyond possible regression, necessitating euthanasia.

[d] Significantly less than placebo.

[e] Significantly less than free doxorubicin.

TABLE 5
Tumor MC2A Incidence[a] with Different Treatments on Days 3, 10, 17

Treatment[b]	Day 13	20	27	34	69[c]
Placebo	8/20	14/20	14/20	14/20	13/20
F.Dox 6 mg/kg	11/20	17/20	16/20	14/20	13/20
F.Dox 9 mg/kg	6/20	18/20	17/20	16/20	14/20
L.Dox 6 mg/kg	11/20	18/20	16/20	15/20	12/20
L.Dox 9 mg/kg	10/20	18/20	18/20	17/20	13/20
Doxil 6 mg/kg	0/20	7/20	5/20	4/20	4/20[d,e]
Doxil 9 mg/kg	0/20	2/20	0/20	0/20	0/20[d,e]

[a] Incidence of tumors per group of 10 mice. Each mouse carried 2 tumor pieces implanted s.c. in the right and left posterior flanks on day 0.

[b] Placebo = saline; F.Dox = free doxorubicin in saline; L.Dox = doxorubicin in standard liposomes; Doxil = doxorubicin in Stealth liposomes.

[c] Final incidence. By this time the tumor had either regressed, or had grown to a size beyond possible regression, necessitating euthanasia.

[d] Significantly less than placebo.

[e] Significantly less than F.Dox and L.Dox.

of each is stored in liquid N_2. The tumors were used here in their third to seventh transplant generations. Tumor MC19 implants have an *in vivo* doubling time of 10 d, and an average 90% incidence of spontaneous metastasis from intramammary implants in untreated mice. Tumor MC65 has a doubling time of 5.5 d, and a 50% incidence of spontaneous metastasis from intramammary implants in untreated mice.

TABLE 6
Tumor MC65 Incidence[a] with Different
Treatments on Days 7, 14, 21

Treatment[b]	Day				
	20	27	34	41	69[c]
Placebo	20/20	20/20	20/20	20/20	20/20
F.Dox 6 mg/kg	15/20	14/20	14/20	14/20	12/20[d]
F.Dox + lip	15/20	14/20	14/20	14/20	14/20[d]
Doxil 6 mg/kg	14/20	11/20	8/20	6/20	3/20[d,e]
Doxil 9 mg/kg	8/20	5/20	4/20	4/20	1/20[d,e]

[a]   Incidence of tumors per group of 10 mice. Each mouse carried
      2 tumor pieces implanted s.c. in the right and left posterior
      flanks on day 0.
[b]   Placebo = saline; F.Dox = free doxorubicin in saline; Doxil =
      doxorubicin in Stealth liposomes.
[c]   Final incidence. By this time the tumor had either regressed, or
      had grown to a size beyond possible regression, necessitating
      euthanasia.
[d]   Significantly less than placebo.
[e]   Significantly less than free doxorubicin.

The intramammary tumor implants were surgically removed as each tumor reached a volume of 1.5 cm^3 (25 ± 1 d for MC65, 39 ± 1.3 d for MC19). The drugs were given in four IV injections over 22 d, with dose levels of 6 and 9 mg/kg body weight. The injections were started 22 d (MC65) or 58 d (MC19) after tumor implantation. Each metastasis experiment was terminated when any of the mice reached an early stage of cachexia, indicating pulmonary metastatic tumor growth. The mice were killed by CO_2 asphyxiation. The lungs were removed and examined for gross and microscopic metastases. The total quantity of metastases found was graded on a scale from 1 to 5, shown in Table 8. The results presented in Table 9 show that the Doxil formulation was significantly more effective than the free drug in preventing the spontaneous development of metastases from intramammary implants of tumor MC19 and tumor MC65.

3. Combination Therapy

The purpose of this study was to compare the therapeutic effects of vincristine in saline and encapsulated in sterically stabilized liposomes (S-VCR), with doxorubicin in saline and in the Doxil formulation, and then to determine whether the combined use of low doses of S-VCR and Doxil, simultaneously or alternately, would result in improved therapeutic effect. The four drug preparations were used to treat s.c. implants of the mouse mammary carcinoma MC2. The single drugs, and the combined drugs, were given IV on days 3, 10, and 17 after tumor implantation. The alternately administered drugs were given on days 3, 6, 10, 13, 17, and 20. Each mouse was killed by CO_2 asphyxiation when one or both of its tumors had grown to a size beyond possible regression, and before showing signs of discomfort. Mice that had received a course of the higher doses of the four drug preparations and were still carrying tumors at the end of the 58- to 79-d observation period were surgically cured and then observed for signs of delayed toxicity for an additional 2 months.

The effects of treatments on the growth rates of tumors are presented in Figures 9 to 13. The effects of treatments on the incidences of measurable tumor development and on the final incidences of progressive tumor growth are presented in Tables 10 to 14. The results show the following:

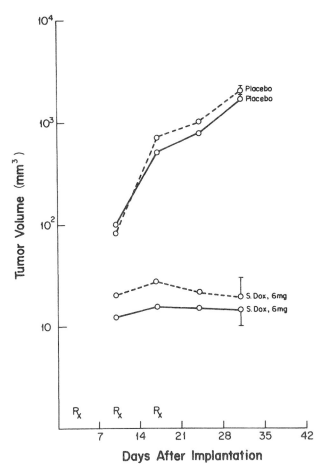

FIGURE 8.   Drug-sensitivity comparison of tissue from an MC2A tumor that grew in a Stealth doxorubicin-treated mouse (dashed line), with tumor tissue from a placebo-treated mouse (solid line). The tumor tissues were from mice in the test presented in Table 5. The values are the mean volumes of 20 tumors.

TABLE 7
Survival with Different Treatments

Treatment[a]	Treatment Started	Alive at 8 weeks
Placebo	Day 34.8 ± 2.5	17/39 (44%)
F.Dox. 9 mg/kg	Day 38.2 ± 1.2	8/21 (38%)
Doxil 9 mg/kg	Day 40.1 + 3.5	22/31 (71%) $p < .05$

Note: Survival of mice carrying long-established implants of the slow-growing mammary carcinoma MC2B. The mice received 4 IV treatments over a 22-d period, starting an average 38 d after the s.c. implantation of tumor tissue.

[a]   Placebo = saline; F.Dox = free doxorubicin in saline; Doxil = doxorubicin in Stealth liposomes.

1.   The improved therapeutic effects of vincristine and doxorubicin in liposomes as compared to the drugs in saline can be seen in the growth inhibition (Figures 9 and 10), and in the higher number of cures when the final tumor incidences were recorded (Tables 10 and 11).

TABLE 8
Quantitative Grading of Pulmonary Metastases[a]

1   1–3 small metastases (≤0.1 mm diameter)
2   4–10 small or medium (0.2–0.3 mm) metastases
3   >10 small, or 2–5 medium, or 1 large (0.4–1.0 mm) metastases
4   >5 medium, or 2–10 large metastases
5   metastases >1 mm visible grossly or with a dissecting microscope, and histologically confirmed

[a]   Total pulmonary metastases found, per mouse.

TABLE 9
Metastases Incidence and Grade with Different Treatments

Treatment[a]	Tumor MC19		Tumor MC65	
	No. of Positive Mice	Average Metastasis Grade[b]	No. of Positive Mice	Average Metastasis Grade[b]
Placebo	8/8	2.8	11/20	2.6
F.Dox 6 mg/kg	6/9	1.8	4/15	2.0[c]
F.Dox + lip.	5/8	2.2	5/16	2.3
Doxil 6 mg/kg	0/9[c,d]		0/13[c]	
Doxil 9 mg/kg	0/9[c,d]		1/20[c]	5

[a]   Placebo = saline; F.Dox = free doxorubicin in saline; F.Dox + lip = 6 mg/kg free doxorubicin + empty Stealth liposomes; Doxil = doxorubicin in Stealth liposomes.
[b]   The incidence of positive lungs and the average grade of metastases were determined by gross and histologic examunations. See Table 8 for derivation of metastasis grades.
[c]   Significantly less than placebo.
[d]   Significantly less than free doxorubicin.

2.   The therapeutic effect of S-VCR was compared at three different dose levels: 0.5, 0.7, and 1.0 mg/kg. Doxil was tested at 1.0, 3.0, and 6.0 mg/kg. The dose-related therapeutic effects can be seen in the results presented in Figures 10 and 11 and in Tables 11 and 12.

3.   The data presented in Figure 12 and in Table 13 show that the combined use of S-VCR and Doxil did not result in improved therapeutic effect, but that S-VCR inhibited the therapeutic efficacy of 6 mg/kg Doxil, a dose which when used alone has had consistent therapeutic effect.[12]

4.   The results presented in Figure 13 and in Table 14 show that when the injections of S-VCR and Doxil were separated by 3 d, the therapeutic effect was significantly improved (1/20 vs. 12/20, $p < .001$). In this test it was again seen that the simultaneous injection of S-VCR reduced the therapeutic effect of Doxil.

The reduction of the therapeutic effect of Doxil when S-VCR was injected simultaneously could have been due to destabilization by vincristine of the microtubules of the mitotic spindle. This would interfere with the progression of tumor cells into S phase, and, consequently, could have precluded doxorubicin's interference with DNA replication.

The aggregation and enhanced excretion of doxorubicin by the polyanion heparin has been reported.[21] There are, however, no indications that the anionic PEG, by itself, or in liposomes, interferes with the therapeutic effect of doxorubicin or Doxil. The 1-mg/kg dose of S-VCR contained one third of the lipids in the 6-mg/kg dose of Doxil. This represents an added

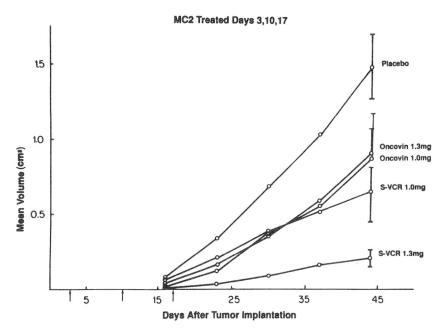

FIGURE 9. The effects of treatments with: saline (Placebo), vincristine in saline (Oncovin), and vincristine in Stealth liposomes (S-VCR), on the growth of s.c. implants of tumor MC2. The arrows indicate the times of treatment. The values are the mean volumes of 20 tumors.

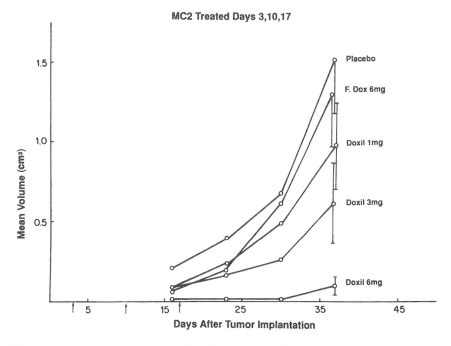

FIGURE 10. The effects of treatments with: saline (Placebo), doxorubicin in saline (F.Dox), doxorubicin in Stealth liposomes (Doxil), on the growth of s.c. implants of tumor MC2. The arrows indicate the times of treatment. The values are the mean volumes of 20 tumors.

TABLE 10
Tumor MC2 Incidence[a] with Treatments on
Days 3, 10, 17

Treatment[b]	Day				
	23	30	37	44	79[c]
Placebo	20/20	20/20	20/20	20/20	20/20
Oncovin 1.0 mg/kg	20/20	20/20	20/20	18/20	18/20
Oncovin 1.3 mg/kg	20/20	20/20	19/20	17/20	16/20
S-VCR 1.0 mg/kg	20/20	20/20	17/20	13/20	10/20[d]
S-VCR 1.3 mg/kg	19/20	18/20	14/20	13/20	12/20[d]

[a] Incidence of tumors per group of 10 mice. Each mouse carried 2 tumor pieces implanted s.c. in the right and left posterior flanks on day 0.

[b] Placebo = saline; Oncovin = free vincristine; S-VCR = vincristine in Stealth liposomes.

[c] Final incidence. By this time the tumor had either regressed or had grown to a size beyond possible regression, necessitating euthanasia.

[d] Significantly less than placebo.

TABLE 11
Tumor MC2 Incidence[a] with Treatments on
Days 3, 10, 17

Treatment[b]	Day				
	23	30	37	44	58[c]
Placebo	19/20	19/20	18/20	18/20	18/20
F.Dox 6.0 mg/kg	20/20	19/20	18/20	18/20	18/20
Doxil 1.0 mg/kg	20/20	20/20	19/20	19/20	19/20
Doxil 3.0 mg/kg	19/20	19/20	18/20	18/20	18/20
Doxil 6.0 mg/kg	12/20	7/20	8/20	9/20	9/20[d]

[a] Incidence of tumors per group of 10 mice. Each mouse carried 2 tumor pieces implanted s.c. in the right and left posterior flanks on day 0.

[b] Placebo = saline; F.Dox = free doxorubicin; Doxil = doxorubicin in Stealth liposomes.

[c] Final incidence. By this time the tumor had either regressed or had grown to a size beyond possible regression, necessitating euthanasia.

[d] Significantly less than placebo.

quantity of PEG in the combined therapy that was unlikely to have interfered with the uptake or effect of Doxil.

C. PRIMARY MOUSE MAMMARY CARCINOMAS

Because breeding C3H/He female mice have a predictable incidence of mammary carcinoma development,[22] these animals present a model to test chemopreventive programs against cancer. During the first pregnancy, which normally starts at an age of 8 weeks, the 10 mammary glands begin to develop premalignant hyperplastic alveolar nodules (HAN). The HAN persist when the normal mammary parenchyma involutes following lactation, and

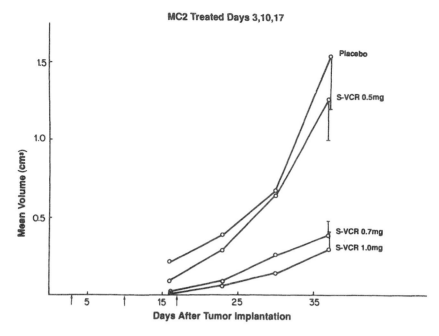

FIGURE 11. The effects of treatments with: saline (Placebo), and vincristine in Stealth liposomes (S-VCR), on the growth of s.c. implants of tumor MC2. The arrows indicate the times of treatment. The values are the mean volumes of 20 tumors.

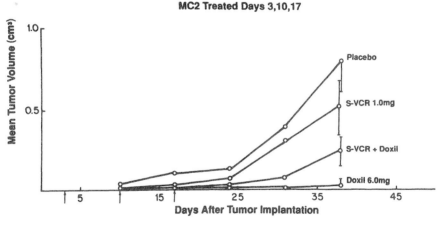

FIGURE 12. The effects of treatments with: saline (Placebo), vincristine in Stealth liposomes (S-VCR), and doxorubicin in Stealth liposomes (Doxil), on the growth of s.c. implants of tumor MC2. The arrows indicate the times of treatment. The values are the mean volumes of 20 tumors.

increase in number with repeated pregnancies and advancing age.[23] The HAN are thought to give rise to most, if not all, mammary carcinomas in C3H/He mice,[24] a situation with a possible parallel in human breast cancer etiology.[25] Most breeding C3H/He mice will develop one or more mammary carcinomas before they are 1 year old.[26] The median age of first tumor development in breeding C3H/He mice is 268 d, but a first tumor has appeared as early as 173 d, and as late as 530 d.[22] Between the first HAN and the first tumor, there may therefore intervene a premalignant phase of a few weeks or of several months in the development of a carcinoma. Genetic factors, nutritional factors, hormonal factors, and reproductive status are known or strongly indicated etiological factors in both human and murine mammary carcinoma development. These points of similarity make mice with primary mammary carcinomas one of the most relevant animal models in cancer research.

TABLE 12
Tumor MC2 Incidence[a] with Treatments on
Days 3, 10, 17

Treatment[b]	Day				
	23	30	37	44	58[c]
Placebo	19/20	19/20	18/20	18/20	18/20
S-VCR 0.5 mg/kg	18/20	17/20	15/20	15/20	15/20
S-VCR 0.7 mg/kg	16/20	13/20	13/20	13/20	13/20
S-VCR 1.0 mg/kg	16/20	18/20	16/20	13/20	9/20[d]

[a] Incidence of tumors per group of 10 mice. Each mouse carried 2 tumor pieces implanted s.c. in the right and left posterior flanks on day 0.
[b] Placebo = saline; S-VCR = vincristine in Stealth liposomes.
[c] Final incidence. By this time the tumor had either regressed or had grown to a size beyond possible regression, necessitating euthanasia.
[d] Significantly less than placebo.

TABLE 13
Tumor MC2 Incidence[a] with Treatments on
Days 3, 10, 17

Treatment[b]	Day				
	17	24	31	38	73[c]
Placebo	20/20	20/20	20/20	19/20	19/20
S-VCR 1.0 mg/kg	20/20	20/20	18/20	18/20	14/20
Doxil 6.0 mg/kg	15/20	2/20	2/20	2/20	2/20[d]
S-VCR 1.0 mg/kg+ Doxil 6.0 mg/kg	17/20	18/20	18/20	13/20	13/20

[a] Incidence of tumors per group of 10 mice. Each mouse carried 2 tumor pieces implanted s.c. in the right and left posterior flanks on day 0.
[b] Placebo = saline; S-VCR = vincristine in Stealth liposomes; Doxil = doxorubicin in Stealth liposomes.
[c] Final incidence. By this time the tumor had either regressed, or had grown to a size beyond possible regression, necessitating euthanasia.
[d] Significantly less than 13/20.

The present investigation had two purposes: (1) to determine the chemopreventive effect of IV injections of Doxil at a low dose (6 mg/kg), and at an easily tolerated treatment schedule (monthly), during the latent phase of mammary carcinoma development; (2) to determine whether tumors that developed in Doxil-treated mice would be resistant or susceptible to further, therapeutic, injections of Doxil.

1. Prophylaxis

Prophylactic injections of 6 mg/kg Doxil were started when the mice were 26 weeks old, an age at which HAN would have developed in all of the mice and when the probability of imminent tumor development was high. The IV injections were repeated every 28th day until a tumor developed, or until the mice were 62 weeks old, at which age the mice were observed until a tumor developed or until the mice were killed at the first sign of age-related poor health.

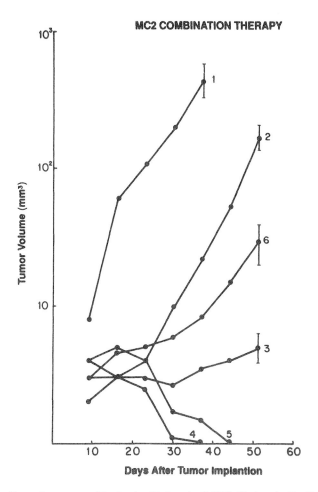

FIGURE 13.   The effects of treatments with: placebo (**1**), 1 mg/kg S-VCR (**2**), 6 mg/kg Doxil (**3**), alternate S-VCR and Doxil (**4**), alternate Doxil and S-VCR (**5**), simultaneous Doxil and S-VCR (**6**). For the treatment schedules, see Table 14. The values are the mean volumes of 18 or 20 tumors. This graph used a semilog scale to distinguish the growth curves for groups with average volumes less than 10 mm³.

Because of the high mortality resulting from repeated IV injections of doxorubicin in the free form (data not shown), a comparison of the effect of Doxil with the effect of the free drug was not possible.

Figure 14 shows the effects of monthly injections of 6 mg/kg Doxil on the tumor incidence, compared to the tumor incidence in control mice that received no treatment. Of the 47 mice given monthly injections, 22 developed a tumor by the time they were 88 weeks of age. These mice had received from two to ten injections (mean 6.8 ± 0.6). Of the 66 untreated mice, 65 developed a tumor by 79 weeks of age (22/47 vs. 65/66, $p <.0001$). Ten treated mice have died tumor-free, at a mean age of 571 ± 18 d, 15 tumor-free mice remain under observation, mean age 560 ± 24 d. Of the 66 untreated mice, one died, tumor-free, of intestinal obstruction, 413 days old. (Ten tumor-free of 32 dead treated mice vs. one tumor-free of 66 dead untreated mice, $p <.001$).

2. Therapy

The treated mice that did develop tumors were then given from 6 to 10 weekly therapeutic injections of Doxil at the same dose level. The number of therapeutic injections that could be given was limited by the development of necrosis due to drug extravasation at the sites of

TABLE 14
Tumor MC2 Incidence[a] with Combination Therapy at Different Schedules

Treatment[b]	Day				
	23	30	37	44	65[c]
Placebo	20/20	20/20	20/20	20/20	19/20[d]
S-VCR 1.0 mg/kg	15/20	15/20	13/20	13/20	13/20
Doxil 6.0 mg/kg	17/18	10/18	10/18	4/18	3/18[e]
S-VCR to Doxil	17/20	4/20	0/20	0/20	0/20
Doxil to S-VCR	18/20	7/20	1/20	0/20	1/20
Doxil with S-VCR	16/20	15/20	12/20	12/20	12/20

[a]   Incidence per group of 9 or 10 mice. Each mouse carried 2 tumor pieces implanted s.c. in the right and left posterior flanks on day 0.

[b]   Placebo = saline; S-VCR = vincristine in Stealth liposomes; Doxil = doxorubicin in Stealth liposomes. Placebo, single drugs, and simultaneous Doxil with S-VCR were injected on days 3, 10, and 17. In the alternating treatments, the first drug was given on days 3, 10, 17, the second drug on days 6, 13, 20.

[c]   Final incidence. By this time the tumor had either regressed or had grown to a size beyond possible regression, necessitating euthanasia.

[d]   Significantly greater than all treatment groups.

[e]   Significantly less than 12/20.

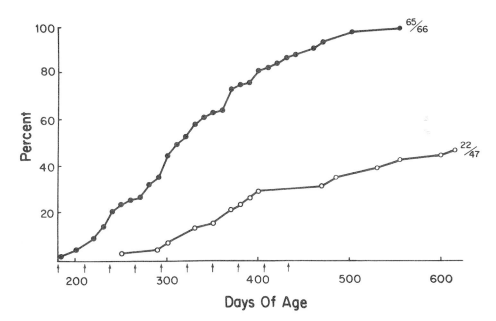

FIGURE 14.   Effect of Doxil chemoprevention. Cumulative curves for mammary tumor development in untreated mice (●), and in treated mice (○). The arrows indicate the times of treatment. Control vs. treated, on day 617, 65/66 vs. 22/47, $\chi^2 = 38.5$, $p < .0001$.

injection. Following the therapeutic injections, the mice were then observed for 1 to 20 more weeks. A mouse was selected for euthanasia by CO_2 asphyxiation according to two criteria: (1) when its tumor became large, (2) if it appeared to be in poor health. All treated mice were necropsied.

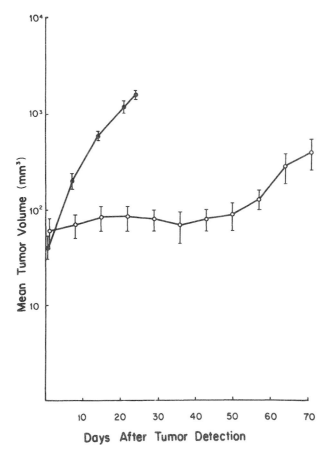

FIGURE 15.   Effect of Doxil therapy on the growth of primary mammary tumors. The mice received 6 mg/kg Doxil IV weekly from the day after a tumor was detected. The mice were killed and necropsied when a tumor reached 1.0 to 1.4 cm³, or when a mouse seemed in poor health. Untreated mice (●), treated mice (○). Control vs. treated, on day 22, t = 7.8, *p* <.0001.

Figure 15 shows the progressive growth, without treatment, of the first 54 tumors that developed in untreated mice, and shows the progressive growth, with 6 to 10 weekly therapeutic injections of 6 mg/kg Doxil, of the first 15 tumors that developed in the treated mice. The untreated tumors grew to a size requiring euthanasia in an average 24 d. On day 22, the last concurrent time point of measurements of tumors in untreated mice and tumors in treated mice, the t value of comparison was *p* <.0001.

Table 15 shows the sizes of the 15 treated tumors during and after therapy. The results show that the average tumor volume did not change significantly during the first 7 weeks of therapy (0.06 ± 0.02 on day 1 vs. 0.09 ± 0.03 on day 50, *p* = 0.42). Three of the tumors were reduced to unpalpable size for the duration of the study. Doxil therapy prolonged the mean survival from 24 ± 1.5 d in untreated mice, to 87 ± 8.9 d in treated mice (*p* <.00001). The mean age at death for the treated mice was 478 ± 16 d, and for the untreated mice, 371 ± 13 d (*p* <.001).

The mice in Table 15 had developed a primary tumor after receiving from 2 to 10 monthly injections of Doxil (median = 7). A comparison of the susceptibility of the tumors to therapy in the seven mice below the median (4 ± 0.7 injections) with the susceptibility in the seven mice above the median (8.7 ± 0.4 injections) (4 ± 0.7 vs. 8.7 ± 0.4, *p* <.001) found that the tumor volumes from day 1 to day 64 were not different (*p* = from 0.1 to 0.6), and the survival times were not different (81 ± 10 d vs. 84 ± 8 d, *p* = 0.8).

TABLE 15

Effect of Weekly Treatments[a] on Primary Mammary Tumor Growth

Mouse i.d.#	Tumor volume (cm^3)							Surv. (d)[c]
	day 1	36[b]	43	50	57	64	71	
12	0.03	0	0	0.01	0.03*	0.03	0.05	71
21	0.09	0	0*	0	0.09	0.21		64
25	0.05	0.21	0.21*	0.21	0.21	0.69	1.40	71
27	0.01	0.01	0.01	0*	0	0	0	71
32	0.03	0.05	0.03	0	0.03	0.05*	0.09	71
37	0.09	0.05	0.09	0.21	0.21	0.21*	0.29	92
41	0.05	0	0	0	0*	0	0	197
45	0.01	0.21	0.29	0.29*	0.40	0.40	1.40	71
47	0.01	0.05	0.05	0.05	0.09*	0.21	0.29	85
52	0.21	0.09	0.05	0.05	0.05*	0.05	0.05	92
59	0.09	0.14*	0.21	0.21	0.40	1.40		64
69	0.01	0.03*	0.03	0.05	0.21	0.40	0.69	71
77	0.21	0.09	0.09*	0.09	0.09	0.09	0.21	127
86	0.01	0.01	0	0*	0	0	0	78
87	0.03	0.03	0.14*	0.14	0.14	0.29	0.69	78
Mean ± SE	0.06 ± 0.02	0.07 ± 0.02	0.08 ± 0.02	0.09 ± 0.03	0.13 ± 0.03	0.29 ± 0.10	0.40 ± 0.14	87 ± 8.9[d]
Untreated[e]	0.04 ± 0.01	1.5 ± 0.03 (day 24 ± 1.5)						24 ± 1.5[d]

[a]　The mice received 6 mg/kg Doxil IV weekly from the day after the tumor was detected. Refer to Figure 2 for the mean tumor volumes during therapy.

[b]　The measurements for the first 4 weeks are left out, to reduce the size of the table.

[c]　Survival from day 1. The mice were killed and necropsied when a tumor reached 1.0 to 1.4 cm^3, or when a mouse seemed in poor health.

[d]　Survival of treated mice vs. untreated mice (p <.0001).

[e]　The mean growth, from detection to euthanasia, of untreated tumors that arose in 65 of the 66 untreated mice (see Figure 1).

*　Last therapeutic injection.

When the tumor carried by mouse number 25 in Table 15 began to grow rapidly, 57 days after its appearance, making euthanasia necessary, the question was asked whether this tumor, that developed after 6 monthly preventive treatments, and had then been treated weekly for 10 weeks, had become drug-resistant. To study the question, tumor tissue removed at the time of euthanasia was transplanted into the right and left number 4 mammary glands of two groups of five mice each. One group of mice received the moderate-to-high, tolerated, dose of 9 mg/kg Doxil IV on days 1, 8, 15, and 22 after the implantation. The second group received four injections of saline. The results of this test, presented in Table 16 and in Figure 16, show that the treatments with Doxil had not resulted in the emergence of a drug-resistant population of cancer cells.

The prolonged treatments with 6-mg/kg doses were well tolerated by the mice, which experienced no weight loss during prophylactic treatment, and only a transient average weight loss of less than 5% during the first 4 weeks of therapeutic injections. None of the treated mice showed other symptoms of systemic toxicity, and most of the mice survived until progressive tumor growth necessitated euthanasia. Two mice, number 12, aged 486 d, and number 32, aged 450 d, were killed because they appeared lethargic. Mouse number 21 died by accident before the end of the observation period. Histological examination revealed no pathologic change in the internal organs or in the leukocyte counts of any of the mice examined. All of the mice eventually developed dermal fibrosis and epidermal necrosis due to drug extravasation at the sites of IV Doxil injections.

TABLE 16
Tumor 25 Incidence[a] with Treatments on
Days 1, 8, 15, 22

Treatment[b]	Day				
	16	23	30	37	44[c]
Placebo	10/10	10/10	10/10	10/10	10/10
Doxil 9.0 mg/kg	6/10	4/10	3/10	4/10	4/10[d]

[a]   Incidence of tumors per group of 5 mice. Each mouse car-
     ried 2 tumor pieces implanted into the right and left number
     4 mammary glands on day 0.
[b]   Placebo = saline; Doxil = doxorubicin in Stealth liposomes.
[c]   Final incidence. By this time the tumor had either regressed
     or had grown to a size beyond possible regression, necessi-
     tating euthanasia.
[d]   Significantly less than placebo.

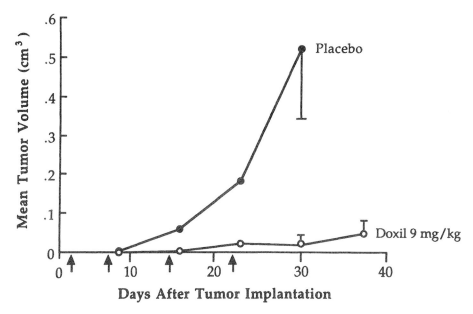

FIGURE 16.   Drug-sensitivity test of transplanted tumor tissue from mouse number 25 in Table 15. Arrows indicate days of Doxil injections. The values are the mean volumes of 10 tumors.

These studies focused on chemoprevention of mammary cancer in a model system in which such cancers were expected to develop with high frequency. Prevention was attempted at a time when malignant cells were likely to be present, and premalignant cells were certainly present. The treatments could have exerted their antineoplastic effect either on the premalignant cells of HAN, or on already transformed cells, or on both. A difference in the tumor incidences in the control group and in the treatment group became apparent soon after the treatments were started (Figure 14). This suggests that occult carcinomas were already formed, and were affected. The tumors that emerged during chemoprevention were, nevertheless, susceptible to intensified therapy. Because monthly and weekly injections of 6 mg/kg Doxil are protocols with only mild systemic toxic side effects (temporary weight loss of less than 5% during the first 4 weeks of therapeutic injections), the significant prophylactic and therapeutic benefits observed in mouse models suggest that doxorubicin in sterically stabilized liposomes may also be clinically useful.

REFERENCES

1. **Olson, F., Mayhew, E., Maslow, D., Rustum, Y., and Szoka, F.,** Characterization, toxicity and therapeutic efficacy of Adriamycin entrapped in liposomes, *Eur. J. Cancer,* 18, 167, 1982.

2. **Mayhew, E. and Papahadjopoulos, D.,** Therapeutic applications of liposomes, in *Liposomes,* Ostro, M., Ed., Marcel Dekker, New York, 1983, 289.

3. **Gabizon, A., Goren, D., Fuks, Z., Meshorer, A., and Barenholz, Y.,** Superior therapeutic activity of liposome-associated Adriamycin in a murine metastatic tumor model, *Br. J. Cancer,* 51, 681, 1985.

4. **Szoka, F.C.,** Liposome drug delivery, in *Membrane Fusion,* Wilschut, J. and Hoekston, R., Eds., Marcel Dekker, New York, 1991, 845.

5. **Mayhew, E., Papahadjopoulos, D., Rustum, Y., and Dave, C.,** Inhibition of cell growth *in vitro* and *in vivo* by cytosine arabinoside entrapped within phospolipid vesicles, *Cancer Res.,* 36, 4406, 1976.

6. **Hwang, K.J., Luk, K.K., and Beaumier, P.L.,** Hepatic uptake and degradation of unilamellar sphingomyelin/ cholesterol liposomes: a kinetic study, *Proc. Natl. Acad. Sci. U.S.A.,* 77, 4030, 1980.

7. **Gabizon, A., Chisin, R., Amselem, S., Druckmann, S., Cohen, R., Goren, D., Fromer, I., Peretz, T., Sulkes, A., and Barenholz, Y.,** Pharmacokinetic and imaging studies in patients receiving a formulation of liposome-associated adriamycin, *Br. J. Cancer,* 64, 1125, 1991.

8. **Gabizon, A. and Papahadjopoulos, D.,** Liposome formulations with prolonged circulation time in blood and enhanced uptake by tumors, *Proc. Natl. Acad. Sci. U.S.A.,* 85, 6949, 1988.

9. **Allen, T.M., Hansen, C., Martin, F.J., Redemann, C., and Yau-Young, A.,** Liposomes containing a synthetic lipid derivative of polyethylene glycol show prolonged circulation half-lives in vivo, *Biochim. Biophys. Acta,* 1066, 29, 1991.

10. **Papahadjopoulos, D., Allen, T., Gabizon, A., Mayhew, E., Matthay, K., Huang, S.K., Lee, K-D., Woodle, M., Lasic, D.D., Redemann, C., and Martin, F.J.,** Sterically stabilized liposomes: improvements in therapy against implanted tumors, *Proc. Natl. Acad. Sci. U.S.A.,* 88, 11460, 1991.

11. **Huang, S.K., Lee, K.-D., Hong, K., Friend, D.S., and Papahadjopoulos, D.,** Microscopic localization of sterically stabilized liposomes in colon carcinoma-bearing mice, *Cancer Res.,* 52, 5135, 1992.

12. **Vaage, J., Mayhew, E., Lasic, D., and Martin, F.,** Therapy of primary and metastatic mouse mammary carcinomas with doxorubicin encapsulated in long circulating liposomes, *Int. J. Cancer,* 51, 942, 1992.

13. **Vichi, P. and Tritton, T.R.,** Adriamycin: protection from cell death by removal of extracellular drug, *Cancer Res.,* 52, 4135, 1992.

14. **Endicott, J.A. and Ling, V.,** The biochemistry of P-glycoprotein-mediated multidrug resistance, *Annu. Rev. Biochem.,* 58, 137, 1989.

15. **Gottesman, M.M.,** How cancer cells evade chemotherapy: Sixteenth Richard and Hilda Rosenthal Foundation Award Lecture, Cancer Res., 53, 747, 1993.

16. **Kojima, N., Ueno, N., Takano, M., Yabushita, H., Noguchi, M., Ishihara, M., and Yagi, K.,** Effect of adriamycin entrapped by sulfatide-containing liposomes on ovarian tumor-bearing nude mice, *Biotechnol. Appl. Biochem.,* 8, 471, 1986.

17. **Nagata, J., Yamauchi, M., Takagi, H., Kojima, N., Hayashi, Y., and Yagi, K.,** Antitumor activity against human gastric cancers of sulfatide-inserted liposomes containing entrapped adriamycin, *J. Clin. Biochem.,* 8, 111, 1990.

18. **Chiarodo, A.,** Meeting Report. National Cancer Institute Roundtable on Prostate Cancer: Future Cancer research Directions, *Cancer Res.,* 51, 2498, 1991.

19. **Vaage, J. and Costanza, M.E.,** Effects of combination drug therapy on the subcutaneous and pulmonary growth of a slow- and a fast-growing C3H/He mammary carcinoma, *Cancer Res.,* 39, 4466, 1979.

20. **Vaage, J. and Costanza, M.E.,** Effects on metastases of drug therapy during developing and established tumor immunity, *Int. J. Cancer,* 32, 759, 1983.

21. **Myers, C.E., Jr. and Chabner, B.A.,** Anthracyclines, in *Cancer Chemotherapy Principles and Practice,* Chabner, B.A. and Collins, J.M., Eds., J.P. Lippincott, Philadelphia, 1990, 356.

22. **Vaage, J., Smith, G.H., Asch, B., and Teramoto, Y.,** Mammary tumorigenesis and tumor morphology in four C3H sublines with or without exogenous mammary tumor virus, *Cancer Res.,* 46, 2096, 1986.

23. **Hui, Y.H., DeOme, K.B., and Briggs, G.M.,** Inhibition of spontaneous development of hyperplastic alveolar nodules and mammary tumors in C3H mice fed phenylalanine deficient diets, *J. Natl. Cancer Inst.,* 47, 687, 1971.

24. **DeOme, K.B., Faulkin, L.J., Bern, H.A., and Blair, P.B.,** Development of mammary tumors from hyperplastic alveolar nodules transplanted into gland free mammary fat pads of female C3H mice, *Cancer Res.,* 19, 515, 1959.

25. **Jensen, H.M., Rice, J.R., and Wellings, S.R.,** Preneoplastic lesions in the human breast, *Science,* 191, 295, 1976.

26. **Vaage, J. and Medina, D.,** Mammary tumor virus oncogenesis and tumor immunogenicity in three sublines of the C3h mouse, *Cancer Res.,* 38, 2443, 1978.

Chapter 15

THERAPEUTIC APPLICATION OF VINCRISTINE IN CONVENTIONAL AND STERICALLY STABILIZED LIPOSOMES

Paul S. Uster

TABLE OF CONTENTS

I. SCOPE OF CHAPTER

The concept and history of both conventional and long circulating Stealth®* liposomes have been reviewed in depth elsewhere in this book, so the focus here is to provide a complete review of the published literature on liposomal-encapsulated vincristine. The ultimate goal of formulating vincristine into liposomes is to make a significant improvement in clinical anticancer therapy. Therefore, this chapter will first summarize the action and pharmacology of vincristine, and then describe strategies that have been employed for formulating liposome-entrapped vincristine. The properties of these formulations will be detailed, and then the activity of these preparations will be compared in animal pharmacokinetics, toxicology, and tumor models.

---

* Stealth® liposomes are a registered trademark of Liposome Technology, Inc.

0-8493-8383-8/95/$0.00+$.50
© 1995 by CRC Press Inc.

II. DRUG SUBSTANCE

A. CLINICAL USE

Vincristine (VCR) is a chemotherapeutic agent isolated from the leaves of the periwinkle *Vinca rosea* Linn, a popular flowering shrub cultivated in gardens throughout the world. The early indigenous uses of the plant for controlling toothache, scurvy, diabetes, and hemorrhage led to the discovery of the antitumor effects of the alkaloid fraction. Out of a group of 30 alkaloids that can be extracted from the plant, vincristine sulfate and vinblastine sulfate have had the greatest clinical evaluation.[1] Vincristine sulfate has a significant spectrum of antitumor activity, being indicated as a single agent for induction therapy of acute leukemia. In combination with other agents, it is used in treating Hodgkin's disease, non-Hodgkin's lymphomas, rhabdomyosarcoma, neuroblastoma, and Wilms' tumor.[2] Vincristine sulfate has been commonly used as part of combination therapy for off-label indications such as metastatic breast cancer, small cell lung carcinoma, "oat" cell lung carcinoma, multiple myeloma, Ewing's sarcoma, metastatic osteogenic carcinoma, metastatic malignant melanoma, advanced colorectal carcinoma, and cervical squamous cell carcinoma.[3] Recently, it has been found a useful part of combination therapy in treating advanced Kaposi's sarcoma.[4]

Vincristine sulfate has been an important component of polychemotherapy because it does not cause significant myelosuppression, nausea, or vomiting when used at maximum tolerated doses. Vincristine's acute toxicity is constipation, which limits the maximum tolerated dose to not more than 2 mg. Doses greater than this can result in severe complications due to paralytic ileus of the intestine. Vincristine's chronic dose-limiting side effect is neurotoxicity. Neuropathy is initially experienced as numbness and tingling in fingers and toes, followed by loss of deep tendon reflexes, and eventual loss of motor function in the hands and feet. Progression to severe toxicity occurs at a cumulative dose of 15 to 20 mg, and requires discontinuing therapy to prevent significant disability.[5] The neurological complications are largely reversible, although sensorimotor loss can require months for recovery.[6]

B. MECHANISM OF ACTION

Despite years of clinical research, the *in vivo* mode of vincristine action is not completely understood. The cytotoxic action of vincristine sulfate is apparently due to interference with microtubule assembly. Treatment of cells *in vitro* with vincristine sulfate blocks mitosis by metaphase arrest. Vinca alkaloid binding to microtubules results in tubule disassembly and 1:1 stoichiometric binding of the alkaloid to tubulin subunits.[7]

C. STRUCTURE AND PHYSICAL CHEMISTRY

The sulfate salt has the empirical formula $C_{45}H_{50}N_4O_{14}S$ and a molecular weight of 824.94 Da (Figure 1). The free base is soluble in organic solvents, and the sulfate salt is extremely water soluble. It has been a misconception that VCR salts are extremely lipophilic and therefore are not hydrophilic. Solutions of 20% (w/v) can be made readily at physiological pH. The free base has two titratable basic groups having pK_as of 5.0 and at 7.4. While stable at pH 3.5 and above, vincristine is rapidly degraded at pHs of 2 and below.[8,31]

III. LIPOSOME FORMULATION METHODS

A. BILAYER ENTRAPMENT STRATEGIES

Layton and Trouet formulated VCR sulfate by dissolving it in the lipid phase of phosphatidylcholine/cholesterol/phosphatidylserine (PC/CH/PS) or PC/CH/stearylamine (7/2/1, mol/mol/mol).[9] Hydrated multilamellar vesicles (MLVs) were down-sized by sonication, but the size distribution of the preparations was not reported. Free vincristine was removed by mixed gel permeation adsorption chromatography. Only 10 to 11% of the input drug remained

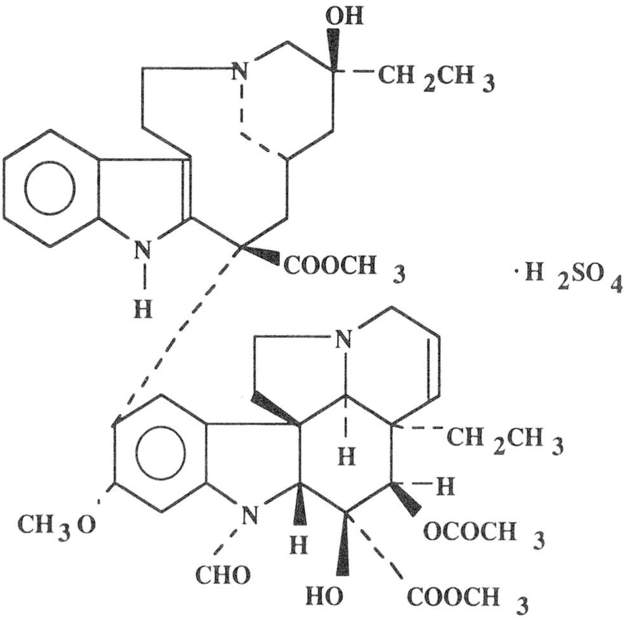

FIGURE 1.   VCR structure.

incorporated, leading to a final lipid-to-drug ratio of the formulations as 667/1, mol/mol. The drug leakage rate in buffer or 50% serum was not more than 0.1% per hour.

Kirby and Gregoriadis prepared 30- to 60-nm small unilamellar vesicles (SUVs) within which aqueous VCR sulfate (11 to 23 µg/mL) at pH 7.4 was passively captured via thin-film hydration of various cholesterol and phospholipid mixtures.[10] Increasing the cholesterol content of the bilayer from 0 to 50 mol% increased the percent entrapment of VCR from 1.8 to 14.4%. An entrapment of 40% was achieved with sphingomyelin/CH SUVs. The presence of CH in VCR-containing SUVs quantitatively slowed the rate of VCR release from PC liposomes incubated in mouse plasma.

The authors hypothesized such entrapment values exceeded the captured volumes of SUVs, making it likely the drug had partitioned into the CH-rich bilayer. This localization of VCR with the bilayer is quite probable. First, a theoretical entrapment of about 2% is expected for SUVs hydrated at 20 mM phospholipid.[11] Fourteen percent would require unilamellar vesicles larger than 400 nm. Second, at pH 7.4 the Henderson-Hasselbach equation indicates that half of the VCR will be uncharged. Thus, the study utilized conditions at which the lipophilicity of VCR is enhanced. Third, the final lipid-to-drug molar ratio of PC:CH VCR vesicles was very small, at best 6000/6000/1 in this study, thereby providing ample lipid phase for VCR partitioning.

B. pH-MEDIATED DRUG ENTRAPMENT

Bilayer entrapment of VCR resulted in having to inject unacceptably large lipid doses. Two methods have been developed which produce formulations with high drug-to-lipid ratio.

It has been known for some time that proton gradients can be used to load biogenic amines into large unilamellar vesicles.[12] The theoretical equilibrium inside/outside concentration ratio for a monoprotic drug (within its solubility limits) is the same as the outside/inside pH ratio, as derived from the Henderson-Hasselbach equation.

Firth and colleagues entrapped VCR inside a mixture of dipalmitoyl phosphatidylcholine, cholesterol, and dipalmitoyl phosphatidic acid (DPPC/CH/DPPA, 7/2/1, mol/mol/mol) 0.1 µm diameter vesicles.[13] They were able to achieve 50 to 60% entrapment of the drug using passive entrapment of VCR solutions with the reverse phase evaporation method.[14] A series of preparations were formulated in which the hydrating buffer pH was varied from pH 5.8 to about 7.8. Each preparation was then dialyzed against pH 7.4 buffer, thereby generating pH gradients of various magnitudes with lower pH inside. VCR release was fastest when there was no pH gradient ($T_{1/2}$ ~2 h at internal pH of 7.4), and slowest at the maximum pH gradient they tested ($T_{1/2}$ ~50 h at internal pH of 5.8). By imposing a ΔpH of 1.6 units (acidic inside), Firth et al. demonstrated the rate constant of VCR release from liposomes was improved 25-fold. Thus, VCR retention inside liposomes can be controlled by imposing a pH gradient.

These observations on the loading of biogenic amines and stability of VCR have been refined into a highly useful strategy for loading a wide variety of pharmacologically active agents.[15] In brief, liposomes are formed in a low pH buffer, and the external aqueous phase is adjusted to a higher pH. Addition of a basic drug to the external buffer above the PC phase transition causes a rapid accumulation of drug inside the liposomes.

Mayer et al. have made use of this property of cationic drugs and pH gradients to efficiently load in excess of 98% of externally added VCR into liposomes.[16,17] In this method 300 mM citric acid at pH 4.0 is entrapped within PC/CH (55/45, mol/mol) LUVs. Then, drug is added to the external phase and the pH is adjusted to 7.5. Essentially complete drug loading can be achieved by the judicious choice of lipid matrix, internal buffer concentration, lipid-to-drug ratio, incubation temperature, and magnitude of the pH gradient.[18]

In contrast to passive encapsulation, vastly improved drug/lipid ratios as high as 0.2/1 (w/w) in 0.2-μm vesicles can be achieved with this method.[16] In later work, it was seen that the internal proton pool available for trapping drug is related to mean liposome diameter. Having a smaller buffer volume, 0.1 μm diameter liposomes typically have reduced stable drug/lipid ratios (0.05, w/w) for comparable loading efficiency and stability.[17] Observations by Firth and colleagues that VCR retention within the liposomes is dependent on maintaining the transmembrane pH gradient were confirmed.[13] Mayer et al. also observed that the pH gradient dissipates much more rapidly with egg PC/CH liposomes than with distearoyl phosphatidylcholine (DSPC/CH).[16] After 24 h at room temperature, 40% of drug leaked from the "fluid" PC formulation, but none had leaked from the "rigid" PC composition. Also, removing the pH gradient by acidifying the external buffer causes a rapid leakage of drug.

C. STERIC STABILIZATION AND AMMONIUM GRADIENT-MEDIATED DRUG ENTRAPMENT

In the late 1980s, several laboratories observed that steric stabilization of the liposome surface with glycolipids[19,20] or poly(ethylene glycol)ated-PE[21] dramatically lengthened the plasma distribution time of liposomes, and favorably altered their biodistribution from the reticuloendothelial organs to other target tissues (for a review of the steric stabilization theory see Reference 22). In our laboratory, we have been interested in determining how steric stabilization using Stealth® liposomes modifies the pharmacokinetics and preclinical efficacy of VCR. To achieve efficient loading and stable encapsulation of VCR, we have used an ammonium gradient-loading method first used for anthracyclines.[23]

Stealth® formulations of VCR were prepared by hydrating a lipid mixture containing about 5 mol% MPEG-DSPE* with an isotonic buffer of ammonium sulfate or ammonium citrate. The hydrated MLVs were extruded to 0.1 μm in diameter, and the external ammonium was removed by dialysis. A VCR solution was combined with the liposomes, and an efflux of ammonium ions was activated by heating the dispersion of liposomes and drug above the PC phase-transition temperature for 10 min. The resulting ammonium efflux powers internalization of the dibasic drug. After rapid cooling, residual unentrapped drug is removed by another

---

* MPEG-DSPE, N-(carbamyl polyethyleneglycol methyl ether)-1,2-distearoyl-*sn*-glycero-3-phospho-ethanolamine, sodium salt. HSPC, hydrogenated soy phosphatidylcholine.

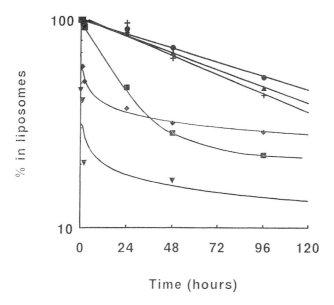

FIGURE 2. Liposomal-encapsulated 3H-VCR formulations were incubated in 25% plasma and aliquots were withdrawn at various time periods. Free drug was separated from encapsulated VCR by column chromatography. (▼) MPEG-DSPE/egg PC; (♦) MPEG-DSPE/HSPC; (▩) PG/egg PC/CH; (●) PG/HSPC/CH; (▲) MPEG-DSPE/egg PC/CH; (+) MPEG-DSPE/HSPC/CH.

dialysis step. Final encapsulations are typically in excess of 97% and drug/total lipid ratios are about 0.05 to 0.10 (w/w). For animal studies discussed below, the lipid composition was MPEG-DSPE/HSPC/CH* (5/56/39, mol/mol/mol), henceforth abbreviated as S-VCR.[24]

With ammonium gradient-loaded liposomes, *in vitro* VCR encapsulation stability in plasma is strongly dependent on vesicle composition (Figure 2).[32] There is a rapid biphasic loss of drug entrapment in MPEG-DSPE/PC liposomes regardless of whether the lecithin is fluid (PC) or rigid (HSPC) at room temperature. If about 40 mol% cholesterol is included in either composition, the *in vitro* leakage is strikingly shifted to monophasic kinetics with a half-life of 4 d. MPEG-DSPE may also be instrumental in the slow monophasic release of drug, because substitution of PG for MPEG-DSPE in MPEG-DSPE/PC/CH liposomes causes a reversion to relatively rapid, biphasic kinetics.

IV. PHARMACOKINETICS

Layton and Trouet observed that the *in vivo* pharmacokinetics of free drug clearance from plasma was biphasic.[9] The first phase cleared well over 90% of tritiated drug with a half-life of 110 s, and the latter phase had a half-life of 26.3 min. With the liposomal preparation having VCR in the bilayer phase, about half of the drug was cleared with a half-life of 105 s. The remainder of the drug was cleared with a half-life of 50 min. The clearance rate of ^{14}C-cholesterol was 50.5 min. Pharmacokinetics were the same for both liposome compositions.

Kirby and Gregoriadis injected SUV-encapsulated VCR at drug doses of 20 to 100 µg/kg and lipid doses of 2 to 3 mg per mouse.[10] VCR clearance was biphasic and independent of CH content. Pharmacokinetic parameters were not calculated, and it is unclear whether the presence of equimolar CH significantly decreased the clearance rate. Inspection of the figures indicates 8% of the dose in PC SUVs and 10 to 15% of the dose in PC/CH (1/1, mol/mol) SUVs remained in the blood 4 h postinjection. One to two percent of the dose in PC/CH (1/1, mol/mol) SUVs remained in the blood 24 h postinjection. Sphingomyelin/CH (1/1, mol/mol) SUVs had 3 to 4% of the dose in the blood 24 h postinjection. It must be noted that the

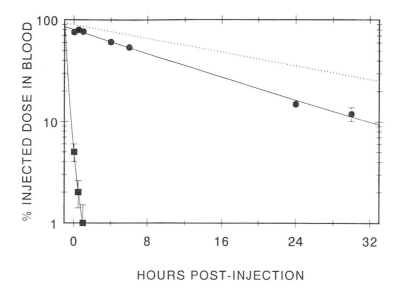

FIGURE 3. Rat *in vivo* plasma pharmacokinetics of VCR as free drug and encapsulated in MPEG-DSPE/HSPC/CH liposomes. (■) free drug; (●) S-VCR; (- - -) 18-h half-life for lipid clearance.

reticuloendothelial system of the mouse is saturated with liposomes at these lipid doses, and clearance would be slower than at lower lipid doses.[25]

Oliver and coworkers studied the pharmacokinetics produced by intracerebral administration of free and liposome-entrapped vincristine.[26] Anesthetized Wistar rats were given single intracerebral injections. Urine and serum were collected at 4 and 24 h. Animals were sacrificed at 4 and 24 h postinjection. Brain tissue, serum, and urine were analyzed by radioimmunoassay for drug (sensitivity not indicated). Vincristine could not be detected in serum at 4 h postinjection. Similar amounts, about 1.4% of the injected dose, were found in the urine at 4 h. About 23% of free vincristine and 43% of the liposomal VCR were distributed in the brain tissue. Twenty-four hours postinjection, VCR (2.5% injected dose) could only be found in the brain of liposome-treated animals.

Mayer et al. found the greatest prolongation of vincristine distribution in the blood using pH gradient-loaded liposomes could be obtained by using 0.1-μm DSPC/CH loaded at a drug/lipid ratio of 0.05/1 (w/w).[17] Drug plasma pharmacokinetics was biphasic over the first 24 h. Approximately 30% of the drug dose was cleared from the plasma during the first hour postinjection. Then, a second distribution phase cleared 62% of the injected dose during the next 23 h leaving about 8% of the injected dose still in plasma at 24 h.

The plasma pharmacokinetics of S-VCR are illustrated in Figure 3. Healthy male Sprague-Dawley rats were given a single bolus dose of 0.25 mg/kg via the tail vein. Plasma levels were monitored by spiking the test articles with tritiated vincristine sulfate. Two percent of the injected dose was the level of radiolabel sensitivity. Free VCR was cleared rapidly, and could not be detected after the first 15 min. This is consistent with an extremely rapid *clinical* initial phase half-life of 5 min. Twenty percent of initial injected dose of S-VCR was cleared within seconds of injection. For the remaining 80%, the distribution half-life was about 10.5 h and fit best as a mono-exponential decay. Fifteen percent of the injected dose of Stealth liposomal vincristine was still in circulation 24 h later.[27]

V. PRECLINICAL TOXICOLOGY

The stearylamine-containing, VCR-loaded liposomes used by Layton and Trouet were extremely toxic in DBA/2 mice inoculated with P388 cells.[9] A severe weight loss compounded

TABLE 1
Relative Weight Loss of ICR Mice
Given Single Doses

Dose Level (mg/kg)	Free VCR	S-VCR
0.9	94.8 ± 2.0%	ND
1.3	83.6 ± 4.9%	93.3 ± 4.1% *
2.0	77.5 ± 5.8%	82.9 ± 6.1% *

ND, not determined; * significantly different from free drug ($p < .05$, T-test).

the toxic effect of the drug. By substituting PS, there was no significant weight loss for free or liposomal VCR at dose levels of 0.9 and 1.8 mg VCR/kg body weight.

Oliver et al. studied the toxicity produced by intracerebral injection of free and liposome-entrapped vincristine.[26] Wistar rats were given single intracerebral injections at comparative dose levels of 5 or 38 g VCR /kg body weight of free or the McKeran liposomal formulation of VCR. Injections of equal quantities of placebo or liposomal formulation were also administered as control. Histological evaluation was by "blinded" observers. Upon necropsy, the placebo liposomes provoked no histological changes in the brain other than needle tract damage. In contrast, animals given liposomal VCR were scored greater in damage than the corresponding dose of free drug. At the 38-μg/kg dose level, free drug caused necrotic damage in 25% of the cerebral hemisphere with marked inflammatory involvement, whereas the liposomal drug caused 90% necrotic damage.

The acute toxicity of single-dose intravenous administration of pH gradient-loaded 0.2 μm diameter DSPC/CH liposomes was compared to free drug.[16] In two strains of mice tested, there was a significant improvement over free drug. CD-1 strain mice had LD_{50}s of 1.9 mg/kg and 4.8 mg/kg for free and liposomal VCR, respectively. Mice of the DBA/2J strain had LD_{50}s of 2.5 mg/kg and 4.2 mg/kg for free and liposomal VCR. Mortality occurred within the first 5 to 10 d post-dosing, with weight loss, not neurotoxicity, being the dose-limiting side effect.

The single-dose toxicity of ammonium gradient-loaded S-VCR was compared to free VCR in male ICR mice with respect to weight loss and survival.[27] The 14-d LD for S-VCR and free VCR was 2.6 ± 0.20 mg/kg and 2.4 ± 0.24 mg/kg, respectively. The 28-day LD_{50} values were similar. The acute dose-limiting toxicity of vincristine in mice was weight loss, not neurotoxicity. At comparable doses of 1.3 and 2.0 mg/kg, S-VCR treated mice lost significantly less weight (Table 1). This is consistent with observations from the colon carcinoma study described below, in which mice tolerated a 50% larger cumulative dose of S-VCR without fatal weight loss. Because of the steep dose response of vincristine, these results suggest that while the LD_{50}s are similar, the maximum tolerated dose of S-VCR may be significantly higher than free VCR.

VI. *IN VITRO* TUMOR MODELS

Luckenbach and Layton studied the effect of passively encapsulated liposomal VCR on cultured L1210 cells.[28] At the only dose studied and 24 h incubation, 180 nM VCR and 260 μM lipid, empty, or VCR-containing liposomes had no effect on L1210 cell viability, while free drug caused a 10% loss of viability. Thus, it seems likely that what little VCR was available in the liposomes was not taken up by the cells.

In a second series of experiments, isolated DBA/2J mouse macrophages were co-cultured with L1210 cells. There was no loss of L1210 viability due to the macrophages. Macrophages and free VCR elicited a 10% loss of L1210 viability at 24 h incubation. Macrophages and empty OLVs produced a 20% loss of L1210 viability. Macrophages and VCR-containing OLVs caused a 60% loss of tumor cell viability. No hypothesis was proposed to explain the synergy between macrophage and drug-loaded liposome.

Multiple drug cross resistance (MDR1) due to P-glycoprotein overexpression in tumor cells is a major focus of interest in oncology. P-glycoprotein is thought to mediate the cellular efflux of drugs like anthracyclines, vinca alkaloids, and taxanes. Thierry et al. measured binding of VCR to P-glycoprotein in membranes from Chinese hamster LZ cells.[29] Cardiolipin/PC liposomes were co-incubated with radiolabeled VCR and LZ membranes. As much as 60% of the specific binding could be inhibited, and the IC_{50} of VCR binding was at a lipid concentration of about 10 μg/ml medium. The authors hypothesize that like MDR1 reversal agents, cardiolipin-containing liposomes could inhibit drug binding to the P-glycoprotein. But in the case of VCR, later work showed that cardiolipin-containing liposomes did not facilitate vincristine's *in vitro* cytotoxicity of drug-resistant LZ cells.[30]

VII. *IN VIVO* TUMOR MODELS

A. P388 LYMPHOCYTIC LEUKEMIA

Bilayer-entrapped liposomal VCR was no more effective than free drug at 0.9 mg/kg given on the schedule of days 1, 5, 9 post-tumor cell injection. At double the dose under the same schedule, liposomal VCR was slightly less effective than free drug. Since there is a drug/lipid ratio of 10^{-3} (w/w), the authors attribute this to the onset of toxicity induced by the total lipid dose of 1.6 g/kg.[9]

A single injection of pH gradient-loaded VCR in 0.2 μm diameter liposomes was able to significantly increase the life span of intraperitoneal tumor-bearing DBA/2J mice relative to the same dose level of free VCR.[16] Mice were given intraperitoneal injections of 10^5 P388 tumor cells, and treated with a single intravenous injection of the test articles 24 h post tumor implantation. Free VCR was able to elicit a median life span increase of 11% (1 mg/kg) and 29%(2 mg/kg) over controls. The DSPC/CH VCR formulation elicited 14% (1 mg/kg), 57% (2 mg/kg), and 96% (3 mg/kg) increased median life span over controls. There were no survivors at 40 d postimplantation.

The effectiveness of ammonium gradient-loaded S-VCR and free VCR was compared in a series of experiments in the P388 leukemia model using B6D2F1 mice.[27] Relative to free drug, encapsulation increased survival times. Single-dose therapy was given 24 h after mice were implanted with 10^6 tumor cells intravenously or intraperitoneally. The site of tumor implantation had a significant effect on antitumor therapy.

When mice were given their tumor burden IV, liposome encapsulation produced significant but modest improvement in mean survival time. Increased mean survival time over controls ranged from 27% (1.3 mg/kg) and 37% (2.0 mg/kg) for free VCR to 31% (1.3 mg/kg) and 44% (2.0 mg/kg) for S-VCR. In contrast, there was a remarkable difference in effectiveness with Stealth liposomal encapsulation when the tumor burden was given intraperitoneally. Increases in mean survival time over controls then ranged from 67% (1.3 mg/kg) and 79% (2.0 mg/kg) for free VCR to 162% (1.3 mg/kg) and 199% (2.0 mg/kg) for S-VCR.

An i.p. injection of tumor cells is likely to delay the migration of tumor cells into other tissues, whereas IV injection disseminates the metastatic cells more rapidly. This is likely why all drug treatments are improved with an i.p. tumor burden. S-VCR may be highly effective in i.p. cases because of favorably altered VCR distribution via Stealth liposome extravasation into the peritoneal cavity.

Intravenous tumor burden represents a more rigorous test of drug treatments. In another series of experiments, mice given an IV tumor burden of 10^5 P388 cells were given three injections of free VCR or S-VCR every 10 d. Animals given a cumulative dose of 3.9 mg/kg had increased mean survival time of 47% (free VCR) and 60% (S-VCR) over controls. Mice given a cumulative dose of 6.0 mg/kg had increased mean survival time of 55% (free VCR) and 87% (S-VCR) over controls. In both cases, encapsulation in Stealth liposomes caused a statistically significant improvement in therapy over free drug.

B. L1210 LEUKEMIA

Single injections of pH gradient-loaded 0.2 μm diameter DSPC/CH liposomes, but not free VCR, displayed dose-dependent anti-L1210 activity.[16] DBA/2J mice were given i.p. inoculations of 10^6 L1210 cells, and a single therapeutic treatment followed 24 h later. Free VCR had no effect on increasing survival time. With liposomal VCR, increases in median life span ranged from 17% at 0.5 mg/kg to 117% at 3 mg/kg. Multiple dose regimens were able to produce 40-d survivors in the liposomal VCR-treated groups. The best results were produced with treatment at days 1, 5, and 9 post-L1210 implantation. Liposomal VCR produced median life span increases of 55% (0.75 mg/kg 3×), 105% (1 mg/kg 3×), and 91% (1.5 mg/kg). Forty-day survivors were observed only in liposomal VCR-treated groups at 1 mg/kg (1/6) and 1.5 mg/kg (2/6).

Mayer et al. studied the effect of liposome composition, drug/lipid ratio, and liposome size on single-dose VCR therapy in the L1210 model.[17] All liposome formulations loaded VCR by the pH gradient method. A tumor burden of 10^5 cells was inoculated intraperitoneally and a single drug dose was given 24 h later. Single injections of egg PC/CH liposomes failed to produce a statistically significant improvement in median life span over equal doses of free VCR. However, single injections of pH gradient-loaded DSPC/CH liposomes at any size were able to improve the percent increase in median life span over the same dose level of free VCR. At subtoxic doses, decreasing mean liposome diameter correlated strongly with increased median life span. Sixty-day long-term survivors were observed at 2 mg/kg dosing of 0.2 μm (1/6 mice) and 0.1 μm (1/12 mice) diameter formulations.

Using 0.12 μm diameter liposomes, therapeutic differences due to drug-to-lipid ratios of 0.1, 0.05, and 0.01 (w/w) were also tested.[17] All liposomal treatments were far superior to free drug. No statistical differences between formulations were observed at the 1-mg/kg dose level. At the other dose level tested, 2 mg/kg, the authors' statistical analysis shows that median life span increases at 0.01 w/w (100% increased life span, ILS) and 0.05 (156% ILS) drug/lipid ratios are both statistically different from 0.1 w/w (133% ILS). While the authors felt that a ratio of 0.01 was significantly less effective, this reviewer notes there is insufficient data in support of a strong trend due to drug/lipid ratio.

Allen et al. investigated Stealth liposome-encapsulated VCR single-dose therapy in B6D2F1 mice.[27] Animals were given 10^6 L1210 cells either intravenously or intraperitoneally, and treated with a single dose 24 h later. Once again, the route of tumor burden had an effect on treatment ability to increase the mean survival time.

The high tumor burden caused the rapid death of the animals, and there were no 70-d survivors in any treatment group. With an IV tumor burden, the only statistical differences between treated and control groups was for 1.3 mg/kg S-VCR. Increased mean survival time over controls ranged from 3% (1.3 mg/kg) and 5% (2.0 mg/kg) for free VCR to 18% (1.3 mg/kg) and 7% (2.0 mg/kg) for S-VCR.

In another series of experiments, the effect of IV or i.p. *drug* administration at 2.0 mg/kg was studied when the tumor burden was given i.p. For both routes of drug treatment, encapsulation in Stealth liposomes resulted in a significant increase in life span. For IV treatment, free drug caused an 8% increase in mean survival time, and S-VCR caused a 46% improvement. When drug was administered i.p., free drug increased mean survival time by 49% and S-VCR increased it by 68%. These results again suggest that with IV drug administration, liposome encapsulation may distribute more drug into the peritoneal cavity.

C. C26 COLON CARCINOMA

The comparative effectiveness of ammonium-gradient-loaded Stealth liposomes with respect to multiple-dose treatment and schedule was studied in the C26 colon carcinoma model.[27] Female BALB/c mice were injected with 4×10^5 tumor cells subcutaneously in the right flank on day zero. Three doses of saline or the test formulations were to be given intravenously 1 week apart starting on day 6 or day 10. Saline and 1.3 mg/kg S-VCR groups were treated as noted, but body weight losses in the 1.3-mg/kg free VCR and 2-mg/kg S-VCR groups caused a 1-week delay in the third dose to avoid drug-induced mortality. Animals could not be given multiple doses of 2.0 mg/kg free VCR without complete mortality by the third dose. Thus, encapsulation in Stealth liposomes appears to increase the tolerable cumulative dose in these tumor-bearing mice.

Tumor size was measured during the first 5 weeks. Multiple doses of 1.3 mg/kg free VCR slightly delayed the growth of C26 tumors implanted 6 or 10 d previously. At the same dose, S-VCR caused tumor volume to regress to unmeasurable levels. Increasing the dose of S-VCR to 2.0 mg/kg caused an earlier tumor regression to unmeasurable levels. With dosing started on day 6, all drug treatment groups had statistically significant increases in median survival time, but if the start of therapy was delayed until day 10 to create a larger tumor burden, only the Stealth treatment groups had statistically significant increases in survival time (Table 2).

Only S-VCR-treated groups had long-term survivors at 120 d post-tumor implant. The day 6 schedule had seven out of ten animals surviving. The day 10 schedule had one survivor out of ten at 1.3 mg/kg and three out of ten at the 2.0-mg/kg dose level. Multiple-dose treatment with S-VCR was markedly more effective than free drug. The Stealth vehicle enabled an apparently ineffective drug to become highly effective in reducing tumor regression and producing cures.

D. MC2 MAMMARY CARCINOMA

The comparative effectiveness of multiple-dose treatment with S-VCR was also studied in the MC2B mammary carcinoma.[24] One cubic millimeter pieces of the tumor were implanted subcutaneously into both posterior flanks of female C3H/HE mice. Starting 3 d postimplantation, mice received three weekly injections of saline, 1.0 or 1.3 mg/kg free VCR, or 1.0 or 1.3 mg/kg S-VCR. Tumor volume and incidence of tumors were monitored.

Tumor growth was significantly reduced by drug treatment. A significant decrease in tumor growth rate was observed for all treatment groups vs. saline control. Also, 1.3 mg/kg S-VCR was significantly more effective than any other drug treatment group. The incidence of tumors at the end of the study was significantly reduced by treatment at any dose of S-VCR over controls, while free VCR produced no significant difference over controls.

Polychemotherapy with Stealth formulations of doxorubicin and VCR was also examined in this model. Suboptimal, multiple doses of S-VCR (1 mg/kg) and Stealth liposomal doxorubicin (6 mg/kg) were selected to provide a delay in tumor growth but not tumor regression. When both liposome formulations were co-injected, therapy was actually inferior to Stealth liposomal doxorubicin alone. However, administering each drug separately and staggered 3 d apart resulted in complete regression of all tumors to unmeasurable levels, regardless of which test article initiated therapy. The large number of cures at subtoxic doses suggests that polychemotherapy with multiple Stealth products may provide highly effective therapy.

VIII. CONCLUSIONS AND PERSPECTIVES

There has been meaningful progress in developing pharmacologically effective liposomal formulations of VCR over the last 5 years, suggesting that the field will provide fruitful research for some time.

TABLE 2
Survival of VCR-Treated Mice Bearing
C26 Carcinomas

Treatment	R_x starting day 6		R_x starting day 10	
	MST	LTS	MST	LTS
Saline	47	0/10	45	0/10
1.3 mg/kg free VCR	49*	0/10	45	0/10
1.3 mg/kg S-VCR	62*	0/10	83*	1/10
2.0 mg/kg S-VCR	>120*	7/10	68*	3/10

MST, median survival time in days; LTS, long-term survivors to 120 d.

* Meier-Kaplan plot statistically significant ($p < .05$, Mann-Whitney test) compared to saline control.

Attempts in the early 1980s to incorporate VCR into the bilayer of liposomes did not produce therapeutically effective therapies. Poor drug loading and rapid VCR leakage from the liposomes were the cause. A key breakthrough of the mid-1980s was recognizing that VCR retention could be improved by imposing a gradient of, for example, protons.[13]

The field remained quiet until 1990 when a new round of publishing activity began. This time around the formulations showed a gratifying improvement of VCR therapeutic activity. This success appears to be due to integration of two liposomal methodologies. First, two successful strategies for achieving high drug loading using either pH gradients[16] or ammonium gradients[24] have been developed. These enable reasonably low lipid doses to be achieved, and have sufficient "battery" to retain VCR for reasonable periods of time. Second, strategies that enable liposomes to persist longer in the plasma compartment have been employed. Either small, rigid liposomes or sterically stabilized liposomes slow VCR plasma clearance significantly.

As a consequence of changing the distribution pattern of the drug, these liposomal vehicles appear to have increased the maximum tolerable dose of drug that can be administered, and have shown significant therapeutic advantages in several murine tumor models. Still, these actively loaded, plasma-persistent formulations are not carrying the maximum drug payload of drug. Drug loss during transit in the central compartment is a current problem which is reducing the maximum potential effectiveness of liposomal VCR therapy. Drug that leaks out of the liposome while in transit is immediately washed out of the blood compartment. Mayer et al. very clearly show total VCR plasma levels dropping much faster than the clearance of the liposome phospholipid.[17] The 10- to 11-h plasma distribution phase of S-VCR is definitely more rapid than the 18 to 20-h clearance rates of several other aqueous-phase markers carried in the same lipid matrix.[22] It may be deduced that in this vehicle as well, some drug is lost during the liposome's transit through the blood compartment. To make full use of the power of long circulating liposomes, a key challenge for future investigations will be to immobilize the drug such that the VCR leakage rate from the liposome is significantly less than the plasma distribution rate.

What, from the clinical perspective, does the future hold for long circulating formulations of vincristine? It is decidedly more effective than free drug in murine leukemia models, but this is not clinically exciting, where current induction therapy has an 80 to 90% response rate. Sterically stabilized liposomes, at least, are markedly more effective than free drug in at least two murine solid tumors. Current clinical practice uses combination therapy in such neoplasms, and to justify moving into the clinic, there will need to be drug interaction toxicology studies. Treatment schedules and combination therapy studies are needed to determine the added value of VCR encapsulation. Efficacy studies in human xenograft models are well warranted.

In conclusion, recent encouraging results have breathed life into the field, and should stimulate new research on the antitumor activity spectrum of liposomal VCR. Perhaps liposome encapsulation has indeed taught an old drug new tricks.

REFERENCES

1. **Johnson, I.S., Armstrong, J.G., Gorman, M., and Burnett, J.P., Jr.,** The vinca alkaloids: a new class of oncolytic agents, *Cancer Res.,* 23, 1390, 1963.
2. **Barnhardt, E.H., Ed.,** *Physicians Desk Reference,* 44th ed., Medical Economics Co., Oradell, NJ, 1990, 1235.
3. **See-Lasley, K. and Ignoffo, R.J.,** *Manual of Oncology Therapeutics,* C.V. Mosby, St. Louis, 1981.
4. **Gill, P., Rarick, M., Bernstein-Singer, M., Harb, M., Espina, B.M., Shaw, V., and Levine, A.,** Treatment of advanced Kaposi's sarcoma using a combination of bleomycin and vincristine, *Am. J. Clin. Oncol.,* 13(4), 315, 1990.
5. **Legha, S.S.,** Vincristine neurotoxicity: pathophysiology and management, *Med. Toxicol.,* 1, 421, 1986.
6. **Postma, T.J., Bernard, B.A., Huijgens, P.C., Ossenkoppele, G.J., and Heimans, J.J.,** Long term effects of vincristine on the peripheral nervous system, *J. Neuro-Oncol.,* 15, 23, 1993.
7. **Zhou, X.-J. and Rahman, R.,** Preclinical and clinical pharmacology of vinca alkaloids, *Drugs,* 44 (Suppl. 4), 1, 1992.
8. **Burns, J.H.,** Vincristine sulfate, *Anal. Profiles Drug Subst.,* 1, 463, 1972.
9. **Layton, D. and Trouet, A.,** A comparison of the therapeutic effects of free and liposomally encapsulated vincristine in leukemic mice, *Eur. J. Cancer,* 16(7), 945, 1980.
10. **Kirby, C. and Gregoriadis, G.,** The effect of lipid composition of small unilamellar liposomes containing melphalan and vincristine on drug clearance after injection into mice, *Biochem. Pharmacol.,* 32(4), 609, 1983.
11. **Deamer, D.W. and Uster, P.S.,** Liposome preparation: methods and mechanisms, in *Liposomes,* Ostro, M.J., Ed., Marcel Dekker, New York, 1983, 27.
12. **Nichols, J.W. and Deamer, D.W.,** Catecholamine uptake and concentration by liposomes maintaining pH gradients, Biochim. Biophys. Acta, 1976.
13. **Firth, G., Oliver, A.S., and McKeran, R.O.,** Studies on the use of antimitotic drugs entrapped within liposomes and of their action on a human glioma cell line, *J. Neurol. Sci.,* 63(2), 153, 1984.
14. **Szoka, F. and Papahadjopoulos, D.,** Procedure for the preparation of liposomes with large internal aqueous space and high capture by reverse-phase evaporation, *Proc. Natl. Acad. Sci. U.S.A.,* 75, 4194, 1978.
15. **Madden, T.D., Harrigan, P.R., Tai, L.C.L., Bally, M.B., Mayer, L.D., Redelmeier, T.E., Loughrey, H.C., Tilcock, C.P.S., Reinish, L.W., and Cullis, P.R.,** The accumulation of drugs within large unilamellar vesicles exhibiting a proton gradient: a survey, *Chem. Phys. Lipids,* 53, 37, 1990.
16. **Mayer, L.D., Bally, M.B., Laughrey, H., Masin, D., and Cullis, P.R.,** Liposomal vincristine preparations which exhibit decreased drug toxicity and increased activity against murine L1210 and P388 tumors, *Cancer Res.,* 50, 575, 1990.
17. **Mayer, L.D., Nayar, R., Thies, R.L., Boman, N.L., Cullis, P.R., and Bally, M.B.,** Identification of vesicle properties that enhance the antitumor activity of liposomal vincristine against murine L1210 leukemia, *Cancer Chemother. Pharmacol.,* 33, 17, 1993.
18. **Boman, N.L., Mayer, L.D., and Cullis, P.R.,** Optimization of the retention properties of vincristine in liposomal systems, *Biochim. Biophys. Acta,* 1152, 253, 1993.
19. **Allen, T.M. and Chonn, A.,** Large unilamellar liposomes with low uptake into the reticuloendothelial system, *FEBS Lett.,* 223(1), 42, 1987.
20. **Gabizon, A. and Papahadjopoulos, D.,** Liposome formulations with prolonged circulation time in blood and enhanced uptake by tumors, *Proc. Natl. Acad. Sci. U.S.A.,* 85, 6949, 1988.
21. **Papahadjopoulos, D., Allen, T.M., Gabizon, A., Mayhew, E., Matthay, K., Huang, S.K., Lee, K.D., Woodle, M.C., Lasic, D.D., Redemann, C., and Martin, F.J.,** Sterically stabilized liposomes: improvements in pharmacokinetics and antitumor therapeutic efficacy, *Proc. Natl. Acad. Sci. U.S.A.,* 88, 11460, 1991.
22. **Woodle, M.C. and Lasic, D.D.,** Sterically stabilized liposomes, *Biochim. Biophys. Acta,* 1113, 171, 1992.
23. **Mayhew, E.G., Lasic, D., Babbar, S., and Martin, F.J.,** Pharmacokinetics and antitumor activity of epirubicin encapsulated in long-circulating liposomes incorporating a polyethylene glycol-derivatized phospholipid, *Int. J. Cancer,* 51, 302, 1993.
24. **Vaage, J., Donovan, D., Mayhew, E., Uster, P., and Woodle, M.,** Therapy of mouse mammary carcinomas with vincristine and doxorubicin encapsulated in sterically stabilized liposomes, *Int. J. Cancer,* 54, 959, 1993.

25. **Allen, T.M., Murray, I., McKeigan, S., and Shah, M.,** *J. Pharmacol. Exp. Ther.,* 229, 267, 1984.

26. **Oliver, A.S., Firth, G., and McKeran, R.O.,** Studies on the intracerebral injection of vincristine free and entrapped within liposomes in the rat, *J. Neurol. Sci.,* 68, 25, 1985.

27. **Allen, T.M., Newman, M.S., Woodle, M.C., Mayhew, E., and Uster, P.S.,** Pharmacokinetics and antitumor activity of vincristine encapsulated in sterically stabilized liposomes, submitted.

28. **Luckenbach, A. and Layton, D.,** Liposomally activated macrophages; subsequent interaction with L1210 leukemic cells, *Int. J. Cancer,* 27, 837, 1981.

29. **Thierry, A.R., Dritschilo, A., and Rahman, A.,** Effect of liposomes on P-glycoprotein function in multidrug resistant cells, *Biochem. Biophys. Res. Commun.,* 187, 1098, 1992.

30. **Thierry, A.R., Vige, D., Coughlin, S.A., Belli, J.A., Dritschilo, A., and Rahman, A.,** Modulation of doxorubicin resistance in multidrug-resistant cells by liposomes, *FASEB J.,* 7, 572, 1993.

31. **Uster, P.,** unpublished data.

32. **Allen, T.,** personal communication.

Chapter 16

STEALTH® LIPOSOMES AS A DRUG SUSTAINED-RELEASE SYSTEM FOR 1-β-D-ARABINOSYLFURANOSYLCYTOSINE (CYTOSINE ARABINOSIDE)

T. M. Allen

TABLE OF CONTENTS

I. HISTORICAL PERSPECTIVE

In the early 1980s we became concerned about the possible adverse consequences of the rapid, repeated localization of large numbers of liposomes in the mononuclear phagocyte system (MPS, also termed reticuloendothelial system). A series of experiments conducted in our laboratory (reviewed in Reference 1) confirmed that MPS impairment could result from repeated injections of liposomes as a result of their high degree of localization in the MPS. These observations prompted us to begin a research program in which we sought means of reducing the MPS uptake of liposomes in order to increase their circulation half-lives so that they would function more effectively as drug sustained-release systems, and also to increase their probability of being delivered to non-MPS tissues.

A clue as to the direction to take came from a review of the red blood cell literature, which revealed that one of the factors associated with the long circulation half-lives of erythrocytes was the presence of sialic acid on the erythrocyte surface. Desialation of erythrocytes led to their rapid removal from circulation into the MPS. After a futile attempt to make long circulating liposomes from total lipids extracted from red blood cells, we turned our attention to the family of sialated glycolipids (gangliosides) present on the surface of red blood cells. It soon became apparent that incorporation of 7 to 10 mol% monosialoganglioside GM1 into egg phosphatidylcholine:cholesterol (PC:CH, 2:1) liposomes increased their circulation half-lives, although only modestly. The first public presentation of this work was at the NATO Advanced Study Institute on "Targeting of Drugs with Synthetic Systems" in 1985. We subsequently observed that inclusion of solid-phase phospholipids in the liposomes further increased their circulation half-lives and the first paper on GM1-containing long circulating liposomes was published in 1987.[2] This was followed closely by an independent publication by Gabizon and Papahadjopoulos showing that not only did GM1-containing liposomes have increased circulation half-lives, but they also resulted in increased uptake into solid tumors.[3] The stage was now set to look at therapeutic applications of these new formulations of long circulating liposomes, which were termed Stealth® liposomes (S-liposomes).[4] The term "Stealth liposomes", implying the ability of the new liposomes to avoid detection by the cells of the MPS, was coined by Dr. Frank Martin of Liposome Technology, Inc., and has an analogy in the military term "stealth bomber" referring to an aircraft which avoids detection by enemy radar.

As reviewed elsewhere in this volume, the near-simultaneous development in several laboratories of second-generation Stealth liposomes, containing lipid derivatives of polyethylene glycol (PEG), also termed sterically stabilized liposomes, resulted in further improvements, most notably in the commercial acceptability of the product,[5-10] which led to further therapeutic experiments. The term "classical" or "conventional" liposomes (C-liposomes) has been used for liposomes which lack the properties of Stealth liposomes, such as long circulating half-lives and dose-independent pharmacokinetics.[9,11]

II. STEALTH LIPOSOMES AS CIRCULATING MICRORESERVOIRS

One of the obvious applications for a liposomal preparation with prolonged circulation half-lives would be as a circulating drug sustained-release system, particularly for those drugs that experience rapid clearance and/or degradation when administered in the free form. Entrapment in liposomes would protect the drugs from degradation and allow for gradual release of free drug into the circulation over the lifetime of the liposomes, which in the case of Stealth liposomes might be for 48 h or longer. A further advantage might result from the alterations, as compared to free drug, in the pharmacokinetics and biodistribution of the S-liposome/drug package, which could lead to reduced toxicity to sensitive tissues and increased therapeutic efficacy.

III. CYTOSINE ARABINOSIDE

The first application for S-liposomes was as a sustained-release system for cytosine arabinoside (1-β-D-arabinofuranosylcytosine, ara-C).[12] Ara-C is a schedule-dependent antineoplastic drug used clinically in the treatment of acute leukemia. The drug kills cells in the DNA-synthesis phase (S phase) of the cell cycle, and consequently it is effective only if the leukemic cells are exposed to the drug sufficiently long for all of the cells to pass through this phase when drug is present. In the clinical setting the free drug is administered as a slow infusion, because ara-C is rapidly degraded with a half-life *in vivo* of only approximately 20 min.[13,14] A large body of literature already existed on ara-C in C-liposomes, to which the results of ara-C in S-liposomes could be compared.[15-21] Previous data showed that C-liposomes containing entrapped ara-C were capable of functioning as a drug sustained-release system in the treatment of murine leukemia with improvements in the therapeutic efficacy of the drug.[21] We were interested in determining if liposomes with longer circulation half-lives, releasing ara-C over longer time periods, would result in greater therapeutic effect than liposomes with equivalent amounts of entrapped drug but shorter half-lives.

IV. THERAPEUTIC EFFICACY OF S-LIPOSOMAL CYTOSINE ARABINOSIDE

All of our experiments on the therapeutic efficacy of ara-C entrapped in S-liposomes were conducted by measuring the mean survival times (MST) or the related percent increased life span (%ILS), following various treatment regimens, of mice bearing L1210 murine leukemia inoculated by either the intravenous (IV) or intraperitoneal (i.p.) routes of injection. In experiments comparing the therapeutic efficacy of drug delivery systems with the efficacy of free drug administration, it is important to compare the results obtained with the drug carrier with those obtained from experiments in which the free drug is given by optimal dosing schedules. Therefore, the efficacy of single or multiple injections of liposomal ara-C was compared with free drug infusion, as well as with single and multiple injections of free drug. Although the mouse leukemia model with IV inoculation of cells is the most difficult to treat because of its aggressive, invasive nature,[22] it bears the closest resemblance to the human clinical situation; therefore, most of our experiments were done with this model, with IV drug treatment.

A. CYTOSINE ARABINOSIDE IN GM1 LIPOSOMES

We initially conducted a series of experiments in which mice received single injections of ara-C entrapped in S-liposomes in which the circulation half-life was prolonged by the addition of 7 mol% monosialoganglioside, GM1.[12] Longer circulating liposomes (0.4 μm extruded, $T_{1/2}$ of approximately 10 h) were composed of PC:sphingomyelin (SM):CH:GM1 at a molar ratio of 1:1:1:0.14 (GM1-araC); control, shorter circulating liposomes (0.4 μm extruded, $T_{1/2}$ of approximately 0.5 h) were composed of PC:CH, 2:1 (C-araC). Mice received 10^6 L1210 leukemia cells by the IV or i.p. route of injection and were treated with single injections of 10 mg/kg GM1-araC, C-araC, or free ara-C, as well as with 24-h infusions of free ara-C given at the same total drug dose. At these low doses of ara-C we demonstrated that single injections of GM1-araC resulted in significant increases in the MST of mice as compared to single injections of C-araC[12] (Table 1), establishing the value of extending the circulation half-lives of the drug delivery system. As would be expected, single injections of 10 mg/kg free ara-C had an insignificant effect in increasing mean survival times due to its rapid rate of degradation. Against leukemia inoculated by the IV route, single IV injections of GM1-araC resulted in MST that were inferior to 24-h infusions of free drug, suggesting that the GM1-araC liposomes were able to sustain minimal therapeutic levels of araC within the circulation for a period of less than 24 h. This may be due, in part, to plasma-induced leakage of ara-C from the liposomes with subsequent rapid degradation of the free drug. However, against i.p.-inoculated leukemia, single IV injections of 10 mg/kg GM1-araC are not significantly different from 24-h drug infusions, suggesting that a dose of 10 mg/kg ara-C, liposomes were sustaining drug levels above the minimal therapeutic concentration in the peritoneal cavity for approximately 24 h. This may be related, in part, to the rapid passage of high levels of drug-containing S-liposomes from the circulation into the peritoneal cavity where the rate of drug release would be slower, more sustained in the absence of high levels of plasma proteins. S-liposomes appear to be able to pass between the circulation and the peritoneal cavity with a half-life of approximately 0.5 to 1 h.[23]

B. EFFECT OF INCREASING DOSE

The effect on MST of increasing dose was studied for GM1-araC liposomes. The maximum tolerated dose of both GM1-araC liposomes and C-araC liposomes was found to be in the range of 100 to 120 mg/kg, which is comparable to the numbers found for optimized multiple, frequent injections of free drug or for 2-d free drug infusions.[22,24] Single injections of GM1-

TABLE 1
A Comparison of the Therapeutic Effects of ara-C
Entrapped in Long Circulating (GM1-araC) and
Short Circulating (C-araC) Liposomes

Liposome Composition (Molar Ratio)	Route of Inoculation	Route of Treatment	%ILS 10 mg/kg
Free ara-C	IV	IV	13.9
	i.p.	IV	6.0
PC:SM:CH:GM1 (1:1:1:0.14)	IV	IV	39.2
	i.p.	IV	49.2
PC:CH (2:1)	IV	IV	19.8
	i.p.	IV	5.2
24-h infusion of free ara-C	IV	IV	68.1
	i.p.	IV	67.5

Note: Mice (8–12/group) received 10^6 L1210 leukemia cells on day
0 and were treated on day 1 with 10 mg/kg ara-C according to
the treatment groups below.

araC were compared to C-araC, free ara-C, and 24-h infusions of free ara-C at moderate and high doses of 20 and 80 mg/kg ara-C, respectively. Mice were injected with either lower or higher tumor burdens (10^5 or 10^6 leukemia cells, respectively) by the IV or i.p. route of injection. It was apparent that the liposomal treatments, at the higher drug doses, were effective across various *in vivo* barriers, i.e., liposomes given IV were effective against leukemia administered i.p. and vice versa[24] (Table 1). With one exception, GM1-araC liposomes were significantly more effective than C-araC liposomes in increasing the MST of mice with both lower and higher tumor burdens, again showing that liposomal drug sustained-release systems with longer circulation half-lives can result in improved therapeutic effects.[24] However, when high doses of C-araC were given i.p. against i.p. tumor, they were equally effective as GM1-araC in increasing MST of the mice.[24] This is not surprising, as one would not expect differences in circulation half-lives of liposomes to have a significant effect on therapeutic outcome when liposomes are given by the i.p. route of injection, against i.p. tumor, as both liposomes and cells are mostly confined within the same anatomical compartment.

When the relationship was examined between the MST obtained in mice bearing IV leukemia, treated with increasing IV doses of GM1-araC, and that obtained in mice receiving 24-h infusions of free ara-C, it was observed that, relative to free drug infusion, liposomal ara-C became more effective with increasing drug dose.[24] Although GM1-araC was significantly less effective than free drug infusion at 10 mg/kg ara-C, it was equivalent to free drug infusion at 20 mg/kg and significantly superior at increasing the MST of mice at doses of 80 mg/kg. In other words, at 10 mg/kg ara-C the liposomes were behaving as though they were sustaining drug release above minimal therapeutic concentrations for periods of time of less than 24 h, at 20 mg/kg ara-C they were sustaining drug release for 24 h, and at 80 mg/kg they were sustaining drug release for longer than 24 h.

Because ara-C is not a very potent drug, relatively high doses of araC are required for a cytotoxic effect (0.01 to 3 µg/ml).[25] In order to entrap sufficient drug for therapeutic experiments, large amounts of lipid are required, particularly at the higher doses. Naturally, as the dose of ara-C increases, the amount of lipid that each animal receives also increases. Because S-liposomes have been shown to have dose-independent pharmacokinetics, there will be a direct relationship between lipid dose and amount of drug that is available.[11] However, for C-

TABLE 2
A Comparison of the Therapeutic Effects of Increasing Doses of ara-C

Liposome Composition (MolarRatio)	Route of Inoculation	Route of Treatment	%ILS, 25 mg/kg	%ILS, 50 mg/kg	%ILS (LTS), 100 mg/kg
Free ara-C	IV	IV	4.8	11.1	15.9
HSPC:CH:PEG (2:1:0.1)	IV	IV	74.6	138.1	228.6 (1/5)
24-h infusion of free ara-C	IV	IV	ND	90.5	122.2

Note: Mice (5/group) received 10^6 L1210 leukemia cells on day 0 and were treated on day 1 with 25, 50, or 100 mg/kg ara-C according to the treatment groups below. Some long-term (70-d) survivors (LTS) occurred.

liposomes, which have dose-dependent pharmacokinetics, as the lipid dose increases the amount of drug that is available will increase disproportionately to the amount of lipid, as a result of MPS saturation.[11] The result will be that, as the dose of C-liposomes increases, they will begin to look more and more like S-liposomes in their therapeutic effects, an observation which we have made from our experiments, and from a comparison of our results with those from the literature, in which high lipid doses of C-liposomes (up to 1 g/kg) were administered.[15–21,24] However, repeated administration of high lipid doses of C-liposomes may result in greater adverse effects on the MPS than would be expected for similar doses of S-liposomes which are taken up into the MPS to a much smaller degree.[11]

C. CYTOSINE ARABINOSIDE IN PEG-DSPE LIPOSOMES

We have also examined the therapeutic effect of ara-C entrapped in long circulating liposomes containing PEG-distearoylphosptatidylcholine (PEG-DSPE).[24] Leukemia-bearing mice were treated with single IV or i.p. injections of 25, 50, or 100 mg/kg ara-C entrapped in liposomes composed of hydrogenated soy PC (HSPC):CH:PEG-DSPE, 2:1:0.1 molar ratio (S-araC). In these experiments the MST of mice bearing IV leukemia, given a single IV injection of 50 mg/kg S-araC, were equivalent to those seen in mice receiving 24-h infusions of free drug at a dose of 100 mg/kg[24] (Table 2). Experimental observations as well as theoretical calculations suggest that a single injection of 50 mg/kg S-araC was maintaining blood levels of ara-C in the therapeutic range for 48 h or longer.[24]

At the highest tested dose (IV 100 mg/kg, IV cells), substantial increases in life spans (%ILS = 229%, Table 2) were observed in treated, as compared to untreated, mice or mice receiving 24-h infusions of 100 mg/kg ara-C. Optimized multiple injection schedules of free ara-C at the same total dose result in %ILS of only 142%.[22] In order to achieve these survival levels with optimized multiple, frequent injections of free drug, the total drug dose would have to be more than double that of S-araC (approximately 240 mg/kg vs. 100 mg/kg).[22] This, therefore, indicates that there has been an increase in the therapeutic index for ara-C when it is administered in S-liposomes. We believe that the increase in therapeutic efficacy of S-araC, as compared to free ara-C, is partly as a result of the sustained period of time over which the free drug is released and partly due to some localization of the drug-containing liposomes in the bone marrow, liver, spleen, and lymph nodes — sites at which significant numbers of leukemia cells can be found as early as 3 h following IV injection of the leukemia cells.[22]

D. EFFECT OF DOSING SCHEDULE

A number of different dosing schedules were examined in order to determine the optimum schedule which would result in long-term survivors among leukemia-bearing mice. Four IV injections of 25 mg/kg S-araC each, given on days 1 to 4 postinoculation (total dose 100/mg kg) did not result in an increase in the MST compared to a single dose of 100 mg/kg given on day 1 postinoculation.[24] As seen above, a 20-g/kg ara-C dose given by IV injection was equivalent to a 24-h free drug infusion at a total dose of 20 mg/kg. Therefore, 4 daily doses of 25 mg/kg might be expected to approximate a 4-d infusion of drug. The observation that a single injection of 100 mg/kg on day 1 gave equivalent increases in MST to 4 daily injections of 25 mg/kg we interpret to indicate that the single dose of 100 mg/kg was sustaining drug levels in the cytotoxic range for approximately 4 d. When 25 mg/kg S-araC was administered as 4 doses (total dose, 100 mg/kg) at intervals of 4 d (days 1, 5, 9, and 13), as would be expected, the %ILS increased from 229% for a single 100-mg/kg dose to 348% for the multiple injections.[24] Interestingly, 2 injections, 48 h apart (days 1 and 3) of 50 mg/kg S-araC (total dose 100 mg/kg) lead to early deaths from toxicity in the mice. Assuming that each dose of 50 mg/kg are maintained ara-C doses in the cytotoxic range for 48 h or more, this would approximate a 4- or 5-d infusion of free drug, and we would be well into the LD_{50} range for infusion of free drug for this period of time (50 mg/kg total dose for a 5-d infusion[26]). Therefore, we administered doses of 50 mg/kg S-araC with a 1-week interval between doses to allow drug levels to fall to zero and some recovery from toxicity to occur between doses. Three IV doses of 50 mg/kg S-araC given on days 1, 8, and 15 resulted in very high MST.[24] Mice given 10^6 IV leukemia cells had a %ILS of 465% following this treatment. When tumor and treatment were given by the i.p. route the %ILS was 493% with 40% long-term survivors.[24] Thus, when the dose and dosing schedule are appropriate to the pharmacokinetics of the drug carrier, excellent results can be obtained.

E. LEAVING FREE DRUG IN THE LIPOSOME PREPARATIONS

Considering the fact that free ara-C is degraded very rapidly, with a $T_{1/2}$ of approximately 20 min,[13,14] we wondered if it would be possible to simplify the preparation of liposomes by leaving some free drug in the preparations. The LD_{50} of free ara-C is approximately 4000 mg/kg in mice. Therefore, a considerable amount of free drug could theoretically be left in the liposome preparation without significantly increasing the overall toxicity of the preparations. We found that it was possible to pellet the liposomes in our preparations by centrifugation, remove the free drug in the supernatant, and resuspend the liposome pellet with drug-free buffer. This left an equal amount of free drug in the preparation to the amount entrapped within the liposomes, i.e., 50 mg/kg liposome-entrapped drug and 50 mg/kg of free ara-C.[24] Fifty milligrams per kilogram of free ara-C has little therapeutic effect on its own because of its rapid degradation, and also has no observable toxicity.[24] Leukemic mice cannot be given three weekly doses of 50 mg/kg of free ara-C because they die of their disease before the second injection can be given. However, mice bearing either IV or i.p. leukemia, injected with three weekly doses with the combination of 50 mg/kg S-araC and 50 mg/kg free ara-C, were all long-term survivors.[24] This suggests that there is a synergistic relationship between the liposome-entrapped drug and the free drug with the results for the combination being superior to either one being given separately. A possible explanation is that the leukemia cells were periodically exposed to high levels of free drug, which would be effective at killing cells going through DNA replication at that time, resulting in reduced tumor burden. These periodic high drug levels may also have led to some synchronization of the cells. After the levels of the injected free ara-C fell below the cytotoxic range, residual (synchronized?) cells would then have been killed by free drug released from the liposome depot. The free drug would also have served as a loading dose, quickly raising drug levels to the therapeutic range, exposing the leukemia cells to cytoxic levels of drug at early time points when the liposomal drug may have been at subtherapeutic levels.

F. EFFECT OF RELEASE RATE OF CYTOSINE ARABINOSIDE

A factor that will have a significant effect on the therapeutic efficacy of any sustained-release drug delivery system is the rate at which the drug is released from the carrier. To take two extremes, liposomes that have a rapid rate of release for ara-C will result in little or no therapeutic improvement because the released drug will be rapidly degraded. At the other extreme, if ara-C is retained indefinitely by the liposomes, or released very slowly, then the levels of free drug may not reach the cytotoxic range, and again no therapeutic effect will result. Obviously there is an optimum rate of drug release that will result in maximum therapeutic effects. For liposomes we would predict this to be on the same order of time as the residence time of the liposomes in the circulation. We examined the relationship between $T_{1/2}$ for leakage of ara-C from liposomes and %ILS in mice bearing 10^6 cells of IV or i.p. leukemia.[24] Mice received single injections of 50 mg/kg liposomal ara-C entrapped in liposomes of various compositions having a wide range of leakage rates ($T_{1/2}$ from a few hours to several hundred hours). Control mice were infused for 24 h with 50 mg/kg total dose of free ara-C. In all cases the liposomal ara-C resulted in significantly greater increases in %ILS than mice receiving infusions of free ara-C (%ILS of 90%, IV cells, IV infusion), but there are also significant differences between the MST for mice receiving liposomes with different release rates for ara-C. The highest increases in MST (%ILS of 197%, IV cells, IV treatment) were seen in mice receiving liposomes with a $T_{1/2}$ for ara-C leakage of 41 h (PC:SM:CH:PEG-DSPE, 1:1:1:0.1), which approximates the residence time in circulation for these liposomes.[24] Liposomes such as those composed of HSPC:CH:PEG-DSPE which had much longer $T_{1/2}$ for ara-C release ($T_{1/2}$ >400 h) resulted in significantly lower MST (%ILS of 138%, IV cells, IV treatment).[24] These experiments demonstrate the importance of matching the release rate of the drug with the residence time of the liposomes in the circulation.

G. SUBCUTANEOUS S-LIPOSOMAL CYTOSINE ARABINOSIDE

Recent experiments from our laboratory have demonstrated that small (<120 nm in diameter) S-liposomes, but not C-liposomes, administered by the subcutaneous (s.c.) route of injection will migrate, together with their entrapped contents, down the lymph node chain draining the site of injection, achieving significant levels in the circulation.[23] Peak blood levels could reach over 30% of injected dose at 12 h postinjection. Larger (>120 nm) S-liposomes and C-liposomes, on the other hand appear to stay at the site of s.c. injection releasing their contents over many days.[23] It was therefore of interest to us to test whether or not the s.c. route of injection of S-araC would result in significant therapeutic activity against L1210 leukemia in mice inoculated by the IV or i.p. routes of injection. Free ara-C administered by the s.c. route has been reported to have a half-life of elimination of approximately 10 min, and is not therapeutically effective.[27,28] As seen above, S-araC administered by the IV or i.p. route of injection had good therapeutic efficacy against murine L1210 leukemia. However, although useful in animal models of disease, the i.p. route has little relevance in human therapy. The IV route is more relevant to human therapy, but generally requires a health professional for administration. The s.c. route of liposomal administration not only provides the possibility of a depot for drug sustained release, but also is an acceptable route for self-administration of therapy in humans.

A number of different compositions and sizes of liposomes, containing entrapped ara-C, were administered by either the IV or the s.c. route in mice inoculated with 10^6 L1210 leukemia cells by the IV or the i.p. route.[29] Liposomes were either small (80 to 90 nm in diameter) or large (150 to 170 nm in diameter) and were either S-liposomes (containing PEG-DSPE) or C-liposomes (lacking PEG-DSPE). Different compositions of S-liposomes with different rates of content release were also tested. Interestingly, for the s.c. injection route, but not the IV injection route, we found that the maximum tolerated dose was only 50 mg/kg S-araC in combination with 50 mg/kg free ara-C.[29] This dose resulted in some degree of gastrointestinal (G.I.) toxicity and weight loss, although there was no apparent toxicity at the site of injection. This indicated that the s.c. route may be releasing free drug over a longer period of time than the IV route, leading to the observed G.I. toxicity, i.e., the longer the period of time over which the animals are exposed to drug, the lower the maximum tolerated dose,

TABLE 3
A Comparison of the IV and s.c. Routes of Injection
of Liposomal ara-C

Liposome Composition (Molar Ratio, Diameter)	$T_{1/2}$ for Leakage of ara-C (h)	Route of Treatment	%ILS
HSPC:CH:PEG-DSPE	463	IV	152.0
(2:1:0.1, 88 nm)		s.c.	148.0
PC40:CH:PEG-DSPE	208	IV	151.9
(2:1:0.1, 89 nm)		s.c.	144.4
PC40:PEG-DSPE	$\alpha = 0.4$	IV	44.4
(1:0.05, 83 nm)	$\beta = 30.4$	s.c.	63.0
PC40:CH:PG	22.2	IV	26.7
(2:1:0.1, 82 nm)		s.c.	88.3

Note: Mice (5/group) received 10^6 L1210 leukemia cells by the IV route of injection on day 0 and treatment according to the various groups below commenced 24 h later. Ara-C dose was 25 mg/kg of free drug in combination with 25 mg/kg of liposome-entrapped drug.

a principle which has been well established for the free drug. We therefore performed all of our s.c. experiments at doses of 25 mg/kg ara-C or less in combination with an equal amount of free drug.

The most notable observation from our results was that the s.c. route of injection was as effective as, or in some cases more effective than, the IV route of injection[29] (Table 3). For liposomes with half-lives for ara-C release from liposomes in plasma of much longer than the residence time of liposomes in the circulation, the IV route and the s.c. route were equally effective[29] (Table 3). For liposomes with much shorter leakage rates (PC40:PEG-DSPE) or much shorter circulation half-lives (PC40:CH:PG) the s.c. route of injection was more effective than the IV route[29] (Table 3). The results can be interpreted as follows. If ara-C is released from liposomes rapidly in plasma, then they can be protected from the action of plasma proteins by administering them s.c. From the s.c. site, the liposomes will serve as a depot to release the drug over a prolonged time period, improving the therapeutic effects. Likewise, if the liposomes are rapidly removed from circulation following IV administration, then administration by the s.c. route will provide for a sustained drug release and improved therapeutic effect. We also observed that s.c. S-araC was as effective as IV S-araC against leukemia inoculated by the i.p. route. This indicates that the S-liposomes and/or their contents were able to cross at least two *in vivo* barriers (s.c. site to blood, and blood to peritoneal cavity) to reach their site of therapeutic action.

When large liposomes containing ara-C were injected s.c. there was only a modest decrease in therapeutic effect as compared to small liposomes. Since the large liposomes are not capable of leaving the s.c. site of injection and reaching the circulation,[23] this indicates that the mechanism of action of the liposomal drug is due, to a significant degree, to the release of free drug from the liposomes at the site of injection. Presumably the free drug is then distributed via the lymph and blood. This observation may have important ramifications for the liposome-mediated delivery of other rapidly degraded drugs, proteins, and peptides from s.c. sites of injection. As with three weekly injections of S-liposomal ara-C given by the IV route of injection, three weekly injections of s.c. S-araC were capable of curing mice, resulting in long-term (70-d) survivors when the leukemia was inoculated either IV or i.p.[29]

V. CONCLUSIONS

We have used leukemic mice as a model system in which to test the hypothesis that long circulating liposomes would be an improved drug sustained-release system for rapidly degraded drugs. The experimental results bear out this hypothesis for lower doses of drugs and lipids, with S-liposomes being more therapeutically effective than C-liposomes. At higher doses of ara-C there is saturation of the MPS by the high lipid doses which are required. This increases the circulating half-life of the C-liposomes considerably.[11] Therefore, the experimental results for ara-C in S-liposomes (which do not experience an increase in circulation half-lives as the lipid dose increases) were approximately the same as for ara-C in C-liposomes at high lipid doses. Relative to C-liposomes, at high lipid doses, the advantages of S-liposomes lie in their dose-independent pharmacokinetics and their decreased tendency to produce adverse effects on the MPS, rather than their improved therapeutic efficacy. S-araC gave results which were compatible with release of free drug from the liposomes at cytotoxic levels for periods of time from 1 d (20 mg/kg ara-C) to 4 d (100 mg/kg ara-C), depending on ara-C dose. When ara-C entrapped in S-liposomes was given by the s.c. route of injection, the therapeutic efficacy was maintained at levels similar, or superior, to those found after IV injection of liposomes. Finally, it could be demonstrated that matching the drug release rate from the liposomes to their circulation half-life could lead to optimization of the therapeutic efficacy. Many of the insights gained from these studies of S-liposomes in a simple model system are applicable in the treatment of other diseases, and to the sustained release of other drugs, particularly those that experience rapid degradation, such as immune modulators, proteins, and peptides.

REFERENCES

1. **Allen, T. M.,** Toxicity and systemic effects of phospholipids, in *Phospholipid Handbook,* Cevc, G., Ed., Marcel Dekker, New York, 1993, 801.
2. **Allen, T. M. and Chonn, A.,** Large unilamellar liposomes with low uptake into the reticuloendothelial system, *FEBS Lett.,* 223, 42, 1987.
3. **Gabizon, A. and Papahadjopoulos, D.,** Liposome formulations with prolonged circulation time in blood and enhanced uptake by tumors, *Proc. Natl. Acad. Sci. U.S.A.,* 85, 6949, 1988.
4. **Allen, T. M.,** Stealth liposomes: avoiding reticuloendothelial uptake, in *Liposomes in the Therapy of Infectious Diseases and Cancer,* New Series, Vol. 89, Lopez-Berestein, G. and Fidler, I., Eds., Alan R. Liss, New York, 1989, 405.
5. **Klibanov, A. L., Maruyama, K., Torchilin, V. P., and Huang, L.,** Amphipathic polyethyleneglycols effectively prolong the circulation time of liposomes, *FEBS Lett.,* 268, 235, 1990.
6. **Blume, G. and Cevc, G.,** Liposomes for the sustained drug release in vivo, *Biochim. Biophys. Acta,* 1029, 91, 1990.
7. **Allen, T. M., Hansen, C., Martin, F., Redemann, C., and Yau-Young, A.,** Liposomes containing synthetic lipid derivatives of poly(ethylene glycol) show prolonged circulation half-lives in vivo, *Biochim. Biophys. Acta,* 1066, 29, 1991.
8. **Senior, J., Delgado, C., Fisher, D., Tilcock, C., and Gregoriadis, G.,** Influence of surface hydrophilicity of liposomes on their interaction with plasma protein and clearance from the circulation: studies with poly(ethylene glycol)-coated vesicles, *Biochim. Biophys. Acta,* 1062, 77, 1991.
9. **Papahadjopoulos, D., Allen, T. M., Gabizon, A., Mayhew, E., Matthay, K., Huang, S. L., Lee, K.-D., Woodle, M. C., Lasic, D. D., Redemann, C., and Martin, F. J.,** Sterically stabilized liposomes: improvements in pharmacokinetics and antitumor therapeutic efficacy, *Proc. Natl. Acad. Sci. U.S.A.,* 88, 11460, 1991.
10. **Lasic, D. D., Martin, F. J., Gabizon, A., Huang, S. K., and Papahadjopoulos, D.,** Sterically stabilized liposomes: a hypothesis on the molecular origin of the extended circulation times, *Biochim. Biophys. Acta,* 1070, 187, 1991.
11. **Allen, T. M. and Hansen, C.,** Pharmacokinetics of stealth versus conventional liposomes: effect of dose, *Biochim. Biophys. Acta,* 1068, 133, 1991.
12. **Allen, T. M. and Mehra, T.,** Recent advances in sustained release of antineoplastic drugs using liposomes which avoid uptake in to the reticuloendothelial system, *Proc. West. Pharmacol. Soc.,* 32, 111, 1989.
13. **Borsa, J., Whitmore, G. R., Valeriote, F. A., Collins, D., and Bruce, W. R.,** Studies on the persistence of methotrexate, cytosine arabinoside, and leucovorin in the serum of mice, *J. Natl. Cancer Inst.,* 42, 235, 1969.
14. **Baguley, B. C. and Falkenhaug, E. M.,** Plasma half-life of cytosine arabinoside (NSC-63878) in patients treated for acute myeloblastic leukaemia, *Cancer Chemother. Rep.,* 55, 291, 1971.
15. **Kobayashi, T., Tsukagoshi, S., and Sakurai, Y.,** Enhancement of the cancer chemotherapeutic effect of cytosine arabinoside entrapped in liposomes on mouse leukaemia L1210, *Gann,* 66, 719, 1975.

16. **Mayhew, E., Papahadjopoulos, D., Rustum, Y. M., and Dave, C.,** Inhibition of tumour cell growth in vitro and in vivo by 1-β-D-arabinofuranosylcytosine entrapped within phospholipid vesicles, *Cancer Res.,* 36, 4406, 1976.

17. **Kobayashi, T., Kataoka, T., Tsukagoshi, S., and Sakurai, Y.,** Enhancement of anti-tumour activity of 1-β-D-arabinofuranosylcytosine by encapsulation in liposomes, *Int. J. Cancer,* 20, 581, 1977.

18. **Rustum, Y. M., Dave, C., Mayhew, E., and Papahadjopoulos, D.,** Role of liposome type and route of administration in the antitumor activity of liposome-entrapped 1-β-D-arabinofuranosylcytosine against mouse L1210 leukaemia, *Cancer Res.,* 39, 1390, 1979.

19. **Ganapathi, R., Krishan, A., Wodinsky, I., Zubrod, C. G., and Lesko, L. J.,** Effect of cholesterol content on antitumor activity and toxicity of liposome-encapsulated 1-β-D-arabinofuranosylcytosine in vivo, *Cancer Res.,* 40, 630, 1980.

20. **Kim, S. and Howell, S. B.,** Multivesicular liposomes containing cytarabine entrapped in the presence of hydrochloric acid for intracavitary chemotherapy, *Cancer Treat. Rep.,* 71, 705, 1987.

21. **Hong, F. and Mayhew, E.,** Therapy of central nervous system leukaemia in mice by liposome-entrapped 1-β-D-arabinofuranosylcytosine, *Cancer Res.,* 49, 50972, 1989.

22. **Skipper, H. E.,** Ara-C and cyclophosphamide; a closer look at the influence of dose intensity and treatment duration on host toxicity and therapeutic response (experimental data), Southern Research Institute, Birmingham, AL, Booklet 13, 1986, 1.

23. **Allen, T. M., Hansen, C. B., and Guo, L. S. S.,** Subcutaneous administration of liposomes: a comparison with the intravenous and intraperitoneal routes of injection, *Biochim. Biophys. Acta,* 1150, 9, 1993.

24. **Allen, T. M., Mehra, T., Hansen, C., and Chin, Y.-C.,** Stealth liposomes: an improved sustained release system for 1-β-D-arabinofuranosylcytosine, *Cancer Res.,* 52, 2431, 1992.

25. **Skipper, H. E.,** Analyses and interpretations of the influence of dose and schedule variables with an S-phase specific drug (ara-C) on (1) toxicity and (2) the degree and duration of response and cure rate of animals bearing known burdens of L1210 leukemia cells, Southern Research Institute, Birmingham, AL, Booklet 4, 1988, 1.

26. **Mayhew, E., Rustum, Y. M., and Szoka, F.,** Therapeutic efficacy of cytosine arabinoside trapped in liposomes, in *Targeting of Drugs, NATO Advanced Study Institutes Series,* Gregoriadis, G., Ed., Plenum Press, New York, 1982, 249.

27. **Finkelstein, J. Z., Scher, J., and Karon, M.,** Pharmacologic studies of tritiated cytosine arabinoside, *Cancer Chemother. Rep.,* 54, 35, 1970.

28. **Kim, S. and Howell, S. B.,** Multivesicular liposomes containing cytarabine for slow-release sc administration, *Cancer Chemother. Rep.,* 17, 447, 1987.

29. **Allen, T. M., Hansen, C. B., and Peliowski, A.,** Subcutaneous administration of sterically stabilized (Stealth) liposomes is an effective sustained release system for 1-β-D-arabinofuranosylcytosine, *Drug Targeting Delivery,* 1, 55, 1993.

Chapter 17

STEALTH® LIPOSOMES AS CARRIERS OF ANTIBIOTICS IN INFECTIOUS DISEASES

Irma A.J.M. Bakker-Woudenberg, G. Storm, and M.C. Woodle

TABLE OF CONTENTS

I. THE ROLE OF LIPOSOMES IN THE TREATMENT OF INFECTIONS

From clinical experience it is well known that in immunocompromised patients, such as patients with underlying malignant disease receiving chemotherapy or radiation therapy, or organ transplant recipients receiving immunosuppressive agents, infections frequently occur and are still a major cause of morbidity and mortality. Failure of antibiotic treatment of these infections occurs despite the availability of new and potent antibiotics. Therefore, there is an evident therapeutic need for improvement of treatment of severe infections. One of the factors that contribute to the lack of success of antibiotic treatment is an impaired host defence system unable to provide adequate support for antibiotic therapy. One way to overcome this problem is stimulation of host defenses by immunomodulating agents. The use of liposomes in the targeting of immunomodulators for this purpose is under investigation. Another way is intensification of antibiotic treatment by developing rational delivery systems of antibiotics. In this regard the use of liposomes as carriers of antibiotics is being investigated.

The applicability of liposomes in the treatment of infections is dependent on the pattern of biodistribution after intravenous administration, and the possibility of manipulating this in order to target the liposomes selectively. After intravenous administration, "classical" liposomes composed primarily of natural phospholipids and cholesterol are unable to leave the general circulation, and rapidly accumulate in cells of the mononuclear phagocyte system (MPS), particularly those in the liver and spleen. The role of these classical liposomes regarding treatment of infectious diseases may be twofold: first, a role for targeting of macrophage modulators to stimulate the cells of the MPS in infections in general (this aspect of enhancing the host defense will not be discussed further here); second, a role as carriers of antibiotic for treatment of infections in MPS tissues. The use of liposomes for delivery of antibiotic to infections in non-MPS tissues requires a reduced uptake by the MPS, enabling the liposomes to circulate long enough to reach the infected tissues or cells. The recent development of liposome formulations with long circulation half-lives opens new therapeutic avenues for improved delivery of antibiotics in infections in general, including infections in non-MPS tissues.

Intensification of antibiotic treatment of infections by the use of liposomes as carriers of antibiotics is under investigation. The use of liposomes as antibiotic delivery systems aims for increased antibiotic concentrations at the site of infection resulting from targeting of antibiotic to the infected tissues; increased intracellular antibiotic concentrations resulting from targeting of antibiotic to the infected cells; and reduced toxicity of potentially toxic antibiotics resulting from targeting of antibiotic to the infectious organisms (away from host cells). Application of liposomes may offer an excellent way to increase the selective delivery of antibiotic in these respects. The first studies on the use of liposomes as carriers of antibiotics were published in 1978.[1,2] Until now most experimental data were derived from studies with classical liposomes encapsulating antibiotics applied in models of infection caused by facultative and obligate intracellular pathogens localized in MPS cells. A variety of protozoal, viral, fungal, and bacterial infections was studied.

II. CLASSICAL LIPOSOMES AS CARRIERS OF ANTIBIOTICS

Due to the rapid localization of classical liposomes in MPS tissues after intravenous administration the encapsulated antibiotic will be targeted to these tissues, resulting in relatively high and prolonged concentrations of the antibiotic. In animal models of intracellular parasitic,[1-5] viral,[6] fungal,[7-16] and bacterial[17-26] infections in MPS tissues the effect of liposomal encapsulation on the therapeutic efficacy of antibiotics has been investigated. As most antibiotics are relatively ineffective for intracellular infections due to poor penetration into the cells or decreased activity intracellularly, in general all these studies demonstrated an increase in therapeutic index of the antibiotic (an increase in efficacy and/or a reduction of toxicity of potentially toxic antibiotic) resulting from encapsulation of the antibiotics in liposomes. Also the data from *in vitro* studies in which phagocytic cells infected with bacterial strains were exposed to antibiotics in the free or liposome-encapsulated form show a superiority of liposome-encapsulated antibiotics.[21,27-33] It was demonstrated that by varying the lipid composition of the liposomes the intracellular degradation of the liposomes could be influenced and thereby the rates at which liposome-encapsulated agents are released and become available to exert their therapeutic action.[30]

The use of liposome-encapsulated antimicrobial agents for treatment of parasitic or bacterial infections in patients was recently reported. In patients suffering from drug-resistant visceral leishmaniasis liposome-encapsulated amphotericin B (AmBisome) was used and appeared to be effective for treatment and secondary prophylaxis.[34,35] The efficacy of liposomal gentamicin for the treatment of *Mycobacterium avium* complex infection in AIDS patients is under investigation.[36] Regarding fungal infections extensive studies in patients have been performed with various preparations of liposomal or lipid-complexed amphotericin B.[37-45]

Animal studies on the treatment of infection by microorganisms not necessarily located intracellularly in MPS tissues show somewhat contradictory results with respect to the efficacy of liposome-encapsulated antibiotics vs. free antibiotics. An explanation for this is that the applicability of classical liposomes in the treatment of infectious diseases caused by extracellular microorganisms not restricted to MPS tissues has been strongly limited by the preferential uptake of intravenously administered liposomes by cells of the MPS.

III. STEALTH® LIPOSOMES AS CARRIERS OF ANTIBIOTICS

The ability to achieve a significantly longer blood residence time of liposomes creates new possibilities for achieving improved delivery of antibiotics to infected tissues in general, including infections in non-MPS tissues. It was shown already many years ago that factors contributing to prolongation of liposome circulation times include reduction of liposome size, and modifications of the bilayer composition yielding an increased rigidity of the bilayer; with such liposomes administration of relatively high lipid doses is needed to obtain a relatively long circulation time.[46-51] Later it was observed that incorporation of specific glycolipids such as monosialoganglioside GM1 and hydrogenated phosphatidylinositol (HPI) resulted in prolonged circulation without the constraint of high lipid doses but still limited to the use of rigid lipids.[52-58] Such liposomes were named Stealth® liposomes[59] or MPS-avoiding liposomes. Importantly, it was shown that such liposomes exhibit enhanced localization in a variety of implanted tumors in mice.[60,61] More recently, many reports have shown that hydrophilic phosphatidylethanolamine derivatives of monomethoxy-polyethyleneglycols (PEG-PE) attached to the liposomes also enhance blood circulation time to an extent equal to or exceeding that found previously but without the requirements of high lipid dose or rigid nature of the bilayers.[52,62-72] The term sterically stabilized liposomes has been proposed for these pegylated liposomes.

In our laboratory we started to investigate the potential of long circulating liposomes for enhanced delivery of antibiotics to sites of bacterial infection.[73,74] We used an experimental model of unilateral pneumonia caused by *Klebsiella pneumoniae* in rats. Bacterial inoculation of the left lung resulted in a progressive infection of the left lung; in the right lung of the same animal signs of an infection were not observed. We examined two types of long circulating liposomes vs. one type of classical (short circulating) liposomes. The degree of localization of liposomes in infected tissue (left lung) was studied. Choice of the two long circulating liposome formulations was based on requirements for extended circulation and on the potential to release therapeutic agents into an infection site once localization has been achieved. Long circulating liposome preparations consisted of either hydrogenated soybean phosphatidylinositol, hydrogenated soybean phosphatidylcholine, and cholesterol (HPI:HPC:Chol, 1:10:5, 100 nm), or polyethyleneglycol (PEG) 1900 derivative of distearoyl phosphatidylethanolamine, partially hydrogenated egg phosphatidylcholine, and cholesterol (PEG-DSPE:PHEPC:Chol, 0.15:1.85:1, 80 nm). Classical (short circulating) liposomes consisted of egg phosphatidylglycerol, egg phosphatidylcholine and cholesterol (PG:EPC:Chol, 1:10:5, 110 nm). Biodistribution of liposomes was detected by using the high-affinity ^{67}Ga-DF complex as aqueous liposomal marker.[75,76] As shown by Gabizon et al.,[75] this complex is appropriate for *in vivo* tracing of intact liposomes because of the advantages of minimal translocation of radioactive label to plasma proteins and the rapid renal clearance rate when the label is released from the liposomes extracellularly.

Both types of the long circulating liposomes indeed showed a prolonged blood residence time in uninfected rats compared to the blood circulation time of PG:EPC:Chol liposomes at the dose of 75 μmol/kg (Figure 1). At 16 h after administration of HPI:HPC:Chol liposomes and at 40 h after administration of PEG-DSPE:PHEPC:Chol liposomes 10% of the dose was still present in the circulation. Circulation half-life for HPI:HPC:Chol liposomes was 5 h, and for PEG-DSPE:PHEPC:Chol liposomes was 20 h. The prolonged residence time in the bloodstream was accompanied by a relatively low hepatosplenic uptake.

In the experimental *Klebsiella pneumoniae* lung infection a progressive development of the infection was observed after inoculation of the left lung with 10^5 bacteria as shown for a group of 80 rats with varying intensity of infection (Figure 2). The increase in bacterial numbers up to 3×10^{10} in the infected left lung was accompanied by an increase in lung weight up to fourfold, reflecting the involvement of the left lung tissue in the infectious process. Infected areas were characterized by serous fluid, polymorphonuclear leukocytes, few macrophages, and large numbers of bacteria. In the right lung bacterial numbers were relatively low and ranged from 10^2 to 10^9. The appearance of bacteria in the right lung was never associated with development of an infectious process, as reflected by the absence of increase in lung weight, and confirmed by histological examination. Comparison of the behavior of HPI:HPC:Chol

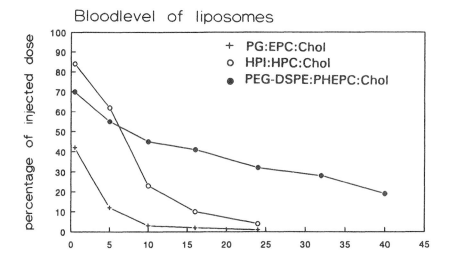

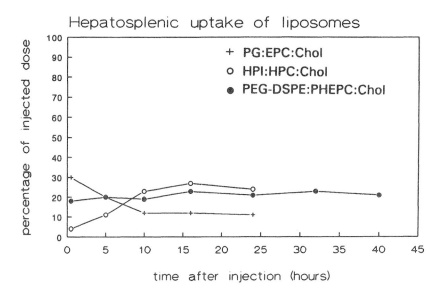

time after injection (hours)

FIGURE 1. Biodistribution of ^{67}Ga-DF-labeled liposomes. Liposomal lipid (75 µmol/kg) was injected intravenously into uninfected rats. Data are mean values for groups of four rats. PG, egg phosphatidylglycerol; EPC, egg phosphatidylcholine; Chol, cholesterol; HPI, hydrogenated soybean phosphatidylinositol; HPC, hydrogenated phosphatidylcholine; PEG-DSPE, polyethyleneglycol 1900 derivative of distearoyl phosphatidylethanolamine; PHEPC, partially hydrogenated egg phosphatidylcholine.

and PEG-DSPE:PHEPC:Chol liposomes in uninfected rats and infected rats (3 d after bacterial inoculation of the left lung) revealed that the bacterial lung infection did not affect the biodistribution substantially (Table 1).

The degree of lung localization of liposomes was determined in groups of rats with varying intensity of infection. This was done at 16 and 40 h after administration of HPI:HPC:Chol and PEG-DSPE:PHEPC:Chol liposomes, respectively, when about 10% of liposomes were still circulating in the blood (Figures 3 and 4). For PG:EPC:Chol liposomes lung localization was determined at 16 h after administration.

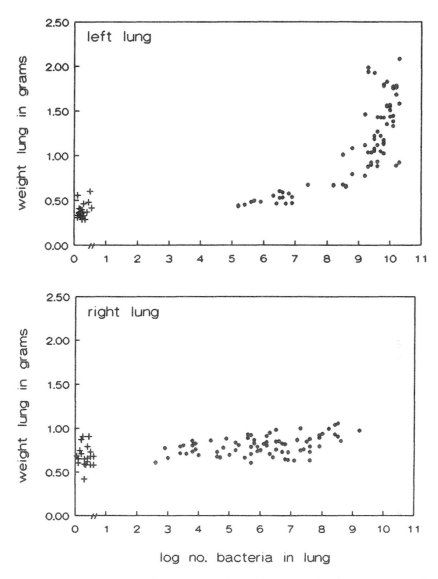

FIGURE 2. Course of lung infection after inoculation of the left lung with 10^5 *Klebsiella pneumoniae*. Increases in bacterial numbers and increases in weight of infected lung tissue were determined in a group of 80 infected rats (●). Weight of uninfected lung tissue was determined in a group of 20 uninfected rats (+). (From Bakker-Woudenberg, I.A.J.M. et al., *J. Infect. Dis.*, 168, 164, 1993. With permission.)

For both types of long circulating liposomes it was observed that in rats with substantially increased bacterial numbers in the left lung the degree of localization of liposomes in infected left lung tissue was up to 10-fold higher compared to the localization of liposomes in the left lung of uninfected rats. At 16 h after injection of HPI:HPC:Chol liposomes about 1% of the injected dose was localized in the most severely infected lungs, whereas at 40 h after injection of PEG-DSPE:PHEPC:Chol liposomes about 9% of the injected dose was localized. The increased localization of either liposome preparation in the infected left lung is likely to be related to the local infectious process, as in the right lung of the same rat (in which infection did not develop) the degree of localization of liposomes was not increased compared with the localization observed in uninfected rats. Interestingly, the increased localization of liposomes in infected left lung tissue appeared to be strongly correlated with the severity of infection as

TABLE 1
Biodistribution of ^{67}Ga-DF Labeled Liposomes in Uninfected and Infected Rats at Various Times After Intravenous Injection

Liposome Composition	0.5 h	5h	10 h	16 h	24 h	32 h	40 h
HPI:HPC:Chol							
Uninfected rats							
Blood	81.1 ± 4.3	61.6 ± 2.5	23.4 ± 0.6	10.2 ± 0.5	4.0 ± 0.7		
Liver	4.8 ± 0.8	10.4 ± 1.7	22.2 ± 1.5	25.8 ± 0.6	22.8 ± 1.4		
Spleen	0.2 ± 0.1	0.6 ± 0.2	1.4 ± 0.3	1.2 ± 0.2	0.8 ± 0.1		
Infected rats							
Blood	72.3 ± 6.4	46.8 ± 5.5	26.4 ± 4.7	7.1 ± 1.8	1.5 ± 0.3		
Liver	11.6 ± 0.8	20.1 ± 2.0	19.4 ± 2.5	22.8 ± 1.2	19.6 ± 3.9		
Spleen	1.2 ± 0.9	2.5 ± 2.8	1.5 ± 0.7	1.6 ± 0.7	2.7 ± 1.7		
PEG-DSPE:PHEPC:Chol							
Uninfected rats							
Blood	70.3 ± 5.1	54.8 ± 3.3	45.0 ± 2.9	40.7 ± 2.0	32.3 ± 2.2	27.8 ± 3.0	19.2 ± 2.2
Liver	15.3 ± 3.0	13.6 ± 2.4	13.2 ± 1.2	15.8 ± 2.4	13.8 ± 0.7	12.5 ± 2.0	
Spleen	2.8 ± 0.1	5.6 ± 2.1	5.5 ± 0.3	7.3 ± 0.8	7.1 ± 1.4	9.1 ± 0.6	7.6 ± 1.7
Infected rats							
Blood	7.9 ± 5.0	61.8 ± 1.5	57.8 ± 5.4	44.1 ± 2.9	33.8 ± 3.1	22.1 ± 2.5	12.6 ± 1.9
Liver	10.7 ± 1.9	13.7 ± 3.2	15.8 ± 1.6	12.8 ± 1.8	15.5 ± 1.8	15.9 ± 2.6	14.9 ± 2.2
Spleen	1.8 ± 1.3	3.7 ± 0.9	4.2 ± 0.6	7.4 ± 0.9	6.3 ± 1.0	6.1 ± 0.7	5.4 ± 0.9

Data are % of injected dose, mean ± SE of 4 rats. Liposomal lipid (75 mmol/kg) was injected intravenously into uninfected rats or in rats with *Klebsiella pneumoniae* lung infection 3 days after inoculation. HPI, hydrogenated soybean phosphatidylinositol;HPC, hydrogenated soybean phosphatidylcholine; Chol, cholesterol; PEG-DSPE, polyethyleneglycol 1900 derivative of distearoylphosphatidylethanolamine; PHEPC, partially hydrogenated egg phosphatidylcholine.

manifested by the increase in weight of the infected left lung, for HPI:HPC:Chol liposomes ($r = 0.93$) as well as PEG-DSPE:PHEPC:Chol liposomes ($r = 0.92$) (Figure 5). The localization of the relatively short circulating PG:EPC:Chol liposomes in infected left lung tissue at 16 h after administration was significantly lower compared to the two long circulating liposome types, and not correlated with the severity of infection (Figure 5).

In conclusion, these experimental data indicate that liposomes with prolonged blood circulation time show substantial localization in infected lung tissue after intravenous administration. The data suggest that the degree of localization of liposomes in the infected tissue is favored by a prolonged residence time of liposomes in the blood compartment, enabling the liposomes to reach the infected tissue more efficiently. The degree of lung localization of liposomes correlated with the intensity of the infection is significantly increased from the moment that bacterial multiplication in the lung results in a local inflammatory response, and appeared to be correlated with increased involvement of the lung tissue in the infectious process. Localization of classical (short circulating) liposomes in the infected lung tissue was significantly lower.

The two types of long circulating liposomes evaluated differed substantially with respect to degree of localization in infected lung tissue: the PEG-containing liposomes were clearly superior over the HPI-containing liposomes and reached a level of 9% of the injected liposomal dose in severely infected rats. A valuable asset of the PEG-containing liposomes is that they show dose-independent pharmacokinetics. High lipid doses are not needed for

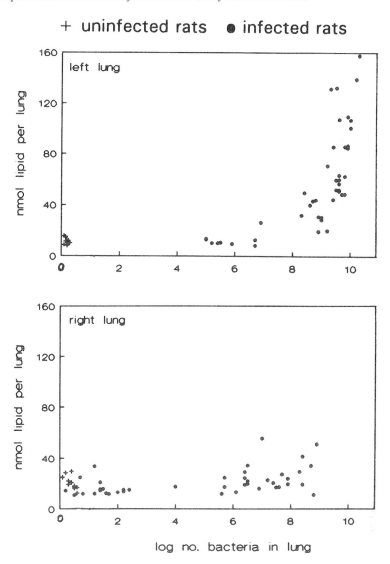

FIGURE 3. Localization of ^{67}Ga-DF-labeled liposomes in lung of rats with *Klebsiella pneumoniae* lung infection, by number of bacteria. Liposomal lipid (75 µmol/kg) was injected intravenously into 44 rats with varying intensity of infection (●) or 10 uninfected rats (+). Localization was determined 16 h later. Liposomal preparations are described in the legend to Figure 1. (From Bakker-Woudenberg, I.A.J.M. et al., *Biochim. Biophys. Acta,* 1138, 318, 1992. With permission.)

prolonged circulation. Therefore, MPS saturation resulting in impaired ability to clear bacteria from the blood can be avoided. This is important, as for immunocompromised patients in which generalization of the infection frequently occurs, maximal blood clearance capacity of the MPS is needed and can be maintained. The activity of PEG in prolonging the blood residence time of liposomes has been reported to be independent of the lipid composition (phase transition temperature, charge).[66] In view of the difference in pharmacodynamics of different classes of antibiotics, the rate of release of encapsulated antibiotic from the liposomes is important. Manipulation of the release rate can be achieved by variation of the lipid composition, which in the case of these PEG-containing formulations can be done without compromising the prolonged circulation properties. Therefore, these PEG-containing liposomes show great promise for increased delivery of antibiotics to sites of bacterial infection.

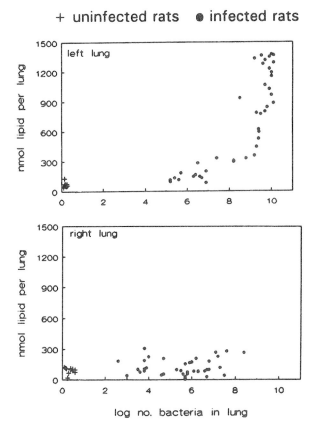

FIGURE 4. Localization of ^{67}Ga-DF-labeled liposomes in lung of rats with *Klebsiella pneumoniae* lung infection, by number of bacteria. Liposomal lipid (75 μmol/kg) was injected intravenously into 40 rats with varying intensity of infection (●) or 10 uninfected rats (+). Localization was determined 40 h later. Liposomal preparations are described in the legend to Figure 1. (From Bakker-Woudenberg, I.A.J.M. et al., *J. Infect. Dis.*, 168, 164, 1993. With permission.)

The mechanism and factors that account for the extravasation of liposomes in the infected lung tissue have yet to be elucidated. Liposomes probably extravasate in areas of inflammation as a result of locally increased capillary permeability or through injured endothelial linings secondary to the infectious process. In addition, phagocytosis of liposomes by inflammatory cells infiltrating the infected tissue may also account for enhanced localization of liposomes in the infected lung. The present results derived from studies in the rat model of *Klebsiella* lung infection indicate that long circulating liposomes are "passively" targeted towards the infected tissue. This passive targeting of long circulating liposomes to sites of infection is certainly of great importance with respect to clinical application, as in immunocompromised patients the localization as well as the etiology of the infection are often unknown.

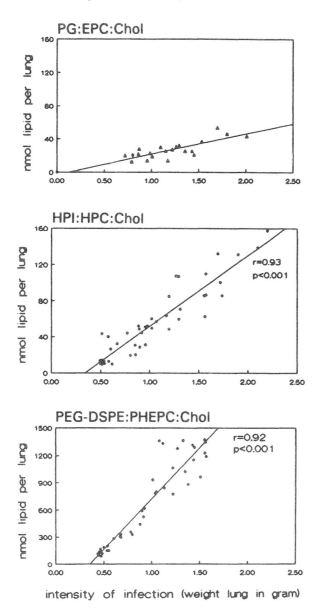

FIGURE 5. Localization of ^{67}Ga-DF-labeled liposomes in lung of rats with *Klebsiella pneumoniae* lung infection, by weight of infected left lung. The increase in weight of infected left lung was taken as a measure of the intensity of the infection. Liposomal lipid (75 μmol/kg) was injected intravenously. Localization was determined 16 h later in a group of 20 rats (PG:EPC:Chol liposomes), 44 rats (HPI:HPC:Chol liposomes), or 40 rats (PEG-DSPE:PHEPC:Chol liposomes). Liposomal preparations are described in the legend to Figure 1. (From Bakker-Woudenberg I.A.J.M. et al., *J. Infect. Dis.*, 168, 164, 1993. With permission.)

REFERENCES

1. **Alving, C. R., Steck, E. A., Chapman, W. L., Waits, V. B., Hendricks, L. D., Swartz, G. M., and Hanson, W. L.,** Therapy of leishmaniasis: superior efficacies of liposome-encapsulated drugs, *Proc. Natl. Acad. Sci. U.S.A.*, 75, 2959, 1978.
2. **New, R. R. C., Chance, M. L., Thomas, S. C., and Peters, W.,** Antileishmanial activity of antimonials entrapped in liposomes, *Nature*, 272, 55, 1978.
3. **Berman, J. D., Ksionski, G., Chapman, W. L., Waits, V. B., and Hanson, W. L.,** Activity of amphotericin B cholesterol dispersion (Amphocil) in experimental visceral leishmaniasis, *Antimicrob. Agents Chemother.*, 36, 1978, 1992.

4. **Croft, S. L., Davidson, R. N., and Thornton, E. A.,** Liposomal amphotericin B in the treatment of visceral leishmaniasis, *J. Antimicrob. Chemother.,* 28 (Suppl. B), 111, 1991.

5. **Croft, S. L.,** Liposomes in the treatment of parasitic diseases, *Pharm. Int,* 7, 229, 1986.

6. **Kende, M., Alving, C. R., Rill, W. L., Swartz, G. M., and Canonico, P. G.,** Enhanced efficacy of liposome-encapsulated ribavirin against Rift Valley fever virus infection in mice, *Antimicrob. Agents Chemother.,* 27, 903, 1985.

7. **Lopez-Berestein, G., Mehta, R., Hopfer, R. L., Mills, K., Kasi, L., Mehta, K., Fainstein, V., Luna, M., Hersh, E. M., and Juliano, R.,** Treatment and prophylaxis of disseminated infection due to *Candida albicans* in mice with liposome-encapsulated amphotericin B, *J. Infect. Dis.,* 147, 939, 1983.

8. **Lopez-Berestein, G., Hopfer, R. L., Mehta, R., Mehta, K., Hersh, E. M., and Juliano, R. L.,** Liposome-encapsulated amphotericin B for treatment of disseminated candidiasis in neutropenic mice, *J. Infect. Dis.,* 150, 278, 1984.

9. **Clark, J. M., Whitney, R. R., Olsen, S. J., George, R. J., Swerdel, M. R., Kuuselman, L., and Bonner, D. P.,** Amphotericin B lipid complex therapy of experimental fungal infections in mice, *Antimicrob. Agents Chemother.,* 35, 615, 1991.

10. **Clemons, K. V. and Stevens, D. A.,** Efficacies of amphotericin B lipid complex (ABLC) and conventional amphotericin B against murine coccidioidomycosis, *J. Antimicrob. Chemother.,* 30, 353, 1992.

11. **Gondal, J. A., Swartz, R. P., and Rahman, A.,** Therapeutic evaluation of free and liposome-encapsulated amphotericin B in the treatment of systemic candidiasis in mice, *Antimicrob. Agents Chemother.,* 33, 1544, 1989.

12. **Adler-Moore, J. P., Chiang, S. M., Satorius, A., Guerra, D., McAndrews, B., McManus, E. J., and Proffitt, R. T.,** Treatment of murine candidosis and cryptococcosis with unilamellar liposomal amphotericin B formulation (AmBisome), *J. Antimicrob. Chemother.,* 28 (Suppl. B), 63, 1991.

13. **Patterson, T. F., Miniter, P., Dijkstra, J., Szoka, F. C., Ryan, J. L., and Andrioli, V. T.,** Treatment of experimental invasive aspergillosis with novel amphotericin B/cholesterol-sulfate complexes, *J. Infect. Dis.,* 159, 717, 1989.

14. **Clemons, K. V. and Stevens, D. A.,** Comparative efficacy of amphotericin B colloidal dispersion and amphotericin B deoxycholate suspension in treatment of murine coccidioidomycosis, *Antimicrob. Agents Chemother.,* 35, 1829, 1991.

15. **Hostetler, J. S., Clemons, K. V., Hanson, L. H., and Stevens, D. A.,** Efficacy and safety of amphotericin B colloidal dispersion compared with those of amphotericin B deoxycholate suspension for treatment of disseminated murine cryptococcosis, *Antimicrob. Agents Chemother.,* 36, 2656, 1992.

16. **Van Etten, E. W. M., Van den Heuvel-de Groot, C., and Bakker-Woudenberg, I. A. J. M.,** Efficacies of amphotericin B-desoxycholate (Fungizone), liposomal amphotericin B (AmBisome), and Fluconazole in the treatment of systemic candidosis in immunocompetent and leukopenic mice, *J. Antimicrob. Chemother.,* 32, 723, 1993.

17. **Karlowsky, J. A. and Zhanel, G. G.,** Concepts on the use of liposomal antimicrobial agents: applications for aminoglycosides, *Clin. Infect. Dis.,* 15, 654, 1992.

18. **Ladigina, G. A. and Vladimirsky, M. A.,** The comparative pharmacokinetics of ^{3}H-dihydrostreptomycin in solution and liposomal form in normal and *Mycobacterium tuberculosis* infected mice, *Biomed. Pharmacother.,* 40, 416, 1986.

19. **Desiderio, J. V. and Campbell, S. G.,** Liposome-encapsulated cephalothin in the treatment of experimental murine salmonellosis, *J. Reticuloendothel. Soc.,* 34, 279, 1983.

20. **Fierer, J., Hatlen, J., Liu, J. P., Estrella, D., Mihalko, P., and Yau-Young, A.,** Successful treatment using gentamicin liposomes of *Salmonella dublin* infections in mice, *Antimicrob. Agents Chemother.,* 34, 343, 1990.

21. **Fountain, M. W., Weiss, S. J., Fountain, A. G., Shen, A., and Lenk, R. P.,** Treatment of *Brucella canis* and *Brucella abortus* in vitro and in vivo by stable plurilamella vesicle-encapsulated aminoglycosides, *J. Infect. Dis.,* 152, 529, 1985.

22. **Bakker-Woudenberg, I. A. J. M., Lokerse, A. F., Roerdink, F. H., Regts, D., and Michel, M. F.,** Free versus liposome-entrapped ampicillin in treatment of infection due to *Listeria monocytogenes* in normal and athymic (nude) mice, *J. Infect. Dis.,* 151, 917, 1985.

23. **Düzgünes, N., Perumal, V. K., Kesavalu, L., Goldstein, J. A., Debs, R. J., and Gangadharam, P. R. J.,** Enhanced effect of liposome-encapsulated amikacin on *Mycobacterium avium-M.intracellulare* complex infection in beige mice, *Antimicrob. Agents Chemother.,* 32, 1404, 1988.

24. **Bermudez, L. E., Yau-Young, A. O., Lin, J. P., Cogger, J., and Young, L. S.,** Treatment of disseminated *Mycobacterium avium* complex infection of beige mice with liposome-encapsulated aminoglycosides, *J. Infect. Dis.,* 161, 1262, 1990.

25. **Klemens, S. P., Cynamon, M. H., Swenson, C. E., and Ginsberg, R. S.,** Liposome-encapsulated gentamicin therapy of *Mycobacterium avium* complex infection in beige mice, *Antimicrob. Agents Chemother.,* 34, 967, 1990.

26. **Düzgünes, N., Ashtekar, D. R., Flasher, D. L., Ghori, N., Debs, R. J., Friend, D. S., and Gangadharam, P. R. J.,** Treatment of *Mycobacterium avium-intracellulare* complex infection in beige mice with free and liposome-encapsulated streptomycin: role of liposome type and duration of treatment, *J. Infect. Dis.,* 164, 143, 1991.

27. **Bonventre, P. F. and Gregoriadis, G.,** Killing of intraphagocytic *Staphylococcus aureus* by dihydrostreptomycin entrapped within liposomes, *Antimicrob. Agents Chemother.,* 13, 1049, 1978.

28. **Desiderio, J. V. and Campbell, S. G.,** Intraphagocytic killing of *Salmonella typhimurium* by liposome-encapsulated cephalothin, *J. Infect. Dis.,* 148, 563, 1983.

29. **Stevenson, M., Baillie, A. J., and Richards, R. M. E.,** Enhanced activity of streptomycin and chloramphenicol against intracellular *Escherichia coli* in the J774 macrophage cell line mediated by liposome delivery, *Antimicrob. Agents Chemother.,* 24, 742, 1983.

30. **Bakker-Woudenberg, I. A. J. M., Lokerse, A. F., and Roerdink, F. H.,** Antibacterial activity of liposome-entrapped ampicillin in vitro and in vivo in relation to the lipid composition, *J. Pharmacol. Exp. Ther.,* 251, 321, 1989.

31. **Al-Awadhi, H., Stokes, G.V., and Reich, M.,** Inhibition of *Chlamydia trachomatis* growth in mouse fibroblasts by liposome-encapsulated tetracyline, *J. Antimicrob. Chemother.,* 30, 303, 1992.

32. **Ashtekar, D., Düzgünes, N., and Gangadharam, P. R. J.,** Activity of free and liposome-encapsulated streptomycin against *Mycobacterium avium* complex (MAC) inside peritoneal macrophages, *J. Antimicrob. Chemother.,* 28, 615, 1991.

33. **Roesler, J., Hockertz, S., Vogt, S., and Lohmann-Matthes, M. L.,** Staphylococci surviving intracellularly in phagocytes from patients suffering from chronic granulomatous disease are killed in vitro by antibiotics encapsulated in liposomes, *J. Clin. Invest.,* 88, 1224, 1991.

34. **Davidson, R. N., Croft, S. L., Scott, A., Maini, M., Moody, A. H., and Bryceson, A. D. M.,** Liposomal amphotericin B in drug-resistant visceral leishmaniasis, *Lancet,* 337, 1061, 1991.

35. **Lazanas, M. C., Tsekes, G. A., Papandreous, Harhalakis, N., Scandali, A., Nikiforakis, E., and Saroglou, G.,** Liposomal amphotericin B for leishmaniasis treatment of AIDS patients unresponsive to antimonium compounds, *AIDS,* 7, 1018, 1993.

36. **Nightingale, S. D., Saletan, S. L., Swenson, C. E., Lawrence, A. J., Watson, D. A., Pilkiewicz, F. G., Silverman, E. G., and Cal, S. X.,** Liposome-encapsulated gentamicin treatment of *Mycobacterium avium-Mycobacterium intracellulare* complex bacteremia in AIDS patients, *Antimicrob. Agents Chemother.,* 37, 1869, 1993.

37. **Lopez-Berestein, G., Fainstein, V., Hopfer, R., Mehta, K., Sullivan, M. P., Keating, M., Rosenblum, M. G., Mehta, R., Luna, M., Hersh, E. M., Reuben, J., Juliano, R. L., and Bodey, G. P.,** Liposomal amphotericin B for the treatment of systemic fungal infections in patients with cancer: a preliminary study, *J. Infect. Dis.,* 151, 704, 1985.

38. **Sculier, J. P., Coune, A., Meunier, F., Brassinne, C., Laduron, C., Hollaert, C., Collette, N., Heymans, C., and Klastersky, J.,** Pilot study of amphotericin B entrapped in sonicated liposomes in cancer patients with fungal infections, *Eur. J. Cancer Clin. Oncol.,* 24, 527, 1988.

39. **Lopez-Berestein, G., Bodey, G. P., Fainstein, V., Keating, M., Frankel, L. S., Zeluff, B., Gentry, L., and Mehta, K.,** Treatment of systemic fungal infections, *Arch. Intern. Med.,* 149, 2533, 1989.

40. **Sculier, J. P., Klastersky, J., Libert, P., Ravez, P., Brohee, D., Vandermoten, G., Michel, J., Thiriaux, J., Bureau, G., Schmerber, J., Sergysels, R., and Coune, A.,** Cyclophosphamide, doxorubicin and vincristine with amphotericin B in sonicated liposomes as salvage therapy for small cell lung cancer, *Eur. J. Cancer,* 26, 919, 1990.

41. **Chopra, R., Blair, S., Strang, J., Cervi, P., Patterson, K. G., and Goldstone, A. H.,** Liposomal amphotericin B (AmBisome) in treatment of fungal infections in neutropenic patients, *J. Antimicrob. Chemother.,* 28 (Suppl. B), 93, 1991.

42. **Ringdén, O., Meunier, F., Tollemar, J., Ricci, P., Tura, S., Kuse, E., Viviani, M. A., Gorin, N. C., Klastersky, J., Fenaux, P., Prentice, H. G., and Ksionski, G.,** Efficacy of amphotericin B encapsulated in liposomes (AmBisome) in the treatment of invasive fungal infections in immunocompromised patients, *J. Antimicrob. Chemother.,* 28 (Suppl. B), 73, 1991.

43. **Meunier, F., Prentice, H. G., and Ringdén, O.,** Liposomal amphotericin B (AmBisome): safety data from a phase II/III clinical trial, *J. Antimicrob. Chemother.,* 28 (Suppl. B), 83, 1991.

44. **Chopra, R., Fielding, A., and Goldstone, A. H.,** Successful treatment of fungal infections in neutropenic patients with liposomal amphotericin B (AmBisome) — a report on 40 cases from a single centre, *Leukemia Lymphoma,* 7 (Suppl.), 73, 1992.

45. **Coker, R. J., Viviani, M., Gazzard, B. G., Du Pont, B., Pohle, M. D., and Murphy, S. M.,** Treatment of cryptococcosis with liposomal amphotericin B (AmBisome) in 23 patients with AIDS, *AIDS,* 7, 829, 1993.

46. **Hwang, K. J., Juen-Fai, S. L., and Beaumier, P. L.,** Hepatic uptake and degradation of unilamellar sphingomyelin/cholesterol liposomes: a kinetic study, *Proc. Natl. Acad. Sci. U.S.A.,* 77, 4030, 1980.

47. **Allen, T. M. and Everest, J. M.,** Effect of liposome size and drug release properties on pharmacokinetics of encapsulated drug in rats, *J. Pharmacol. Exp. Ther.*, 226, 539, 1983.

48. **Senior, J. H.,** Fate and behavior of liposomes in vivo: a review of controlling factors, *Crit. Rev. Ther. Drug Carrier Syst.*, 3, 123, 1987.

49. **Allen, T. M.,** A study of phospholipid interactions between high-density lipoproteins and small unilamellar vesicles, *Biochim. Biophys. Acta*, 640, 385, 1981.

50. **Senior, J. and Gregoriadis, G.,** Stability of small unilamellar liposomes in serum and clearance from the circulation: the effect of the phospholipid and cholesterol components, *Life Sci.*, 30, 2123, 1982.

51. **Hwang, K. J., Padki, M. M., Choro, D. D., Essien, H. E., Lai, J. Y., and Beaumier, P. L.,** Uptake of small liposomes by non-reticuloendothelial tissues, *Biochim. Biophys. Acta*, 901, 88, 1987.

52. **Allen, T. M.,** Stealth liposomes: five years on, *J. Liposome Res.*, 2, 289, 1992.

53. **Wasseff, N. M., Matyas, G. R., and Alving, C. R.,** Complement-dependent phagocytosis of liposomes by macrophages: suppressive effects of "stealth" lipids, *Biochem. Biophys. Res. Commun.*, 176, 866, 1991.

54. **Allen, T. M. and Chonn, A.,** Large unilamellar liposomes with low uptake into the reticuloendothelial system, *FEBS Lett.*, 223, 42, 1987.

55. **Papahadjopoulos, D. and Gabizon, A.,** Targeting of liposomes to tumor cells in vivo, *Ann. NY Acad. Sci.*, 507, 64, 1987.

56. **Allen, T. M., Hansen, C., and Rutledge, J.,** Liposomes with prolonged circulation times: factors affecting uptake by reticuloendothelial and other tissues, *Biochim. Biophys. Acta*, 981, 27, 1989.

57. **Allen, T. M., Ryan, J. L., and Papahadjopoulos, D.,** Gangliosides reduce leakage of aqueous-space markers from liposomes in the presence of human plasma, *Biochim. Biophys. Acta*, 818, 205, 1985.

58. **Gabizon, A. and Papahadjopoulos, D.,** The role of surface charge and hydrophilic groups on liposome clearance in vivo, *Biochim. Biophys. Acta*, 1103, 94, 1992.

59. **Allen, T. M.,** Stealth liposomes: avoiding reticuloendothelial uptake, in *Liposomes in the Therapy of Infectious Diseases and Cancer, New Series,* 89, Lopez-Berestein, G. and Fidler, I., Eds., Alan R. Liss, New York, 1989, 405.

60. **Gabizon, A., Price, D. C., Huberty, J., Bresalier, R. S., and Papahadjopoulos, D.,** Effect of liposome composition and other factors on the targeting of liposomes to experimental tumors: biodistribution and imaging studies, *Cancer Res.*, 50, 6371, 1990.

61. **Huang, S. K., Mayhew, E., Gilani, S., Lasic, D. D., Martin, F. J., and Papahadjopoulos, D.,** Pharmaco-kinetics and therapeutics of sterically stabilized liposomes in mice bearing C-26 colon carcinoma, *Cancer Res.*, 52, 6774, 1992.

62. **Klibanov, A. L. and Huang, L.,** Long-circulating liposomes: development and perspectives, *J. Liposome Res.*, 2, 321, 1992.

63. **Allen, T. M., Hansen, C., Martin, F., Redemann, C., and Yau-Young, A.,** Liposomes containing synthetic lipid derivatives of poly(ethylene glycol) show prolonged circulation half-lives in vivo, *Biochim. Biophys. Acta,* 1066, 29, 1991.

64. **Allen, T. M. and Hansen, C.,** Pharmacokinetics of Stealth versus conventional liposomes: effect of dose, *Biochim. Biophys. Acta,* 1068, 133, 1991.

65. **Klibanov, A. L., Maruyama, K., Beckerleg, A. M., Torchillin, V. P., and Huang, L.,** Activity of amphipathic poly(ethylene glycol) 5000 to prolong the circulation time of liposomes depends on the liposome size and is unfavorable for immunoliposome binding to target, *Biochim. Biophys. Acta,* 1062, 142, 1991.

66. **Woodle, M. C., Matthay, K. K., Newman, M. S. et al.,** Versatility in lipid compositions showing prolonged circulation with sterically stabilized liposomes, *Biochim. Biophys. Acta,* 1105, 193, 1992.

67. **Blume, G. and Cevc, G.,** Liposomes for the sustained drug release in vivo, *Biochim. Biophys. Acta,* 1029, 91, 1990.

68. **Senior, J., Delgado, C., Fisher, D., Tilcock, C., and Gregoriadis, G.,** Influence of surface hydrophilicity of liposomes on their interaction with plasma protein and clearance from the circulation: studies with poly(ethylene glycol)-coated vesicles, *Biochim. Biophys. Acta,* 1062, 77, 1991.

69. **Woodle, M. C. and Lasic, D. D.,** Sterically stabilized liposomes, *Biochim. Biophys. Acta,* 1113, 171, 1992.

70. **Lasic, D. D., Martin, F. J., Gabizon, A., Huang, S. K., and Papahadjopoulos, D.,** Sterically stabilized liposomes: a hypothesis on the molecular origin of the extended circulation times, *Biochim. Biophys. Acta,* 1070, 187, 1991.

71. **Chonn, A. and Cullis, P. R.,** Ganglioside GM1 and hydrophilic polymers increase liposome circulation times by inhibiting the association of blood proteins, *J. Liposome Res.*, 2, 397, 1992.

72. **Blume, G. and Cevc, G.,** Molecular mechanism of the lipid vesicle longevity in vivo, *Biochim. Biophys. Acta,* 1146, 157, 1993.

73. **Bakker-Woudenberg, I. A. J. M., Lokerse, A. F., Ten Kate, M. T., and Storm, G.,** Enhanced localization of liposomes with prolonged blood circulation time in infected lung tissue, *Biochim. Biophys. Acta,* 1138, 318, 1992.

74. **Bakker-Woudenberg, I. A. J. M., Lokerse, A. F., Ten Kate, M. T., Mouton, J. W., Woodle, M. C., and Storm, G.,** Liposomes with prolonged blood circulation and selective localization in *Klebsiella pneumoniae*-infected lung tissue, *J. Infect. Dis.,* 168, 164, 1993.

75. **Gabizon, A., Huberty, J., Straubinger, R. M., Price, D. C., and Papahadjopoulos, D.,** An improved method for in vivo tracing and imaging of liposomes using a gallium-67-deferoxamine complex, *J. Liposome Res.,* 1, 123, 1988.

76. **Woodle, M.C.,** [67]Gallium-labeled liposomes with prolonged circulation: preparation and potential as nuclear imaging agents, *Nucl. Med. Biol.,* 20, 149, 1993.

Chapter 18

STERICALLY STABILIZED LIPOSOME-ENCAPSULATED HEMOGLOBIN

Shuming Zheng, Yaoming Zheng, Richard L. Beissinger, and Frank J. Martin

TABLE OF CONTENTS

I. INTRODUCTION

There is an urgent need for a universally transfusable (nonallergenic), oxygen-carrying blood replacement fluid that can be used in emergency situations to provide temporary life support until a transfusion of whole blood can be administered. Numerous approaches have been taken in the attempt to develop materials that can deliver oxygen effectively and are safe for use as a red blood cell substitute; liposome-encapsulated Hb (LEH) has been developed more recently.[1-9] Liposome technology provides a mechanism for encapsulation and *in vivo* delivery of drugs, proteins, etc., which probably would otherwise be degraded, cleared rapidly, or toxic to the host.[10]

The overall goal of our work is the development of a safe, efficacious, and commercially viable oxygen-carrying red blood cell substitute composed of hemoglobin solution encapsulated in a liposome. In these studies "Stealth®" liposomes (liposomes containing phosphatidylinositol, PI, or polyethyleneglycol distearoyl phosphatidylethanolamine, PEG-PE) that are designed to evade recognition and rapid uptake by the reticuloendothelial system (more recently referred to as the mononuclear phagocytic system, MPS),[11-15] are used. Hemoglobin encapsulated in more conventional liposomes have been shown to provide an effective means of oxygen delivery *in vitro* and in experimental animals.[7,16] However, it was recently found that phosphatidylglycerol (from egg) liposomes bound to rat platelets, which was mediated by complement.[17] Although LEH containing PI have also been shown to be efficacious,[18] recent experiments[19] suggest that such systems, by overloading the MPS,[20,21] cause alterations in phagocytic activity and increase host susceptibility to infectious challenge.[22] Our recent results with LEH containing PEG-PE suggest that they are less immunotoxic as they cause less adverse effects when treated animals were tested by infectious challenge.[19] It is the PEG-PE lipid component that is believed to be responsible for producing liposomes with surfaces that are sterically stabilized such that, e.g., plasma protein uptake is greatly reduced.[23,36,38] This study encompasses the further development, characterization, and efficacy in life support of LEH using PEG-PE lipids.

II. MATERIALS AND METHODS

Human stroma-free hemoglobin solutions were prepared at 4°C following aseptic techniques as described elsewhere.[3,5] All equipment used in processing Hb and LEH was depyrogenated and all water used (for example in the washing and lysing steps and in the preparation of phosphate-buffered saline) was sterile and pyrogen-free. To maximize oxygen-carrying capacity high concentrations of Hb solution up to 35 g% (i.e., 35 g Hb per 100 ml solution) were used in some experiments to prepare liposomes. Pyridoxal-5-phosphate (P-5-P) (Sigma Chemical Co.) was added to the hemoglobin solutions to control oxygen affinity of the LEH to a value similar to that of fresh red blood cells. The antioxidant catalase, which acts as a scavenger of free radicals, was added to the Hb solution.[24] Also it is known that other constituents of the red blood cell hemolysate from which the hemoglobin solution is prepared protect against oxidation of hemoglobin and phospholipid.[24]

The membrane lipids used to encapsulate Hb solution include partially hydrogenated egg phosphatidylcholine with an "iodine value" of 40 (PC, IV 40), cholesterol (CHOL), phosphatidylinositol (PI) or PEG-PE, and α-tocopherol (α-Tc). PC was from Asahi Chemical Company, Ltd., Tokyo, Japan and PEG-PE was obtained from Liposome Technology, Inc. (Menlo Park, CA). Other lipids were obtained from Sigma Chemical Co. (St. Louis, MO). All these materials were used as obtained without further purification. The liposome membrane was formulated to contain the lipid molar ratios for PC:CHOL:PI or PEG-PE:α-Tc of 1.0:1.0:0.1:0.02. Both the PI and PEG-PE formulations were prepared at a lipid to Hb loading of about 150 μmol per milliliter of precursor Hb solution.

An aseptic double emulsion technique was followed in the preparation of LEH.[18] This method first involves the formation of a water-in-oil-in-water type multiple emulsion[25] using PC and cholesterol as primary and secondary emulsion surfactants, respectively, followed by organic solvent removal in a rotary evaporator operating under partial vacuum. As the organic solvents are removed, LEH spontaneously form in the excess lipid system. The evaporation procedure is continued until dryness using a single-stage vacuum pump to maximize removal of all organic solvent and water so that the Hb concentration within the LEH is as high as possible. This results in the deposition of an apparently dry Hb/lipid thin film on the walls of the round bottom flask. Concentrated Hb solution is then added under agitation to rehydrate the lipid and form a coarse LEH suspension. Size reduction was achieved by homogenization using a Microfluidizer.[18] The LEH so formed were washed at least three times in isotonic phosphate-buffered saline (PBS) and centrifuged at 30,000 $\times$ g for 30 min to remove all unencapsulated Hb solution and any residual organic solvent. Ideally, a finishing membrane extrusion step was used to ensure the final LEH suspension would pass through a 0.45-μm sterilizing filter. Approximately 200 ml of the susension was placed in a 400-ml, 76 mm diameter pressure filtration cell (Nucleopore, Pleasanton, CA) fitted with the appropriate pore size filter. Prior to addition of the LEH sample to the filtration cell, it was first prefiltered and then filtered under vacuum through 5-, 3-, and 1-μm CF filters (Nucleopore). The resulting LEH samples were then reloaded into the pressure cell and using nitrogen as the pressure source (at pressures well below 100 psi) they were filtered through 0.8-, 0.6-, and 0.45-μm CF filters; the last filter used is a sterile-grade 0.45-μm filter. If desired all filtration steps can be done in the filtration cell and the LEH can

Oxygen content in terms of ml O$_2$ per ml of the LEH suspension sample was determined using a Lex-O$_2$-Con (Hospex Fiberoptics, Chestnut Hill, MA).[28] The oxy-hemoglobin equilibrium dissociation and association curves for the two LEH preparations, measured as a function of oxygen partial pressure, were obtained using a Hemox-Analyzer (TCS Medical Products Co., Huntingdon Valley, PA). Oxygen affinity (P$_{50}$) and cooperativity (Hill exponent n) were determined from the generated curves. Encapsulated Hb concentration was determined by dissolving the liposome membrane with *n*-octyl β-D-glucopyranoside detergent solution as described elsewhere.[3,18] The resulting Hb solution concentration was measured for oxy, reduced, and met-Hb components by the method of Benesch et al.,[26] modified using the extinction coefficient values provided by Van Assendelft and Zijlstra.[27]

Steady shear viscosity of the suspension samples was measured in a uniform shear field with a Wells-Brookfield Syncro-Lectric Microviscometer (Model LVT) equipped with a 0.80° cone (Model CP-40, Soughton, MA). Shear rates from 45 to 450 s^{-1} at 37°C were evaluated. The cone-and-plate geometry is very useful as it gives a good approximation of viscometric flow with constant shear rate throughout the flow field.[29] The stability of the LEH to shear, i.e., leakage of Hb, was evaluated by shearing the LEH samples for 30 min in the viscometer as a function of shear rate. The effect of shear rate on leakage of encapsulated Hb was obtained for freshly prepared LEH samples in either 7.5 g% egg albumin/PBS or human plasma, both at 30% by volume. Following centrifugation (13,600 × g, Microcentrifuge Model 235C, Fisher Scientific) the concentration of total Hb in the supernatant of the sheared sample was compared with that of the unsheared sample. The benzidine method,[30] which is known to be accurate down to concentrations of 1 mg/dl, was used for determining plasma Hb concentration at low concentration levels (i.e., in the mg/dl range) as a result of leakage of Hb from LEH. The effect of 1-d incubation at 37°C on leakage of encapsulated Hb was obtained for freshly prepared LEH samples in either 7.5 g% egg albumin/PBS or human plasma (both at 30% by volume LEH). Also, the effect of cholesterol content in the liposome membrane on leakage of encapsulated Hb was evaluated.

Circulation half-life of LEH and efficacy in life support was evaluated in unconscious rats in Illinois Institute of Technology's Small Animal Lab. Female rats (Harlan Sprague, Indianapolis, IN) weighing 225 to 275 g (8 to 12 weeks of age) were anesthetized using ketamine. Cannulation of the femoral artery and vein was carried out based on the model developed by Keipert and Chang.[31] Using a doubly cannulated rat and a peristaltic pump (Manostat, New York), exchange transfusions were performed by removing blood at a constant rate of about 0.2 ml/min, coupled with its simultaneous isovolemic replacement with either LEH suspension or control. The decrease in hematocrit levels was recorded during the exchange-transfusion.

III. RESULTS AND DISCUSSION

Hemoglobin concentrations of up to 25 g% were achieved in both PEG-PE and PI liposomes, representing greater than 80% of precursor Hb solution concentration. Possibly the other apparent 20% can be accounted for by the volume occupied by the lipid phase. LEH processing resulted in LEH containing about 0.9 μmol encapsulated Hb per 82 μmol of lipid (i.e., total lipid including cholesterol). These results were determined using a total phosphorus analysis in conjunction with Bligh-Dyer extraction of 1-ml samples of PEG-PE-based LEH at a lipocrit of 30% containing an encapsulated Hb solution concentration of 20 g%. The phosporus analysis was performed by LTI's quality control unit. Met-Hb generation accompanying LEH processing (for either lipid formulation) appeared to be small with only a 3% increase for encapsulated over precursor. These results correspond to an oxygen content for an LEH suspension sample (50% by volume LEH) of 15 volume% oxygen. Storage of PEG-PE-based LEH for up to 1 month at –20°C resulted in percent met-Hb concentrations to levels of about 9%. Additions of various components to the lipid phase of LEH systems appeared to reduce the oxidative interactions between hemoglobin and membrane lipid. Some of these included the addition of cholesterol to the membrane phase to protect Hb from oxidation.[32,33] Also oxidation of Hb to met-Hb may have been inhibited by using partially hydrogenated PC instead of natural unsaturated egg-PC.[3] Partial hydrogenation of egg-PC to an iodine value of 40 as used here is known to convert the polyunsaturated fatty acids to monosaturated species, which are far less susceptible to oxidation.[34]

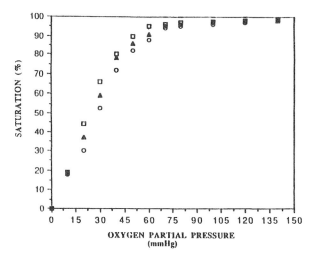

FIGURE 1.   Oxygen dissociation curve of precursor hemoglobin solution ▲, whole blood ○, and PEG-PE-LEH □, measured at 37°C.

The size of the LEH containing PEG-PE in which thin-section electron micrographs were prepared was similar to that reported previously for liposomes containing PI.[18] LEH were passed 10 times through a Microfluidizer™ and resulted in a particle size range of a 3-μm filtered sample from 50 nm to a little greater than 1 μm with a median particle size of about 300 to 400 nm. The oxygen saturation curve obtained for LEH containing PEG-PE, the precursor Hb solution, and a whole blood sample using a Hemox Analyzer is shown in Figure 1. Oxygen affinity (based on P_{50}) and cooperativity (as characterized by the Hill coefficient) for the LEH suspensions appeared to be near the normal values seen for whole blood.

Steady shear viscosity results were obtained for PEG-PE- and PI-based LEH suspension samples (in PBS containing 7.5 g% albumin at isooncotic levels) for shear rates to about 500 s⁻¹. Although viscosity results for both formulation suspension samples prepared at 50% lipocrit were higher than those obtained for human and rat whole blood (both at 45 to 46% hematocrit), viscosity results obtained for either LEH suspension sample of 30% lipocrit were slightly lower than that measured for the whole blood samples. All measurements were made at 37°C. The effect of shear rate on leakage of encapsulated Hb was obtained for freshly prepared LEH samples in either 7.5 g% egg albumin/PBS or human plasma, both at 30% by volume. Although leakage was highest for either formulation suspended in plasma, only 0.5% leakage of the encapsulated Hb was measured at the highest shear rate value tested, i.e., about 500 s⁻¹. The effect of storage of either LEH containing PEG-PE or LEH containing PI at 37°C on leakage of encapsulated Hb was obtained for freshly prepared LEH samples in either 7.5 g% egg albumin/PBS or human plasma (both at 30% by volume LEH). Again, less than 0.5% leakage of the encapsulated Hb was observed for either LEH sample in plasma even after 24 h of incubation (see Figure 2A). As shown in Figure 2B, cholesterol content lower than that used in the standard formulation, which is a 1:1 molar ratio of cholesterol to egg-PC, resulted in substantial leakage of Hb from LEH into 7.5% egg albumin/PBS. For example, a molar ratio of 0.4:1.0 is accompanied by 35% leakage of hemoglobin. It is well known that high cholesterol content in the membrane reduces membrane permeability in serum and plasma.[35]

Circulation half-life following 50% isovolemic exchange-transfusion typically was about 15 to 20 h for both of the formulation samples tested (see Figure 3). These times are desirably long and compare very favorably to results reported in another study for LEH containing dimyristoyl phosphatidylglycerol.[5] Also, high cholesterol content in the membrane, as is the case for both PEG-PE- and PI-LEH formulations, prolongs liposome stability in circulation.[35] Other recent studies have also found that PEG-PE significantly increased the blood circulation time of liposomes formulated with it.[11-15] Also those studies have shown PEG-PE liposomes

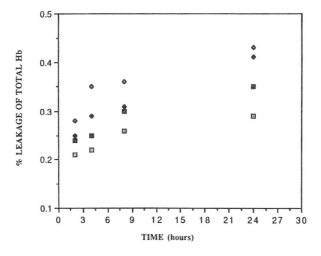

FIGURE 2A. Effect of incubation time of PI-LEH in human plasma ◆, in 7.5 wt% albumin-PBS □, PEG-PE-LEH in human plasma ◇, and in 7.5 wt% albumin-PBS ▩ on hemoglobin leakage at 37°C.

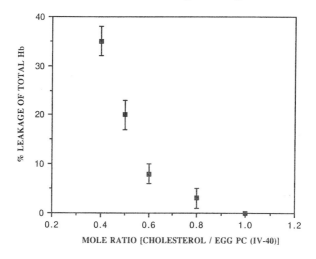

FIGURE 2B. Effect of cholesterol molar ratio in PEG-PE-LEH in 7.5 wt% albumin-PBS on hemoglobin leakage following 1-d storage at 4°C.

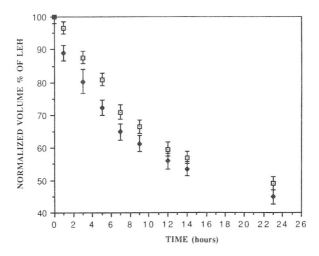

FIGURE 3. Clearance of PI-LEH ◆ and PEG-PE-LEH ⊡ in rats following 50% isovolemic exchange transfusion.

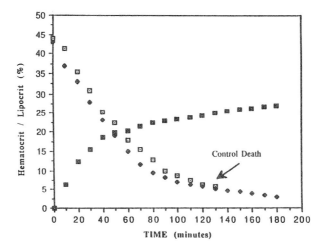

FIGURE 4A.   Hematocrit in rats during isovolemic exchange transfusion with 7.5 wt% albumin-PBS ☐ and PE-LEH ◆. The lipocrit of PEG-PE ▨ during the exchange is also shown.

to have greatly decreased uptake by the reticuloendothelial system (RES) and thus enhanced circulation time compared to many other conventional phospholipid formulations. A possible mechanism being developed to explain this behavior is that the PEG-PE polymer-phospholipid component sterically stabilizes the LEH[23] like that previously reported for the nonionic surfactant coating of colloidal particles.[37] This results in "limited accessibility",[13] i.e., because of steric stabilization of the liposome surface to adsorption by plasma proteins[37] such as immunoglobulins and high-density lipoproteins, which could cause vesicular breakdown[39] and opsonization followed by RES uptake.[13] Also, it was shown recently that the binding of phosphatidylglycerol liposomes to rat platelets was mediated by complement.[17] Recently it was found that replacement of PI with PEG-PE in LEH may provide a mechanism for *in vivo* oxygen delivery with less adverse impact on host resistance and immunity.[19] Groups of CD-1 mice were given an intravenous infectious challenge with a 20% lethal dose of *Listeria monocytogenes*. Mice then dosed 1 d later with LEH containing PI died rapidly from Listeria infection, whereas mice dosed with LEH containing PEG-PE lived significantly longer.

As shown in Figures 4A and 4B, nearly total (97%) isovolemic exchange transfusion demonstrates efficacy of PI- and PEG-PE-based LEH suspension samples, since administration of LEH supported life in rats whose hematocrit had been reduced to levels below 5%, which are incompatible with survival when performing exchange transfusion with isotonic/isooncotic PBS containing 7.5 g% albumin. These results confirming the efficacy of PI are consistent with those found in other recent studies for terminal hematocrit obtained for control and LEH-exchanged animals[7,16,17] and show the PEG-PE liposomes are equally efficacious.

IV. CONCLUSIONS

The finding that PEG-PE is a suitable substitute for PI in this setting improves the prospects of developing a commercially viable liposome-based blood substitute for several reasons: PEG-PE is synthetic and thus has the potential to be produced in commercial quantities at reasonable cost, PEG-PE exhibits favorable solubility properties (similar to PC), LEH containing PEG-PE are efficacious and, perhaps most importantly, recent evidence suggests that liposomes containing PEG-PE may have less adverse impact than PI on the capacity of the MPS to clear pathogens from the bloodstream.[19]

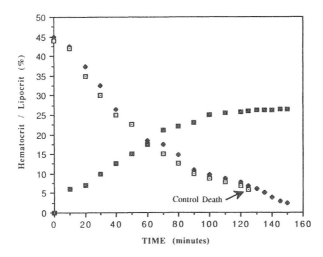

FIGURE 4B. Hematocrit in rats during isovolemic exchange transfusion with 7.5 wt% albumin-PBS □ and PI-LEH ◆. The lipocrit of PI-LEH ▥ during the exchange is also shown.

ACKNOWLEDGMENTS

We would like to thank the Microfluidics Corporation for aid in adapting the Microfluidizer™ M110 equipment for our purpose and Dr. John Hess of Blood Research at Letterman Army Institute of Research for the pyrogen- and stroma-free solutions of hemoglobin. This work was supported by a grant from the Army Medical Research and Development Command (DAMD17-89-Z-9007) and from funds provided by Liposome Technology, Inc.

REFERENCES

1. **Djordjevich D. and Miller, I.F.,** *Exp. Hematol.* (Copenhagen), 8, 584-92.
2. **Gaber, B.P., Yager, P., Sheridan, J.P., and Chang, E.L.,** *FEBS Lett.,* 153, 285, 1983.
3. **Farmer, M.C. and Gaber, B.P.,** *Methods Enzymol.,* 149, 184, 1987.
4. **Brandl, M., Becker, D., and Bauer, K.H.,** *Drug Dev. Ind. Pharm.,* 15, 655, 1989.
5. **Beissinger, R.L., Farmer, M.C., and Gossage, J.L.,** *Trans. Am. Soc. Artif. Intern. Organs,* 32, 58, 1986.
6. **Jopski, B., Pirkl, V., Jaroni, H.W., Schubert, R., and Schmidt, K.H.,** *Biochim. Biophys. Acta,* 978, 79, 1989.
7. **Hunt, C.A., Burnett, R.R., MacGregor, R.D., Strubbe, A.E., Lau, D.T., Taylor, N., and Kawada, H.,** *Science,* 230, 1165, 1985.
8. **Kato, A. and Kondo, T.,** in *Advances in Biomedical Polymers,* Gebelein, C.G., Ed., Plenum Press, New York, 1987.
9. **Hayward, J.A., Levine, D.M., Neufeld, L., Simon, S.R., Johnston, D.S., and Chapman, D.,** *FEBS Lett.,* 187, 261, 1982.
10. **Allen, T.M.,** *Adv. Drug Del. Rev.,* 2, 55, 1988.
11. **Allen. T.M., Austin, G.A., Chonn, A., Lin, L., and Lee, K.C.,** *Biochim. Biophys. Acta,* 1061, 56, 1991.
12. **Klibanov, A.L., Maruyama, K., Torchilin, V.P., and Huang, L.,** *FEBS Lett.,* 268, 235, 1990.
13. **Allen, T.M., Hansen, C., Martin, F., Redemann, C., and Yau-Young, A.,** *Biochim. Biophys. Acta,* 1066, 29, 1991.
14. **Blume, G. and Cevc, G.,** *Biochim. Biophys. Acta,* 1029, 91, 1990.
15. **Papahadjopoulos, D., Allen, T.M., Gabizon, A., Mayhew, E., Matthay, K., Huang, S.K., Lee, K.-D., Woodle, M.C., Lasic, D.D., Redemann, C., and Martin, F.J.,** *Proc. Natl. Acad. Sci. U.S.A.,* 88, 11460, 1991.
16. **Djordjevich, L., Mayoral, J., Miller, I.F., and Ivankovich, A.D.,** Crit. Care Med., 15, 318, 1987.
17. **Loughrey, H.C., Bally, M.B., Reinish, L.W., and Cullis, P.R.,** Thromb. Haemostasis, 64, 172, 1990.
18. **Zheng, S., Zheng, Y., Beissinger, R.L., and Fresco, R.,** *Biomater. Artif. Cells Artif. Organs,* in press.
19. **Sherwood, R.L., McCormick, D.H., Zheng, S., and Beissinger, R.L.,** *Toxicologist,* 12, 176a, 1992.
20. **Allen, T.M., Murray, L., Alving, C.R., and Moe, J.,** *Can. J. Physiol. Pharmacol.,* 65, 185, 1987.
21. **Merion, R.M.,** *Transplantation,* 40, 86, 1985.

22. **Nugent, K.M.,** *J. Leuk. Biol.,* 36, 123, 1984.
23. **Lasic, D.D., Martin, F.J., Gabizon, A., Huang, S.K., and Papahadjopoulos, D.,** *Biochim. Biophys. Acta,* 1070, 187, 1991.
24. **Szebeni, J., Winterbourn, C.C., and Carrell, R.W.,** *Biochem. J.,* 220, 685, 1984.
25. **Zheng, S., Beissinger, R.L., and Wasan, D.T.,** *J. Colloid Interface Sci.,* 144, 72, 1991.
26. **Benesch, R.E., Benesch, R., and Fury, S.,** *Anal. Biochem.,* 55, 245, 1973.
27. **Van Assendelft, O.W. and Zijlstra, W.G.,** *Anal. Biochem.,* 69, 43, 1975.
28. **Kusumi, F., Butts, W.C., and Ruff, W.L.,** *J. Appl. Physiol.,* 35, 229, 1973.
29. **Beissinger, R.L. and Williams, M.C.,** *AIChE J.,* 30, 569, 1984.
30. **Crosby, W.H. and Furth, F.W.,** *Blood,* 11, 380, 1956.
31. **Keipert, P.E. and Chang, T.M.S.,** *Vox Sang.,* 53, 7, 1987.
32. **Chapman, D.,** *Q. Rev. Biophys.,* 8, 185, 1975.
33. **Szebeni, J., Hauser, H., Eskelson, C.C., Waston, R.R., and Winterhalter, K.H.,** *Biochemistry,* 27, 6425, 1988.
34. **Lang, J., Vigo-Pelfrey, C., and Martin, F.,** *Chem. Phys. Lipids,* 53, 91, 1990.
35. **Damen, J., Regts, G., and Scherphof, G.,** *Biochim. Biophys. Acta,* 665, 538, 1981.
36. **Milner S.T.,** *Science,* 251, 905, 1991.
37. **Illum, L., Hunneyball, I.M., and Davis, S.S.,** *Int. J. Pharm.,* 29, 53, 1986.
38. **Jeon, S.I., Lee, J.H., Andrage, J.D., and de Gennes, P.G.,** *J. Colloid Interface Sci.,* 142, 149, 1991.
39. **Tall, A.R. and Small, D.M.,** *Nature,* 265, 163, 1977.

Chapter 19

PEG-MODIFIED LIPOSOMES FOR GAMMA- AND MAGNETIC RESONANCE IMAGING

V.P. Torchilin, V.S. Trubetskoy, J. Narula, and B.A. Khaw

TABLE OF CONTENTS

I. INTRODUCTION

In this chapter two novel practical applications of PEG-containing liposomes in the field of biomedical research are presented. Both of them are connected with possible use of liposomes for the delivery of diagnostic (imaging) agents. The corresponding animal experiments will be described.

The use of liposomes as carriers for different pharmaceuticals, including both therapeutic and diagnostic agents, is now a well-established branch of drug delivery research.[1-3] Numerous attempts have been made to make liposomes target-specific by attaching organ- or tissue-specific monoclonal antibodies (mAb) to their surface.[4-6] Despite evident success in the development of mAb-to-liposome coupling techniques and improvements in the targeting efficacy, the majority of mAb-modified liposomes still ends in the liver, which is usually a consequence of insufficient time for the interaction between the target and targeted liposome. This is especially true in cases when a target of choice has diminished blood supply (ischemic or necrotic areas). Even good liposome binding to the target could not provide high liposome accumulation because of the small quantity of liposomes passing through the target with the blood during the time period when liposomes are still present in the circulation. The same lack of targeting can happen if the concentration of the target antigen is very low, and even sufficient blood flow (and consequently, liposome passage) through the target still does not result in good accumulation effect due to the small number of productive collisions between antigens and immunoliposomes. It is quite evident that in both cases much better accumulation can be achieved if liposomes can stay in the circulation long enough. This will increase the total quantity of immunoliposomes passing through the target in the first case, and the number of productive collisions between immunoliposomes and target antigen in the second.

The situation has changed with the discovery of so-called long circulating liposomes.[7-10] These are the liposomes containing ganglioside GM1 or poly(ethylene glycol) (PEG) on their surface. We shall concentrate here only on PEG-coated liposomes. Prolonged circulation times of such liposomes might facilitate their nonspecific or specific (in the case of mAb-modified liposomes) accumulation in the target due to the sharply increased probability of liposome interaction with the target. However, some doubt has been expressed that the co-immobilization of PEG and mAb on the same liposome can result in the decrease of targeting effect because of the steric hindrances created by the presence of protective polymer.[11] In Chapter 6 we have discussed the approach that can be used to overcome this difficulty and to prepare long circulating immunoliposomes. The experimental proof for this approach is described here, using as an example targeted delivery of ^{111}In-antimyosin-PEG-liposomes into the infarcted myocardium in rabbits and dogs, and subsequent visualization of necrotic areas on gamma-camera following infarct accumulation of liposome-bound [111]In γ-radioactivity.

The acute myocardial infarction model is convenient for studying *in vivo* liposome delivery toward the tissue antigen directly exposed into the bloodstream. In a model like this, we can escape problems connected with liposome extravasation, and the relationship between immunoliposome circulation time and its target accumulation can be directly investigated. As we have already mentioned, the time during which the contact of circulating immunoliposome with the antigen is possible, is quite important for the target localization. Co-immobilization of mAb and PEG on the liposome surface can modulate this parameter. Besides, the model suggested permits performance of some optimization experiments, controlling such parameters as liposome size and PEG or/and mAb quantity.

The presence of PEG on the liposome surface can affect not only liposome longevity, but also some other important characteristics of liposomal preparations intended for diagnostic application. Thus, for example, taking into account the ability of the PEG molecule to bind water molecules tightly, we can expect that liposome-bound PEG might affect relaxivity properties of liposomal membrane-incorporated paramagnetic labels used for magnetic resonance (MR) imaging. Thus, coating liposomes with polymers can help to determine optimal performance composition in the delivery of paramagnetic label-loaded liposomes to the lymph nodes after subcutaneous administration for the purpose of MR visualization. This application is quite important from a practical point of view, because the imaging of lymph nodes plays a major role during the early detection of neoplastic involvement in cancer patients.[12] Liposomes, as any other particulates, have been shown to accumulate in lymph nodes with the lymph flow upon subcutaneous injection.[13] Tagging of the vesicles with different reporting groups might detect the abnormalities within the node architecture and, hence, be helpful in diagnosis of cancer.

II. GAMMA-VISUALIZATION OF INFARCTED MYOCARDIUM WITH ^{111}In-LABELED PEG-COATED IMMUNOLIPOSOMES

To prove that long circulating PEG-coated liposomes can be made targeted by co-incorporation of an antibody onto the liposome surface, we have studied *in vivo* PEG-liposomes with anti-myosin antibody[14] for targeting of experimental myocardial infarction in rabbits. The antibody used (mAb R11D10) effectively binds myosin inside ischemic or/and necrotic cardiomyocytes with affected or destroyed cellular membranes, but does not interact with normal cells, being unable to penetrate the intact plasmic membrane.[15,16] This forms the basis for the targeted delivery of radiolabeled PEG-coated long circulating liposomes in the region of ischemically compromised myocardium.

Infarcts in rabbits were generated as described in Reference 14. Briefly, rabbits (New Zealand white rabbits, 3.3 kg) were anesthetized with ketamin/xylazine. A femoral artery cut-down was performed to establish a blood pressure line and for arterial blood sampling. An ear vein was catheterized to allow intravenous injections. An endotracheotomy was performed, followed by ventilation with a Harvard Rodent Ventilator, model 683. A left thoracotomy was performed and the mid-left anterior descending coronary artery was occluded with a silk

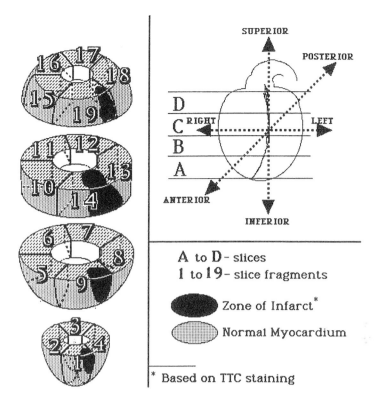

FIGURE 1. Diagrammatic representation of sample preparation. The postmortem heart was cut into 5-mm slices (A-D), stained with TTC (dark areas show infarct zone), and each slice was further cut into numbered fragments (**1** to **19**, referred to as left ventricle - LV - samples in Figure 2), that were then counted for ^{111}In radioactivity. The actual number of fragments from each heart varied, depending on the size of the heart. (Adapted from Reference 14.)

suture. After 40 min, the snare was released and removed. Different radiolabeled (^{111}In) liposome preparations were injected intravenously after 30 to 60 min of reperfusion. Blood samples were taken after specified time intervals to measure liposomal radioactivity. Five to six hours after liposome injection, animals were killed by an overdose of pentobarbital. The heart was excised and cut into 5-mm slices, stained with 2% triphenyl tetrasolium chloride, and each slice was further divided into smaller segments (see the scheme of sample preparation on Figure 1). Samples of normal and infarcted myocardium were weighed and counted in a gamma-counter. The data on the liposome accumulation in the heart were expressed as infarct-to-normal myocardium radioactivity ratio.

Liposomes were prepared by detergent dialysis method from a mixture of phosphatidyl choline and cholesterol in 3:2 molar ratio. Additionally liposomes contained 1 mol% of ^{111}In-labeled diethylenetriamine pentaacetic acid conjugated with stearylamine (DTPA-SA) and, when necessary, 4 or 10 mol% of PEG-phosphatidyl ethanolamine (PEG-PE).[8,14] Total liposome-associated radioactivity was usually around 200 μCi. For the incorporation into the liposomal membrane anti-myosin antibody was preliminary modified with hydrophobic "anchor", *N*-glutaryl phosphatidyl ethanolamine (Avanti Polar Lipids) as described in Reference 17. All liposomes were extruded through 0.4- and 0.2-μm Nuclepore filters. According to the measurements on Coulter Submicron Particle Size Analyzer N-4, liposomes in all preparations were monodisperse with size between 150 and 190 nm.

The half-life of PEG-free immunoliposomes in rabbits was 40 min, and increased to 200 min for liposomes with 4 mol% PEG and to about 1000 min for liposomes with 10 mol% PEG. Antibody-free liposomes with 4 mol% of PEG had a lifetime of about 300 min. The lifetime

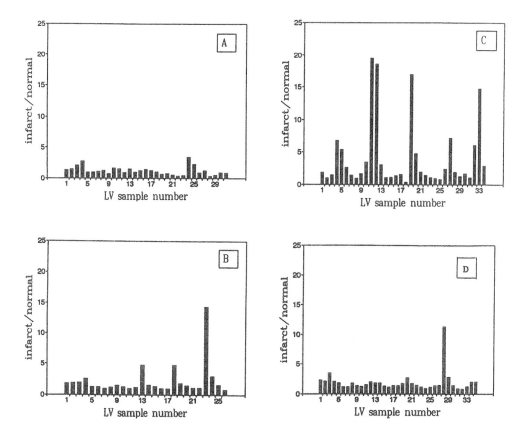

FIGURE 2. The distribution of ^{111}In-labeled liposomes in rabbit infarcted myocardium 6 h after intravenous injection. Typical patterns of radioisotope distribution in the left ventricle (LV) fragments for individual animals are presented as infarct-to-normal ratio. (A) Liposomes with 4 mol% of PEG; (B) liposomes with antimyosin antibody; (C) liposomes with 4 mol% PEG and antimyosin antibody; (D) liposomes with 10 mol% PEG and antimyosin antibody. TTC staining (infarct areas) coincides with radioactivity (liposomes) accumulation. M_w (PEG) = 5000. (From Torchilin, V.P. et al., *FASEB J.*, 6, 2716, 1992. With permission.)

of antibody-free liposomes with 10 mol% of PEG did not differ from that of immunoliposomes with 10 mol% of PEG. Two conclusions can be drawn from these data. First, the increasing quantity of PEG increasingly protects liposomes from the clearance; second, co-immobilization of an antibody and PEG decreases the half-life of liposomes only at the lower PEG concentration. High PEG concentration blocks the recognition of antibody by liver cells. The results agree well with our hypothesis on the mechanism of PEG action (see Chapter 6, Figure 8).

Figure 2 presents the typical patterns of liposome-associated ^{111}In radioactivity distribution over the infarcted heart in rabbits (expressed as infarct-to-normal ratios). mAb-free PEG-coated liposomes showed only slight and probably nonspecific accumulation in the infarct (Figure 2A). This observation might reflect the known phenomenon of "plain" liposome accumulation in the infarcted heart region, described in Reference 18. The highest uptake ratio with antimyosin liposomes without PEG was 14:1 (Figure 2B). The highest uptake ratio was achieved for PEG-immunoliposomes and reached 20:1 (Figure 2C). Very high PEG concentration (10 mol%) on immunoliposomes diminished the uptake ratio back to 12:1 (Figure 2D).

All acute myocardial infarctions have been confirmed by histochemical staining with triphenyl tetrazolium chloride (TTC), which is specific for delineation of normal tissues following dehydrogenase activity, whereas infarcted myocardium remains unstained.[15] It appeared that lack of TTC staining coincided with increased ^{111}In radioactivity accumulation, and the uptake of PEG-immunoliposomes really corresponded to the areas of infarction.

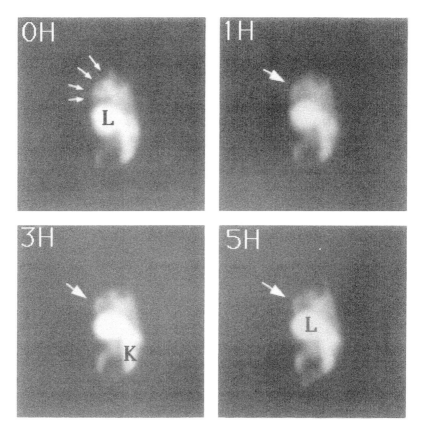

FIGURE 3. Radioimmunoscintigraphy of experimental myocardial infarction in dog with ^{111}In-PEG-antimyosin-liposomes. Images obtained in increasing time intervals (0, 1, 3, 5 h) gradually reveal the infarct localization, which can be seen already as early as 1 h (single arrow). Multiple arrows show the area of the initial blood flow pattern in the heart area; L, liver; K, kidney.

Thus, PEG-coated antimyosin-immunoliposomes can be used for the specific delivery of pharmaceuticals into the necrotic areas of the reperfused infarcted myocardium. Besides, long circulating immunoliposomes with the properly chosen PEG-to-mAb ratio provide the most effective accumulation in the regions where not all cells are dead, and the target antigen concentration might be relatively low (see Figure 2C). Myocardial infarction seems to be a good target for the liposomal delivery of imaging and therapeutic agents (radiometals, thrombolytics, superoxide dismutase, proteases). Moreover, the method developed can be easily extended onto other species. Thus, using similar protocol[19] we have performed experimental visualization of the myocardial infarction in dog using PEG-antimyosin-^{111}In-labeled liposomes. The imaging data are presented in Figure 3. The results prove that gamma-labeled PEG-immunoliposomes can effectively accumulate in the target area and permit its rather fast visualization. Generally speaking, the principal behavior of PEG-coated liposomes *in vivo* agrees well with our hypothetical model of PEG action on liposomes (Chapter 6).

To investigate the relative importance of such parameters as liposome size and antibody or/and PEG presence on the liposome surface for liposome biodistribution and targeting, we have studied the behavior of different liposome preparations in rabbits with experimental myocardial infarction.[20] Liposomes of two sizes (110 to 150 nm, small liposomes; and 330 to 400 nm, large liposomes) containing 4 mol% PEG have been used in our experiments.

In the case of small liposomes the fastest blood clearance was observed for plain liposomes. Antibody slightly increases circulation time probably because of making part of the surface less accessible for opsonins, which interacts mostly with the "empty" surface of the liposome.[21] As far as we have used Fab, there was no immunoliposome capture via Fc fragment, which usually occurs in the case of whole IgG. PEG (4 mol%) sharply increases circulation time: liposome half-life increases from 40 for plain liposomes to 400 min for PEG-liposomes. Simultaneous incorporation of Fab and PEG somewhat decreases the circulation time of PEG-liposomes (probably because of more pronounced interaction between Fab and plasma proteins than between PEG and these proteins), but they still circulate long enough — half-life of about 300 min — to perform effective binding to target. Circulation times for small liposomes are very similar to those for 150- to 200-nm liposomes. The picture for large liposomes in principle is similar to that for small ones, despite the half-life of large PEG-liposomes (around 320 min) being somewhat smaller than for small PEG-liposomes, and Fab incorporation slightly increases the circulation time of PEG-liposomes. This can be explained by the size difference — the surface area of large liposomes is about six times larger than that of small liposomes. The possible irregularities in PEG location or PEG cluster formation can expose some part of liposome surface for the opsonization. This is why the circulation time for large PEG-liposomes is less than for small PEG-liposomes. Additional incorporation of Fab onto the surface of large liposomes does not change the whole picture much, or can even additionally protect the liposome surface. Thus, the circulation time is strongly influenced by all three factors studied — liposome size and the presence of Fab, PEG, or both on the liposome surface.

Both small PEG-liposomes and Fab-liposomes accumulate in the necrotic area almost identically if expressed in absolute quantities — 0.13 and 0.14% dose/g respectively. This points to two different ways for liposome accumulation — the specific one, which requires the presence of antibodies on the surface of short circulating liposomes to make them capable of sufficient binding to the target even after small number of passages through it,[22] and the nonspecific one, which proceeds via impaired filtration mechanism in affected tissues and requires many passages of liposomes through the target, i.e., prolonged circulation. At the same time, infarct-to-normal ratio (or relative targeting) is higher for small Fab-liposomes than for PEG-liposomes — 22.5 against 7.5, respectively. The reason for this interesting phenomenon might be that the accumulation of Fab-liposomes in normal tissues is very low (the time of Fab-liposomes residence in the blood is insufficient for nonspecific accumulation), whereas long circulating small PEG-liposomes can slowly accumulate in any vascular defects which are present even in normal tissues. The combination of Fab and PEG on the liposome surface gives excellent results in absolute terms, almost 0.25% dose/g (twice as high as for Fab-liposomes!) because both accumulation mechanisms are working in this particular case, resulting in an additive effect.

The increase in the liposome size affects their ability for nonspecific accumulation in the necrotic tissues via the mechanism of impaired filtration. It can also affect the efficacy of Fab-liposome interaction with the target, for example in the case when a single Fab-antigen bond is not sufficient to anchor a large liposome in the target, and multiple bonds cannot be formed for each random immunoliposome collision with the target. Besides, repeated passages of these "unsuccessful" liposomes through the target is very improbable because of their fast clearance. So, the number of productive collisions in this case might be less than for small immunoliposomes. These considerations permit us to understand the results observed. Plain large liposomes do not give any noticeable accumulation in the infarction similar to small liposomes — the residence time is too short to reveal the differences. The accumulation of both Fab-liposomes and PEG-liposomes is quite good. Moreover, PEG-liposomes demonstrate absolute accumulation, even slightly better than Fab-liposomes (0.14 and 0.1% dose/g, respectively). Probably, prolonged circulation can be more efficient for gradual accumulation than short-term specific interactions, part of which can be nonproductive (see above). For infarct-to-normal ratio we observe the same picture as for small liposomes. Very low nonspecific capture of Fab-liposomes makes their relative accumulation higher than that for long-circulating PEG-liposomes (16 vs. 7.5 times for infarct-to-normal ratio). Coincorporation of

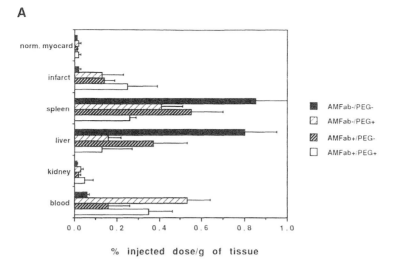

A

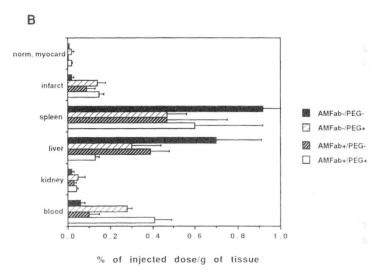

B

FIGURE 4. Biodistribution pattern of small (A) and large (B) ^{111}In-labeled liposomes in rabbits 5 h after intravenous injection. Signs "+" and "–" designate presence or absence of mAb or PEG on the liposome surface.

Fab and PEG into the same liposome does not improve absolute accumulation of large liposomes (nonspecific accumulation via impaired filtration mechanism is the limiting step of the whole process, which does not leave place for additive effect with Fab). Infarct-to-normal ratio for this particular case is somewhat in between those values for Fab-liposomes and PEG-liposomes. Maximal accumulation of large liposomes achieved is somewhat less than that for small liposomes.

Thus, the maximal absolute delivery can be achieved in the case of small liposomes with co-immobilized Fab and PEG, which makes them potentially useful for the delivery of pharmaceutical agents.

General biodistribution patterns of both small and large liposomes are partially presented in Figure 4. The data obtained demonstrate that both plain and Fab-coated small liposomes demonstrate the usual organ distribution and tend to accumulate in the liver and spleen, with minor accumulation in kidneys and lungs. PEG modification, of liposomes, as well as double Fab and PEG modification, sharply decrease liposome accumulation in the liver and spleen, while increasing it in kidneys and lungs. Lung concentration can be increased by eightfold (not shown).

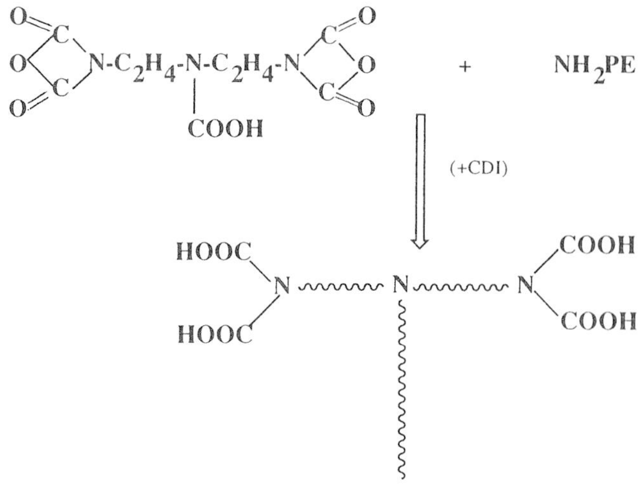

FIGURE 5.   Scheme of DTPA derivatization with hydrophobic "tail" for incorporation into liposomes.

Principal regularities in liposome biodistribution are preserved for different types of large liposomes. However, the large size of liposomes results in many nonspecific phenomena, which makes the whole picture less vivid. Still the maximal accumulation was observed for the liver and spleen in the case of plain and Fab-liposomes. The protective effect of PEG is less pronounced than for small liposomes. Lung accumulation is relatively high even for plain liposomes, but still increases for long circulating liposomes. The kidney pattern for large liposomes remains the same as for small liposomes.

III. DELIVERY OF IMAGING AGENTS BY PEG-MODIFIED Gd-LIPOSOMES. MR VISUALIZATION OF LYMPH NODES

Another possible approach in the use of PEG-liposomes is connected with the area of targeted delivery of MR imaging agents. "Plain" liposomes have been used already as carriers for imaging agents[23,24] both for gamma-scintigraphy and for MR-tomography. For the latter purpose liposomes are loaded with corresponding imaging agents, such as Gd complex with the strong chelating agent diethylenetriamine pentaacetic acid (DTPA).[24] The loading can be performed by the entrapment of the corresponding agent into the water interior of the liposome or by the attachment of the agent (usually heavy metal ion) to the liposome surface via different chelating groups. For better incorporation into the liposomal membrane a chelator can be preliminarily modified with fatty acid or phospholipid residue,[25] see Figure 5 for the scheme of diethylenetriamine pentaacetic acid (DTPA) modification with the hydrophobic tail.

In many cases liposomes used for the delivery of diagnostic agents do not need to be administered via the circulation system. Thus, the major part of intramuscularly or subcutaneously injected liposomes is delivered through the lymphatic channels to the nearest lymph nodes and accumulates there, being mainly taken up by lymph node macrophages.[26] As lymph nodes are frequent sites for tumor metastases, this ability of liposomes is of great interest for the lymphatic delivery of diagnostic and therapeutic agents. In this particular case, the coating of the liposome surface aims to improve its accumulation in lymph nodes or/and to enhance the signal of liposome-associated label.

In experiments on enhanced lymph node visualization we have used NMR-spectroscopy with Gd-containing liposomes. The coating of MR-active Gd-liposomes with PEG can permit alteration of the Gd-water surroundings due to the presence of water molecules tightly

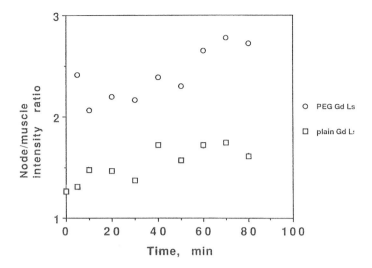

FIGURE 6. The kinetics of the relative axillary lymph node-to-muscle MR signal intensity ratio after the subcutaneous administration of different Gd-containing liposomes in rabbits. Software "Khoros" (University of New Mexico) was used for image analysis. From 10 to 15 pixels were selected as the region of interest within both lymph node and muscle image boundaries for the assessment of average signal intensity.

associated with the PEG molecule, and, thus, to increase the possible signal. With this in mind, Gd-labeled PEG-modified liposomes (PEG molecular weight 5000) were used for MR-imaging of lymph nodes in rabbits and compared with nonmodified Gd-liposomes. For incorporation into the liposomal membrane, PEG was modified (hydrophobized) using phosphatidyl ethanolamine (PE).[8] Liposomes were prepared from egg phosphatidyl choline, cholesterol, Gd-DTPA-PE and PEG-PE in 60:25:10:5 molar ratio. To prepare liposomes the lipid suspension was extruded through polycarbonate filters with consequently reducing pore size until vesicles reached mean diameters of 200 nm, as determined by laser light scattering. Liposome relaxation parameters were measured using Praxis II proton spin analyzer operated at 10.7 MHz. *In vivo* imaging of axillary/subscapular lymph node area in rabbits was performed using 1.5 Tesla GE Signa MRI scanner (T_1 weighed pulse sequence, fat suppression mode) during 2 h after the subcutaneous administration of a liposomal preparation into the paw of the anesthetized rabbit (n = 3 for each group). The raw data from the MRI instrument were analyzed by the image processing software in order to determine the relative target/nontarget (lymph node/muscle) pixel intensity.

In vivo imaging after the subcutaneous injection of different preparations (20 mg of total lipid in 0.5 ml of saline) has demonstrated that all Gd-liposomes are able to visualize axillar/subscapular lymph nodes within minutes (see signal accumulation kinetics in Figure 6). Target area relative signal intensity has been found to be noticeably higher for Gd-PEG-liposomes than for plain liposomes. Signal from plain (nonmodified) liposomes develops only slightly in both lymph nodes, the node-to-muscle intensity ratio being around 1.5 even after 80 min of observation. PEG-coated Gd-containing vesicles rapidly and substantially increase lymph node signal: the node-to-muscle ratio reaches about 2.5 within 5 to 10 min. To illustrate the remarkable ability of PEG-Gd-liposomes to visualize lymph nodes within minutes upon administration, Figure 7 shows a typical transverse slice image of rabbit axillary and subscapular lymph nodes in 30 min.

Surprisingly, the measurements of the actual delivery of liposomes using the surface-bound ^{111}In radiolabel demonstrated the decreased accumulation of PEG-Gd-liposomes with 5 mol% PEG in the lymph nodes under study (Figure 8). The lymph node delivery experiments were carried out essentially under the same conditions as the MR imaging studies. This may reflect

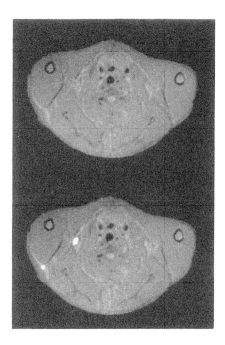

FIGURE 7. Transverse scan of axillary and subscapular lymph node area in rabbit. *Upper image,* before injection; *lower image,* 30 min postinjection of PEG-Gd-liposomes.

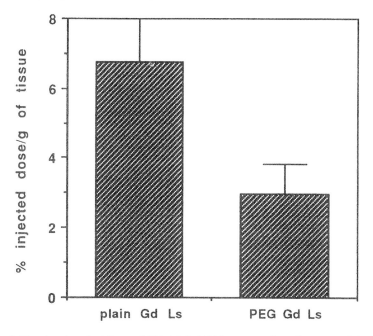

FIGURE 8. Axillary lymph node delivery of ^{111}In-labeled Gd-liposomes after subcutaneous injection in rabbit. Liposomes were labeled with trace amounts of ^{111}In-DTPA-stearylamine. Rabbits were sacrificed 2 h postinjection, lymph nodes were removed, weighed, and counted for ^{111}In radioactivity.

the macrophage-evading properties of PEG-modified liposomes. Yet, the lesser amount of the contrast material caused greater MR signal enhancement. The relaxivity ($1/T_1$) measurements of our liposome preparations resolve this apparent paradox. Figure 9 demonstrates that $1/T_1$ values of PEG-Gd-liposomes are about two times higher than the corresponding parameters for plain Gd-liposomes. This fact might be explained by the presence of increased amounts

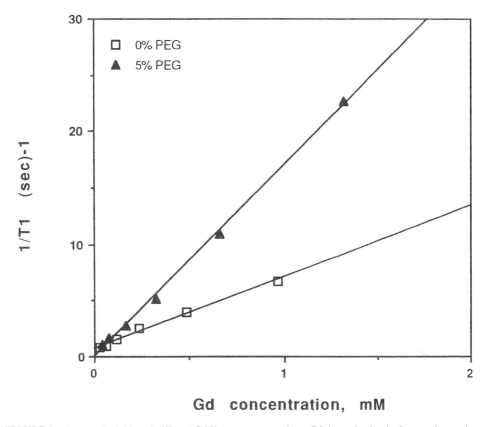

FIGURE 9. Molar relaxivities of different Gd-liposome preparations. Gd determination in the samples used was performed on a commercial basis by Galbraith Laboratories, Inc. (Knoxville, TN). Details of relaxivity measurements are given in text.

of PEG-associated water protons in the close vicinity of chelated Gd ions located on the liposomal membrane.

In order to create an MR contrast medium for lymphography, the use of liposomes as targeting agent for effective paramagnetic chelates seems to be the optimal choice. It has already been shown that the particles with a size of about 200 nm are the most likely to be retained within the lymph node.[26] Like no other nanoparticulate carriers, liposomes can be easily surface-modified in order to modulate their delivery properties. The most appropriate candidates for such modification are different polymers which can be easily derivatized (hydrophobized) and incorporated into the liposome bilayer. The modification of liposomes with PEG had a potent effect on nodal MR images. The relaxivity of Gd ions associated with liposome preparation was increased more than twofold compared to plain liposomes while the actual liposome accumulation in the target was found to be two times less.

It is also worth noting again that the visualization of lymph nodes using PEG-Gd-liposomes takes place very rapidly (within minutes after subcutaneous injection). This circumstance is in sharp contrast with other imaging modalities, where it takes substantially longer times to get a good lymph node image. For example, it takes 24 to 48 h when using the X-ray contrast agent Ethiodol.[27] Evidently, this is just a minor example of modified liposome applicability for the delivery of diagnostic agents, including those for MR imaging. Besides, lymph nodes are natural targets for subcutaneously administered liposomes, like organs of the reticuloendothelial system are for intravenously injected liposomes. The approach could be generalized using PEG-liposomes targeted to different areas of interest. Such liposomes can be prepared by co-immobilization of PEG and mAb on the liposome surface.

IV. CONCLUSION

PEG-coated liposomes and immunoliposomes can be considered as promising diagnostic (imaging) agents. They can be easily loaded with sufficient quantities of radioactive (for gamma-imaging) or MR-active (for MR imaging) agents. PEG can perform two different functions on liposomes. It can serve as a capture-avoiding agent, permitting effective accumulation of the diagnostic label in the target and high target-to-normal ratio, as we have seen with ^{111}In-PEG-immunoliposome use for gamma-imaging of myocardial infarction. On the other hand, surface modification of Gd-liposomes might improve their properties as MR imaging contrast agents. Visualization of lymph nodes with Gd-liposomes can be achieved within minutes after subcutaneous administration. The covering of the liposomal surface with PEG increases the "region of interest" pixel intensity by 2.5- to 2.8-fold compared with the plain Gd-liposomes.

REFERENCES

1. **Gregoriadis, G., Ed.,** *Liposome Technology*, 2nd ed., CRC Press, Boca Raton, 1993.
2. **Gregoriadis, G., Ed.,** *Liposomes as Drug Carriers*, John Wiley & Sons, Chichester, 1988.
3. **Torchilin, V.P.,** *Immobilized Enzymes in Medicine*, Springer-Verlag, Heidelberg, 1991, pp. 60–70, 100–123.
4. **Torchilin, V.P.,** Liposomes as targetable drug carriers, *Crit. Rev. Ther. Drug Carrier Syst.*, 2, 65, 1985.
5. **Torchilin, V.P.,** Immobilization of specific proteins on liposome surface: systems for drug targeting, in *Liposome Technology*, Vol. 3, 1st ed., Gregoriadis, G., Ed., CRC Press, Boca Raton, 1984, 75.
6. **Weissig, V. and Gregoriadis, G.,** Coupling of aminogroup-bearing ligands to liposomes, in *Liposome Technology*, 2nd ed., Gregoriadis, G., Ed., CRC Press, Boca Raton, 1992, 231.
7. **Allen, T.M. and Chonn, A.,** Large unilamellar liposomes with low uptake into the reticuloendothelial system, *FEBS Lett.*, 223, 42, 1987.
8. **Klibanov, A.L., Maruyama, K., Torchilin, V.P., and Huang, L.,** Amphipathic polyethyleneglycols effectively prolong the circulation time of liposomes, *FEBS Lett.*, 268, 235, 1990.
9. **Woodle, M.C., Matthay, K.K., Newman, M.S., Hidayat, J.E., Collins, L.R., Redemann, C., Martin, F.J., and Papahadjopoulos, D.,** Versatility in lipid compositions showing prolonged circulation with sterically stabilized liposomes, *Biochim. Biophys. Acta*, 1105, 193, 1992.
10. **Mori, A., Klibanov, A.L., Torchilin, V.P., and Huang, L.,** Influence of steric barrier activity of amphipathic poly(ethyleneglycol) and ganglioside GM1 on the circulation time of liposomes and on the target binding of immunoliposomes in vivo, *FEBS Lett.*, 284, 263, 1991.
11. **Klibanov, A.L., Maruyama, K., Beckerleg, A.M., Torchilin, V.P., and Huang, L.,** Activity of amphipathic poly(ethyleneglycol) 5000 to prolong the circulation time of liposomes depends on the liposome size and is unfavorable for immunoliposome binding to target, *Biochim. Biophys. Acta*, 1062, 142, 1991.
12. **Charman, W.N. and Stella, V.J., Eds.,** *Lymphatic Transport of Drugs*, CRC Press, Boca Raton, 1992.
13. **Patel, H.M., Boodle, K.M., and Vaughan-Jones, R.,** Assessment of the potential uses of liposomes for lymphoscintigraphy and lymphatic drug delivery, *Biochim. Biophys. Acta*, 801, 76, 1984.
14. **Torchilin, V.P., Klibanov, A.L., Huang, L., O'Donnell, S., Nossiff, N.D., and Khaw, B.A.,** Targeted accumulation of polyethylene glycol-coated immunoliposomes in infarcted rabbit myocardium, *FASEB J.*, 6, 2716, 1992.
15. **Khaw, B.A., Mattis, J.A., Melnicoff, G., Strauss, H.W., Gold, H.K., and Haber, E.,** Monoclonal antibody to cardiac myosin: imaging of experimental myocardial infarction, *Hybridoma*, 3, 11, 1984.
16. **Khaw, B.A., Yasuda, T., Gold, H.K., Strauss, H.W., and Haber, E.,** Acute myocardial infarct imaging with indium-111-labeled monoclonal antimyosin Fab, *J. Nucl. Med.*, 28, 1671, 1987.
17. **Weissig, V., Lasch, J., Klibanov, A.L., and Torchilin, V.P.,** A new hydrophobic anchor for the attachment of proteins to liposomal membranes, *FEBS Lett.*, 202, 86, 1986.
18. **Caride, V.J. and Zaret, B.L.,** Liposome accumulation in regions of experimental myocardial infarction, *Science*, 198, 735, 1977.
19. **Khaw, B.A., Klibanov, A., O'Donnell, S.M., Saito, T., Nossif, N., Slinkin, M.A., Newell, J.B., Strauss, H.W., and Torchilin, V.P.,** Gamma imaging with negatively charge-modified monoclonal antibody: modification with synthetic polymers, *J. Nucl. Med.*, 32, 1742, 1991.
20. **Torchilin, V.P., Narula, J., and Khaw, B.A.,** submitted for publication.
21. **Senior, J.H.,** Fate and behavior of liposomes in vivo: a review of controlling factors, *Crit. Rev. Ther. Drug Carrier Syst.*, 3, 123,1987.

22. **Torchilin, V.P., Khaw, B.A., Smirnov, V.N., and Haber, E.,** Preservation of antimyosin antibody activity after covalent coupling to liposomes, *Biochem. Biophys. Res. Commun.*, 89, 1114, 1979.
23. **Mauk, M.R. and Gamble, R.C.,** Preparation of lipid vesicles containing high levels of entrapped radioactive cations, *Anal. Biochem.*, 94, 302, 1979.
24. **Kabalka, G., Buonocore, E., Hubner, K., Moss, T., Norley, N., and Huang, L.,** Gadolinium-labeled liposomes: targeted MR contrast agents for the liver and spleen, *Radiology*, 163, 255, 1987.
25. **Kabalka, G.W., Davis, M.A., Moss, T.H., Buonocore, E., Hubner, K., Holmberg, E., Maruyama, K., and Huang, L.,** Gadolinium-labeled liposomes containing various amphiphilic Gd-DTPA derivatives: targeted MRI contrast enhancement agents for the liver, *Magn. Reson. Med.*, 19, 406, 1991.
26. **Takakura, Y., Hashida, M., and Sezaki, H.,** Lymphatic transport after parenteral drug administration, in *Lymphatic Transport of Drugs*, Charman, W.N. and Stella, V.J., Eds., CRC Press, Boca Raton, 1992, 256.
27. **Swanson, D.P. and Shetty, P.C.,** Lymphographic contrast media, in *Pharmaceuticals in Imaging*, Swanson, D.P. and Thrall, J., Eds., Macmillan, New York, 1991, 236.

Chapter 20

ANTIBODY-TARGETED STEALTH® LIPOSOMES

Theresa M. Allen, Christian B. Hansen, and Samuel Zalipsky

TABLE OF CONTENTS

I. INTRODUCTION

Following the first description of liposomes 30 years ago,[1] the use of liposomes as delivery systems to selectively deliver drugs to specific cells or tissues has received considerable attention. After the initial burst of enthusiasm for the use of liposomal drug carriers, it was realized that the mononuclear phagocyte system (MPS, also termed the reticuloendothelial system) presented a barrier that had to be overcome before liposomal drug therapy could be used effectively in the treatment of a wide range of diseases involving tissues other than the MPS.[2] The pronounced tendency of "classical" liposome formulations (C-liposomes), particularly those containing phosphatidylserine, to target cells of the MPS, presents some important therapeutic opportunities.[3,4] However, the short circulation half-lives of C-liposomes has frustrated attempts to use them for the selective delivery of drugs *in vivo* to other sites.

The advent of new formulations of long circulating liposomes, first described by Allen and Chonn,[5] has renewed activity in the area of liposomal drug delivery systems, with several promising applications currently receiving preclinical testing in animal models,[6–12] and one formulation in clinical trials in humans.[13,14] The first long circulating liposome formulations resulted from attempts to mimic some of the properties of the outer surface of red blood cells, and contained monosialoganglioside GM1 as the hydrophilic, opsonin-repelling surface component thought to be responsible for their survival in the circulation.[5,15] Gabizon and Papahadjopoulos demonstrated increased tumor uptake of these long circulating liposomes in experiments which, for the first time, demonstrated that site-specific delivery of liposome-entrapped drug to sites outside the MPS was a realistic possibility.[16] Several reviews describing the dose-independent pharmacokinetics, the biodistribution and the mechanisms of action of these "first-generation" long circulating liposomes have been published.[17–20] Potential problems with the commercial acceptability of products containing GM1 led to a search for safe, inexpensive substitutes with increased clinical acceptability. Second-generation formulations, containing lipid derivatives of polyethylene glycol (PEG), such as PEG-distearoyl phosphatidylethanolamine, have subsequently been described by us and others,[21–25] and their properties have been explored.[26–31] Long circulating liposomes have been termed Stealth® liposomes (Stealth® is a registered trademark of Liposome Technology, Inc.), because of their poor affinity for the cells of the MPS, i.e., their ability to avoid detection.[32] These liposomes are also referred to as sterically stabilized liposomes (S-liposomes).[25,26] S-liposomal drug delivery systems are suitable for a number of therapeutic applications involving, primarily, sustained drug release, and selective delivery of drugs to specific target tissues such as cancers.

One of the underlying principles of chemotherapy is that of selective toxicity, i.e., the concept that chemotherapeutic drugs are more toxic to diseased or invading cells than to normal cells. One of the ways in which selective toxicity can be achieved, at least in principle, is by targeting drugs to diseased cells through use of target-specific ligands associated with the drug (the "magic bullet" concept). However, attachment of ligands such as antibodies to C-liposomes greatly increased the already rapid rate with which they were removed from circulation.[33] Therefore, the development of liposomes with long survival times in circulation was a necessary first step before ligand-mediated targeting of liposomes to specific cells *in vivo* could be attempted. The next challenge was to develop methodologies that could be used to couple ligands to the surface of S-liposomes, with both long survival times in circulation and target recognition being retained.

II.   WHAT CONSTITUTES AN "IDEAL" IMMUNOLIPOSOME?

There are a number of important properties that are desirable in S-immunoliposomes. Our search for an ideal coupling method was dictated by a number of considerations, outlined in Table 1, which are discussed briefly below.

A. RAPID SIMPLE COUPLING METHOD

The method should involve minimal manipulation to either the liposomes or the antibody (Ab). For example, methods that involve several dialysis or column chromatography steps may increase the possibilities of contamination, degradation, or dilution of the product. We believe that, because liposomes are inherently more stable than Abs (i.e., proteins), the coupling procedure should be designed so that most of the steps are performed on the liposomes prior to antibody addition, and that manipulation of the Ab should be kept to a minimum.

TABLE 1

Properties of Ideal Immunoliposomes

1. Rapid, simple coupling method
2. High coupling efficiency of Ab to liposome
3. Optimum Ab density at liposome surface
4. Retention of long circulation half-lives
5. Retention of target recognition
6. Appropriate drug loading and release characteristics
7. Ingredients compatible with use in humans
8. *In vivo* efficacy

B. HIGH COUPLING EFFICIENCY OF Ab TO LIPOSOME

Because production of pure monoclonal antibody (mAb) tends to be an expensive process, it is frequently very difficult for researchers to get ready access to the large amounts (tens of milligrams or more) of mAb required for *in vivo* targeting experiments. Therefore, an efficient coupling procedure is essential. Ideally, close to 100% of added mAb should be coupled to liposomes. A method in which only 10 or 20% of added mAb is coupled would result in either a large degree of wastage of the mAb or in time-consuming procedures to recover the uncoupled mAb.

C. OPTIMUM Ab DENSITY AT THE LIPOSOME SURFACE

The Ab density required for high levels of specific binding of immunoliposomes to epitopes at the target cell surface will undoubtedly vary with a number of factors, such as the density of the epitope at the cell surface, the binding affinity of the Ab to its epitope, and the nature of the epitope, e.g., if epitope expression varies with cell confluence or changes in the cellular environment. There is evidence that binding of immunoliposomes to their target cells increases with increasing Ab density on the liposomes, given preparations of equal liposome concentration.[34] However, it is not possible to make general predictions about optimum Ab densities; they will have to be determined for each individual system.

D. APPROPRIATE DRUG LOADING AND RELEASE CHARACTERISTICS

With the description of remote loading techniques, which rely on a pH[35] or chemical gradient[36] across the liposome bilayer to load drug into the liposomes, it is now possible to rapidly and efficiently entrap drugs such as doxorubicin (DOX) and vincristine in liposomes. Ab coupling methods which interfere with either the maintenance of the gradient across the liposome bilayer, or with the loading of drug into liposomes in response to the gradient, would be undesirable as it would lead to either reduced loading of drugs into the liposomes or too rapid rates of leakage of drugs from the liposomes.

E. RETENTION OF LONG CIRCULATION HALF-LIVES

Obviously, we would want S-immunoliposomes to have equally long survival times in circulation as S-liposomes in order that they will have sufficient time to find and bind to their target. The coupling of Ab to C-liposomes has been found to greatly increase their rate of removal from circulation,[33] so it is important to determine for each coupling method whether or not prolonged circulation times can be maintained. For example, does it affect circulating half-lives if the Ab is coupled to the surface of an S-liposome vs. to the terminus of PEG? Because S-liposomes achieve maximum accumulation in solid tumors at approximately 24 h postinjection,[25] a circulation half-life for S-immunoliposomes of 12 to 24 h would be desirable.

F. RETENTION OF TARGET RECOGNITION *IN VITRO* AND *IN VIVO*

If S-immunoliposomes are to bind to their targets it is important that Ab binding is not sterically hindered by the PEG molecules present on the liposome surface. The steric barrier which PEG imparts to liposomes has been found to be directly proportional to the chain length of PEG,[37,38] with PEG 5000 imparting a strong steric barrier which inhibited target recognition to Ab coupled at the liposome surface.[37,38] A good compromise between retention of both circulation half-lives and targeted recognition for liposomes with Ab coupled at the liposome surface is PEG 2000. Another strategy which could be employed to overcome the PEG steric barrier is to couple Ab to the PEG terminus (see below).

G. INGREDIENTS COMPATIBLE WITH USE IN HUMANS

A variety of chemicals and proteins (avidin, streptavidin) are used in the various coupling methods. Some of these ingredients will be more acceptable for use in humans than others. In particular, the proteins avidin and streptavidin are potentially immunogenic and it may be difficult to get approval for their use in humans.

H. *IN VIVO* EFFICACY

Even if a perfect coupling method were devised for attaching Ab to drug-containing liposomes, according to the above criteria, there remains the problem of *in vivo* efficacy, i.e., will the S-immunoliposomes bind to target cells *in vivo* in sufficient numbers in order to release drug in the therapeutic concentration range over a satisfactory period of time so that an improved therapeutic effect over that achievable for free drug or liposomal drug in the absence of Ab can be achieved? The answer will depend on many factors, including the nature of the target (small vs. large tumor having poor vs. good blood supply of leaky vs. non-leaky capillaries), the occurrence of barriers which hinder access to the target (blood-brain barrier, capillary endothelium), etc. The answers to the question of *in vivo* efficacy may have to be derived empirically on a case-to-case basis.

III. COUPLING METHODS FOR ATTACHING ANTIBODIES TO S-LIPOSOMES

To date we have used five different methods for coupling antibodies to S-liposomes, and Huang and colleagues have used a sixth method. The six methods are discussed briefly below along with their advantages and disadvantages.

A. NONCOVALENT BIOTIN-ADVIDIN METHOD

The first method that we tried for attaching antibodies (Ab) to S-liposomes was the noncovalent biotin-avidin method[39] of Urdal and Hakomori,[40] which relies on the strong binding between biotin and avidin (or streptavidin) (Figure 1). This method, adapted by Loughrey et al.,[41,42] has been used for attaching Ab to C-liposomes. When used without modification, the method resulted in low Ab densities and low coupling efficiencies when applied to S-liposomes, possibly due to steric interference by PEG with the approach of the large avidin and/or Ab molecules to the biotinylated phosphatidylethanolamine (PE) at the liposome surface. Ab densities at the liposome surface could be modestly improved (Table 2) by increasing the mol% of biotin-PE in the S-liposomes to 0.5 mol%, a 5- to 10-fold increase over that used for C-liposomes. Biotin-PE cannot easily be used at this level in C-liposomes as it causes extensive aggregation and cross-linking of the liposomes,[41] but the presence of PEG in the S-liposomes inhibits this aggregation and allows increased levels of avidin binding to the S-liposomes.[38] High percentages of DOX could be loaded into immunoliposomes formed by this method.

B. COVALENT ATTACHMENT OF Ab TO THE LIPOSOME SURFACE

A number of methods have been described for coupling Ab to the surface of C-liposomes, and some of these methods can be adapted for use with S-liposomes.

A Biotin-avidin

DOPE–BIOTIN

↓ ← AVIDIN

DOPE–BIOTIN–AVIDIN   NHS–BIOTIN

↓ ← BIOTIN-mAb ← NH₂–mAb

DOPE–BIOTIN–AVIDIN–BIOTIN-mAb
Immunoliposomes

B₁ MPB-PE

SPDP

↓ ← mAb–NH₂

mAb–NH–C–(CH₂)₂–S–S–N
PDP-mAb

↓ ← DTT

mAb–NH–C–(CH₂)₂–SH
thiolated mAb

↓ ← MPB-PE   (CH₂)₃–C–NH–DOPE

mAb–NH–C–(CH₂)₂–S
Immunoliposomes

B₂ PDP-PE

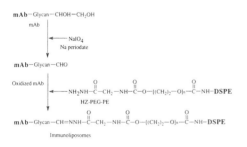

SMPB   PDP-PE

↓ ← NH₂–mAb   ↓ ← DTT

mAb–NH–C–(CH₂)₃   HS–(CH₂)₂–C–NH–DOPE
MPB-mAb   thiolated PE

mAb–NH–C–(CH₂)₃
S–(CH₂)₂–C–NH–DOPE
Immunoliposomes

C₁ HZ-PEG-PE

mAb–Glycan–CHOH–CH₂OH
mAb

↓ ← NaIO₄
Na periodate

mAb–Glycan–CHO
Oxidized mAb

↓ ← NH₂NH–C–CH₂–NH–C–O–[(CH₂)₂–O]ₙ–C–NH–DSPE
HZ-PEG-PE

mAb–Glycan–CH=NNH–C–CH₂–NH–C–O–[(CH₂)₂–O]ₙ–C–NH–DSPE
Immunoliposomes

C₂ PDP-PEG-PE

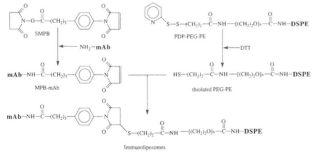

SMPB   PDP-PEG-PE

↓ ← NH₂–mAb   ↓ ← DTT

mAb–NH–C–(CH₂)₃   HS–(CH₂)₂–C–NH–[(CH₂)₂O]ₙ–C–NH–DSPE
MPB-mAb   thiolated PEG-PE

mAb–NH–C–(CH₂)₃
S–(CH₂)₂–C–NH–[(CH₂)₂O]ₙ–C–NH–DSPE
Immunoliposomes

D N-Glutaryl-PE

Octylglucoside + HO–C–NGPE
N-glutaryl-PE

Sulfo-NHS   C₂H₅–N=C=N–(CH₂)₃N(CH₃)₂
EDC

↓

mAb–NH₂

↓

mAb–NH–C–NGPE
mAb-glutaryl-PE

Phospholipids → dialysis

↓

Immunoliposomes

FIGURE 1. Strategies for linking mAb to S-liposomes to form immunoliposomes. The abbreviations used are as follows: mAb, monoclonal antibody; Biotin-DOPE, *N*-biotinyl dioleoylphosphatidylethanolamine; DOPE, dioleoylphosphatidyl ethanolamine; DSPE, distearoyl phosphatidylethanolamine; DTT, dithiothreitol; EDC, 1-ethyl-3-(3-dimethylaminopropyl)-carbodiimide; HZ-PEG-DSPE, *N*-hydrazide-poly(ethylene glycol) distearoyl phosphatidylethanolamine; NGPE, *N*-glutarylphosphatidylethanolamine; NHS-biotin, *N*-hydroxysuccinimidobiotin; MPB-DOPE, *N*-maleimidophenylbutyrate dioleoylphosphatidylethanolamine; PDP-PE, *N*-pyridyldithiopropionate dioleoylphosphatidylethanolamine; PDP-PEG-DSPE, *N*-pyridyldithiopropionamide poly(ethylene glycol) distearoyl phosphatidylethanolamine; SMPB, succinimidyl-4-(*p*-maleimidophenyl)butyrate; SPDP, *N*-succinimidyl-3-(2-pyridyldithio)propionate; Sulfo-NHS, *N*-hydroxysulfosuccinimide.

TABLE 2
A Comparison of Ab Coupling Methods for the Formation of S-Immunoliposomes

Coupling Method	mAb Density μg Ab/μmol PL	Coupling Efficiency (% of Initial mAb)	% Drug Loading
Biotin-avidin	8–20	10	95–100
MPB-PE	50–200	50–100	30–50
HZ-PEG-PE	20–40	10–25	95–100
PDP-PEG-PE	50–200	50–85	95–100
PDP-PE	15–30	10	80–95

Note: Range of monoclonal antibody (mAb) densities and coupling efficiencies for a variety of coupling methods used for attaching mAb at the surface of Stealth liposomes. The numbers given are for a practical range of antibody concentrations (5–20 mg/ml) and mAb:PL starting ratios (1:500 to 1:2000 molar ratio). Liposomes were 0.1 μm in diameter and composed of HSPC:CH:PEG 2000-DSPE, 2:1:0.1 molar ratio. After Stealth immunoliposome formation, liposomes were loaded with doxorubicin using the ammonium sulfate gradient method[36] for 1 h at 65°C.

1. MPB-PE

The *N*-[4-(*p*-maleimidophenyl)butyrate]-PE (MPB-PE) method, first described for Ab attachment to C-liposomes,[43] has been tested by us for use with S-liposomes. MPB-PE (1 mol%) incorporated into liposomes reacts with thiolated antibodies, in which thiol groups are introduced by the amine reactive heterobifunctional reagent *N*-succinimidyl 3-2-pyridyl-dithiol)-propionic acid (SPDP), to form thioether bonds linking the Ab to the liposome surface. This method results in high coupling efficiencies and high Ab densities at the liposome surface when PEG 2000 is present in the liposomes (Table 2). When the coupling was done in the presence of PEG 5000, the coupling efficiency, Ab density, and Ab recognition decreased substantially,[38] likely due to the strong steric barrier imparted to liposomes by high-molecular-weight PEG.[37,38]

The major problem that we encountered with this method was an alteration of the kinetics of drug loading by the somewhat hydrophobic phenylbutyrate group of MPB at the liposome surface. At 1 mol% of MPB-PE, DOX loading was retarded, with 50% or less of DOX being loaded into solid-phase liposomes during 1 h at 65°C (Table 2). Higher percentages of drug loading could be achieved by reducing the mole percent of MPB-PE in the liposomes (which reduced coupling efficiency and Ab densities), by increasing the incubation time to 6 h at 65°C (which, however, could increase liposome degradation), or by using fluid-phase liposomes (although these had high rates of DOX leakage). Even under conditions where the liposomes could be loaded with drug, MPB appeared to cause substantially increased rates of leakage of the drug from the liposomes.

2. PDP-PE

In a variation of the above method, *N*-[3-(2-pyridyldithio)propionate]-PE (PDP-PE, 1 mol%) was incorporated into liposomes, activated to produce thiol-PE, and then coupled to an MPB-Ab, which is formed by incubation of the Ab with succinimidyl 4-(*p*-maleimidophenyl)butyrate (SMPB) (Figure 1). This method has the potential advantage of reducing the amount of MBP in the resulting immunoliposomes to much lower levels than those found in the MPB-PE method, which would be predicted to cause less interference with drug remote loading. As can be seen from Table 2, the percent drug loading of DOX increased for PDP-PE liposomes. However, the coupling efficiency and the Ab density on liposomes were both low using this method (Table 2), again likely a result of the steric barrier imparted to the liposome surface by PEG, particularly given the small size of the thiol-PE headgroup.

C. COVALENT ATTACHMENT OF ANTIBODY TO THE TERMINUS OF PEG

An alternative approach is available for coupling Ab to S-liposomes, which cannot be used for C-liposomes: attachment of the Ab to the PEG terminus. This would have the advantage of overcoming the PEG steric barrier which not only limits access of free Ab to the liposome surface in the above methods but also causes reductions in Ab recognition of its target epitope.

1. PEG-HZ

A reactive hydrazide group has been added to the terminus of PEG 2000-DSPE (PEG-HZ).[38,44,45] When 1 to 5 mol% of PEG-HZ is incorporated into liposomes, Ab can be coupled to the PEG terminus following oxidation of the carbohydrate groups of the oligosaccharide portion of the Ab with periodate (Figure 1). Because the oligosaccharide portion of the Ab is in the Fc portion of the Ab, the binding of Ab to liposomes occurs exclusively through this region of the molecule, unlike some other methods in which thiolation can occur in several regions of the Ab, including the Fab region, which can lead to some degree of interference with the Ab binding to its epitope. The Ab densities and coupling efficiencies with this method were intermediate between the MPB-PE method and the biotin-avidin and PDP-PE methods. No interference was observed with either drug remote loading or drug leakage properties of S-immunoliposomes made by this method.

2. PDP-PEG

Because we wanted to take advantage of the highly efficient formation of thioether bonds, combined with the advantages of linking Abs to the PEG terminus, we synthesized and tested a PDP-PEG molecule which could be used to attach MPB-Ab to liposomes (Figure 1). The chemistry in this method is the same as for the PDP-PE method above, and is the "reverse" of the MPB-PE method. When PDP-PEG is incorporated in S-liposomes at a concentration of 0.5 to 2 mol%, high Ab densities and high coupling efficiencies could be obtained (Table 2). Furthermore, high drug remote loading efficiencies could be obtained (Table 2) with good retention of drug contents. Of all the above methods, this one comes closest to ideal.

D. *N*-GLUTARYL-PE METHOD

Huang and coworkers have taken a different approach to incorporation of Ab into S-liposomes. Ab was covalently attached to *N*-glutaryl-PE (Ab-NGPE) in the presence of octyl glucoside detergent using *N*-hydroxysulfosuccinimide as a carboxyl activation reagent.[46] S-immunoliposomes were prepared from Ab-NGPE and lipids by detergent dialysis[34,46,47] or reverse-phase evaporation.[47] Approximately 35 to 50% of the starting levels of Ab-NGPE is incorporated into liposomes.[47] Because some unknown fraction of the Ab-NGPE is oriented towards the liposome interior with this method, the Ab density at the surface of the liposomes is more difficult to calculate, but total Ab density (interior plus exterior) is of the order of 24 to 80 µg IgG/µmol lipid.[34]

IV. TARGET RECOGNITION *IN VITRO*

Once a successful coupling method for the production of S-immunoliposomes has been identified, the next step is to show antibody-mediated specific binding and selective cytotoxicity of drug-containing S-immunoliposomes to cells in culture. In *in vitro* experiments, S-immunoliposomes have been shown to result in threefold increased specific target binding to murine squamous carcinoma (KLN-205) cells[39] and to human ovarian carcinoma (CaOV.3) cells,[38] as compared to S-liposomes lacking antibody. Antibody-mediated targeting of S-liposomes has also been demonstrated *in vitro* in the HER-2 overexpressing breast cancer cell line.[48] In the case of the murine squamous carcinoma cell line, an increase in cytotoxicity of approximately fourfold was observed for S-immunoliposomes, containing entrapped DOX (SI-DOX) as compared to that seen for similar liposomes lacking antibody (S-DOX).[39] DOX entrapped in S-immunoliposomes was approximately fivefold more cytotoxic than free DOX.[39] On the other hand, for human ovarian carcinoma cells *in vitro*, in spite of a threefold increase in antibody-mediated binding of liposomes, the cytotoxicity of SI-DOX, although significantly greater than that seen for S-DOX, was similar to the cytotoxicity for free DOX.[38] One can speculate that the increased cytotoxicity for SI-DOX against KLN-205 cells may be related to internalization of the drug package, but to date there is little published evidence for endocytosis of immunoliposomes by their target cells.

Failure to show increased cytotoxicity for drug-containing S-immunoliposomes *in vitro* may not be a cause for pessimism, as *in vitro* experiments cannot reflect additional factors such as pharmacokinetics, biodistribution, enzymatic degradation, etc., of drug, which will occur *in vivo*. As the pharmacokinetics of the free drug is dramatically different from that of the liposome-entrapped drug,[36] the ability to target a drug package *in vivo* (leading to a higher percentage of the drug localized at the desired site of action) should lead to an improved therapeutic effect over that seen for free drug.

Two other examples of ligand-mediated targeting of S-liposomes have appeared recently in the literature.[38,45,49] In each of these examples, targeting was mediated by ligands bound at the terminus of PEG. We have attached mAb, by means of a hydrazone bond, to the terminus PEG-hydrazide incorporated into S-liposomes and compared the binding of these liposomes by KLN-205 squamous carcinoma cells to the binding of similar liposomes in which the mAb was attached at the liposome surface by means of biotin-avidin.[38] The level of binding of each type of liposomes was similar and, in each case, the binding was approximately threefold higher than that seen for control liposomes lacking mAb.[38] In the other example, a protein, plasminogen, was linked to the terminus of PEG 5000-PE in liposomes.[49] These proteoliposomes were capable of binding to their target molecule, fibrin, to the same extent as proteoliposomes formed by linking plasminogen at the liposome surface in the absence of PEG. Recently we have also been able to demonstrate increased binding of S-immunoliposomes, formed by the PDP-PEG-PE method for attaching Ab to the PEG terminus, to KLN-205 cells *in vitro* (Figure 2), which demonstrates that Abs subject to this coupling method are also capable of retaining their ability to bind to their targeted epitope.

V. TARGET RECOGNITION *IN VIVO*

If Ab-mediated specific binding of S-immunoliposomes can be demonstrated *in vitro*, one would next want to examine the pharmacokinetics and biodistribution of these liposomes *in vivo* and demonstrate their selective uptake into target tissues. Evidence to date suggests that S-immunoliposomes[38,45] retained their prolonged circulation half-lives *in vivo*, with only small decreases in their circulating half-lives.[45] Naturally, as S-immunoliposomes localize in their target tissues *in vivo*, the numbers of circulating liposomes will decrease, which must be borne in mind in interpreting pharmacokinetic data.

In one of the first experiments to show *in vivo* target binding of immunoliposomes, Maruyama et al.[34] demonstrated that GM1-immunoliposomes containing an antibody against lung endothelial cells would localize efficiently in murine lung tissue, with as much as 70% of injected GM1-containing S-immunoliposomes localizing in lung within 15 min of injection. C-immunoliposomes had significantly lower lung uptake at equivalent antibody:lipid ratios as a result of their short circulating half-lives. The uptake of lung-targeted immunoliposomes was dependent on antibody density on the liposomes.[34] More recently, the ability of S-

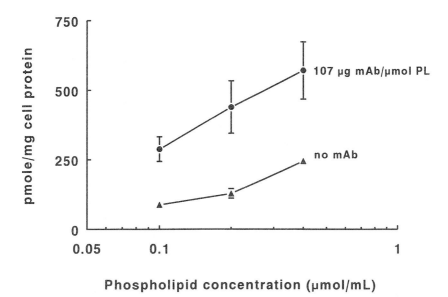

FIGURE 2. Uptake of S-immunoliposomes by KLN-205 squamous cell lung carcinoma. Specific mAb 174H.64 was coupled to liposomes (HSPC:CH:PEG-DSPE:PDP-PEG-DSPE, 2:1:0.08:0.02) by the PDP-PEG-PE method at an mAb:PL molar ratio of 1:500. KLN-205 cells were plated at a density of 1.2×10^5 cells/well and after 3 d of incubation at 37°C in a humidified CO_2 environment liposomes, labeled with ^{3}H-cholesteryl hexadecyl ether, were incubated with the cells for 1 h.

immunoliposomes containing lipophilic antitumor prodrugs to target to lung endothelial cells *in vivo* has been demonstrated.[50] The ability of S-immunoliposomes to be taken up selectively *in vivo* has also been demonstrated in infarcted rabbit myocardium, where anti-myosin antibodies on S-immunoliposomes resulted in significantly increased uptake into rabbit heart.[51] The ability to target S-immunoliposomes *in vivo* was determined to depend on the molecular weight of the PEG. In a number of studies, it has been demonstrated that higher molecular weights of PEG (>2000 Da) interfere with the ability of S-immunoliposomes to recognize their target antigens.[37,38,52] S-immunoliposomes containing PEG in the molecular weight range of 1000 to 2000 Da appeared to be optimum for retaining both antibody recognition and prolonged circulation half-lives.[37,38]

To date, there is only one published example of a therapeutic application of targeted S-immunoliposomes.[53] In these experiments mice bearing the murine squamous carcinoma cell, KLN-205, which localizes in lung within 3 d of injection, were treated with 6 mg/kg of either free DOX, S-DOX, or DOX entrapped in S-DOX liposomes with mAb attached at the liposome surface by the biotin/avidin method (SI-DOX).[53] Mice received single injections of free DOX, S-DOX, or SI-DOX at day 3 after intravenous injection of tumor (2×10^5 cells). At 45 d postinoculation of tumor, the uptake of ^{125}I-deoxyuridine (a measure of cell proliferation) into the lungs of mice receiving SI-DOX was not significantly different from that of normal (tumor-free) controls, suggesting that there was a substantial reduction in tumor burden with the SI-DOX treatment. Histopathology on the lungs of mice receiving SI-DOX showed dramatically lower numbers of tumor foci than that seen for mice receiving either S-DOX or free DOX, and some mice treated with the SI-DOX appeared to have tumor-free lungs.[53] In survival studies, 40 to 60% long-term (170-d) survivors were observed in mice treated with single injections of 6 mg/kg SI-DOX (10 to 12 µg mAb). No long-term survivors were noted in mice treated with single injections of either S-DOX or free DOX. In this solid tumor model, therefore, antibody-mediated targeting of long circulating liposomes resulted in significant improvement in therapeutic effect over that which could be obtained with antibody-free S-liposomes or free drug. More significantly, long-term survivors could be obtained at levels of antibody that are significantly lower than those required to achieve similar results with single injections of doxorubicin immunoconjugates against the same tumor.[54] In therapeutic experiments in which treatment was delayed until the squamous cell lung carcinoma tumor was well established, however, the therapeutic efficacy of the mAb-S-DOX liposomes was only modestly higher than that seen for nontargeted treatments.[55] Thus, the S-immunoliposomal treatment appears to be most effective against small, newly established micrometastases.

The mechanism for the therapeutic effect of immunoliposomes against solid tumors is speculated to be a result of extravasation of the liposomes from the circulation into the tumor interstitial space through leaky capillary ends arising during the process of angiogenesis, followed by binding of the drug-containing immunoliposomes to their targeted epitopes at the surface of tumor cells. This would lead to sustained release of entrapped drug at locally increased drug concentrations in the vicinity of the targeted cells. For some mAb, internalization of the liposome-drug package may occur and also contribute to the mechanism of cytotoxicity. In addition, surrounding cells in a heterogeneous cancer, which lack the specific epitope, may be killed by the "bystander effect", in which there is specific binding of S-immunoliposomes to their target epitopes with release and diffusion of the entrapped drug to nearby cells.

VI. CONCLUSIONS

The technology for attaching Abs to S-liposomes has developed rapidly and it is now possible to achieve high Ab densities on S-immunoliposomes with high coupling efficiencies, and good remote drug loading. Of the methods researched to date the PDP-PEG-PE is the closest to ideal. S-immunoliposomes have been shown to have good levels of *in vitro* target binding, and have sufficiently long survival times in circulation to reach *in vivo* target sites. In at least one case, targeting of S-immunoliposomes, containing entrapped DOX, *in vivo* targeting has been shown to result in superior therapeutic effects in the treatment of a solid tumor. Many therapeutic opportunities exist for S-immunoliposomes, including targeting within the vasculature, targeting to cells within lymph and lymph nodes, and targeting to micrometastases of solid tumors.

REFERENCES

1. **Bangham, A. D., Standish, M. M., and Watkins, J. C.,** Diffusion of univalent ions across the lamellae of swollen phospholipids, *J. Mol. Biol.,* 13, 238, 1965.
2. **Poznansky, M. J. and Juliano, R. L.,** Biological approaches to the controlled delivery of drugs: a critical review, *Pharmacol. Rev.,* 36, 277, 1984.
3. **Kleinerman, E. S., Murray, J. L., Synder, J. S., Cunningham, J. E., and Fidler, I. J.,** Activation of tumoricidal properties in monocytes from cancer patients following intravenous administration of liposomes containing muramyl tripeptide phosphatidylethanolamine, *Cancer Res.,* 49, 4665, 1989.
4. **Fidler, I. J.,** Systemic activation of macrophages by liposomes containing muramyltripeptide phosphatidyl-ethanolamine for therapy of cancer metastases, *J. Liposome Res.,* 1, 461, 1990.
5. **Allen, T. M. and Chonn, A.,** Large unilamellar liposomes with low uptake into the reticuloendothelial system, *FEBS Lett.,* 223, 42, 1987.
6. **Gabizon, A. A.,** Selective tumor localization and improved therapeutic index of anthracyclines encapsulated in long-circulating liposomes, *Cancer Res.,* 52, 891, 1992.
7. **Allen, T. M., Mehra, T., Hansen, C., and Chin, Y.-C.,** Stealth liposomes: an improved sustained release system for 1-β-D-arabinofuranosylcytosine, *Cancer Res.,* 52, 2431, 1992.
8. **Huang, S. K., Mayhew, E., Gilani, S., Lasic, D. D., Martin, F. J., and Papahadjopoulos, D.,** Pharmacokinetics and therapeutics of sterically stabilized liposomes in mice bearing C-26 colon carcinoma, *Cancer Res.,* 52, 6774, 1992.
9. **Mayhew, E. G., Lasic, D., Babbar, S., and Martin, F. J.,** Pharmacokinetics and antitumor activity of epirubicin encapsulated in long-circulating liposomes incorporating a polyethylene glycol-derivatized phospholipid, *Int. J. Cancer,* 51, 302, 1992.

10. **Vaage, J., Mayhew, E., Lasic, D., and Martin, F. J.,** Therapy of primary and metastatic mouse mammary carcinomas with doxorubicin encapsulated in long circulating liposomes, *Int J. Cancer,* 51, 942, 1992.

11. **Woodle, M. C., Storm, G., Newman, M. S., Jekot, J. J., Collins, L. R., Martin, F. J., and Szoka, F. C., Jr.,** Prolonged systemic delivery of peptide drugs by long-circulating liposomes: illustration with vasopressin in the Brattleboro rat, *Pharm. Res.,* 9, 260, 1992.

12. **Huang, S. K., Martin, F. J., Jay, G., Vogel, J., Papahadjopoulos, D., and Friend, D. S.,** Extravasation and transcytosis of liposomes in Kaposi's sarcoma-like dermal lesions of transgenic mice bearing the HIV tat gene, *Am. J. Pathol.,* 143, 10, 1993.

13. **Northfelt, D. W., Martin, F. J., Kaplan, L. D., Russell, J., Anderson, M., Lang, J., and Volberding, P. A.,** Pharmacokinetics, tumor localization and safety of Doxil (liposomal doxorubicin) in AIDS patients with Kaposi's sarcoma, *Proc. Am. Soc. Clin. Oncol.,* 12, 51, 1993.

14. **Gabizon, A., Catane, R., Uziely, B., Kaufman, B., Safra, T., Cohen, R., Martin, F., Huang, A., and Barenholz, Y.,** Prolonged circulation time and enhanced accumulation in malignant exudates of doxorubicin encapsulated in polyethylene-glycol coated liposomes, *Cancer Res.,* 54, 897, 1994.

15. **Allen, T. M., Hansen, C., and Rutledge, J.,** Liposomes with prolonged circulation times: factors affecting uptake by reticuloendothelial and other tissues, *Biochim. Biophys. Acta,* 981, 25, 1989.

16. **Gabizon, A. and Papahadjopoulos, D.,** Liposome formulations with prolonged circulation time in blood and enhanced uptake by tumors, *Proc. Natl. Acad. Sci. U.S.A.,* 85, 6949, 1988.

17. **Mumtaz, S., Ghosh, P. C., and Bachhawat, B. K.,** Design of liposomes for circumventing the reticuloendothelial cells, *Glycobiology,* 1, 505, 1991.

18. **Allen, T. M.,** Stealth liposomes: five years on, *J. Liposome Res.,* 2, 289, 1992.

19. **Klibanov, A. L. and Huang, L.,** Long-circulating liposomes: development and perspectives, *J. Liposome Res.,* 2, 231, 1992.

20. **Allen, T. M. and Papahadjopoulos, D.,** Sterically stabilized ("Stealth") liposomes: pharmacokinetics and therapeutic advantages, in *Liposome Technology,* Vol. 3, 2nd ed., Gregoriadis, G., Ed., CRC Press, Boca Raton, FL, 1993, 59.

21. **Klibanov, A. L., Maruyama, K., Torchilin, V. P., and Huang, L.,** Amphipathic polyethyleneglycols effectively prolong the circulation time of liposomes, *FEBS Lett.,* 268, 235, 1990.

22. **Blume, G. and Cevc, G.,** Liposomes for the sustained drug release in vivo, *Biochim. Biophys. Acta,* 1029, 91, 1990.

23. **Allen, T. M., Hansen, C., Martin, F., Redemann, C., and Yau-Young, A.,** Liposomes containing synthetic lipid derivatives of poly(ethylene glycol) show prolonged circulation half-lives in vivo, *Biochim. Biophys. Acta,* 1066, 29, 1991.

24. **Senior, J., Delgado, C., Fisher, D., Tilcock, C., and Gregoriadis, G.,** Influence of surface hydrophilicity of liposomes on their interaction with plasma protein and clearance from the circulation: studies with poly(ethylene glycol)-coated vesicles, *Biochim. Biophys. Acta,* 1062, 77, 1991.

25. **Papahadjopoulos, D., Allen, T. M., Gabizon, A., Mayhew, E., Matthay, K., Huang, S. L., Lee, K.-D., Woodle, M. C., Lasic, D. D., Redemann, C., and Martin, F. J.,** Sterically stabilized liposomes: improvements in pharmacokinetics and antitumor therapeutic efficacy, *Proc. Natl. Acad. Sci. U.S.A.,* 88, 11460, 1991.

26. **Lasic, D. D., Martin, F. J., Gabizon, A., Huang, S. K., and Papahadjopoulos, D.,** Sterically stabilized liposomes: a hypothesis on the molecular origin of the extended circulation times, *Biochim. Biophys. Acta,* 1070, 187, 1991.

27. **Lasic, D. D., Woodle, M. C., and Papahadjopoulos, D.,** On the molecular mechanism of steric stabilization of liposomes in biological fluids, *J. Liposome Res.,* 2, 335, 1992.

28. **Needham, D., Hristova, K., McIntosh, T. J., Dewhirst, M., Wu, N., and Lasic, D. D.,** Polymer-grafted liposomes: physical basis for the "stealth" property, *J. Liposome Res.,* 2, 411, 1992.

29. **Needham, D., McIntosh, T. J., and Lasic, D. D.,** Repulsive interactions and mechanical stability of polymer-grafted lipid membranes, *Biochim. Biophys. Acta,* 1108, 40, 1992.

29a. **Woodle, M. C., Collins, L. R., Sponsler, E., Kossovsky, N., Papahadjopoulos, D., and Martin, F. J.,** Sterically stabilized liposomes: reduction in electrophoretic mobility but not electrostatic surface potential, *Biophys. J.,* 61, 902, 1992.

30. **Blume, G. and Cevc, G.,** Molecular mechanism of the lipid vesicle longevity in vivo, *Biochim. Biophys. Acta,* 1146, 157, 1993.

31. **Allen, T. M.,** The use of glycolipids and hydrophilic polymers in avoiding rapid uptake of liposomes by the mononuclear phagocyte system, *Adv. Drug Del. Rev.,* 13, 285, 1994.

32. **Allen, T. M.,** Stealth liposomes: avoiding reticuloendothelial uptake, in *Liposomes in the Therapy of Infectious Diseases and Cancer, New Series,* Vol. 89, Lopez-Berestein, G. and Fidler, I., Eds., Alan R. Liss, New York, 1989, 405.

33. **Aragnol, D. and Leserman, L. D.,** Immune clearance of liposomes inhibited by an anti-Fc receptor antibody in vivo, *Proc. Natl. Acad. Sci. U.S.A.,* 83, 2699, 1986.

34. **Maruyama, K., Kennel, S. J., and Huang, L.,** Lipid composition is important for highly efficient target binding and retention of immunoliposomes, *Proc. Natl. Acad. Sci. U.S.A.,* 87, 5744, 1990.

35. **Mayer, L. D., Bally, M. B., and Cullis, P. R.,** Uptake of adriamycin into large unilamellar vesicles in response to a pH gradient, *Biochim. Biophys. Acta,* 857, 123, 1986.

36. **Gabizon, A., Shiota, R., and Papahadjopoulos, D.,** Pharmacokinetics and tissue distribution of doxorubicin encapsulated in stable liposomes with long circulation times, *J. Natl. Cancer Inst.,* 81, 1484, 1989.

37. **Mori, A., Klibanov, A. L., Torchilin, V. P., and Huang, L.,** Influence of the steric barrier of amphipathic poly(ethyleneglycol) and ganglioside GM1 on the circulation time of liposomes and on the target binding of immunoliposomes in vivo, *FEBS Lett.,* 284, 263, 1991.

38. **Allen, T. M., Agrawal, A. K., Ahmad, I., Hansen, C. B., and Zalipsky, S.,** Antibody-mediated targeting of long-circulating (Stealth) liposomes, *J. Liposome Res.,* 4, 1, 1994.

39. **Ahmad, I. and Allen, T. M.,** Antibody-mediated specific binding and cytotoxicity of liposome-entrapped doxorubicin to lung cancer cells in vitro, *Cancer Res.,* 52, 4817, 1992.

40. **Urdal, D. L. and Hakomori, S.,** Tumor-associated ganglio-N-triosylceramide: target for antibody-dependent, avidin-mediated drug killing of tumor cells, *J. Biol. Chem.,* 255, 10509, 1980.

41. **Loughrey, H., Bally, M. B., and Cullis, P. R.,** A noncovalent method of attaching antibodies to liposomes, *Biochim. Biophys. Acta,* 901, 157, 1987.

42. **Loughrey, H. C., Choi, L. S., Cullis, P. R., and Bally, M. B.,** Optimized procedures for the coupling of proteins to liposomes, *J. Immunol. Methods,* 132, 25, 1990.

43. **Heath, T. D.,** Covalent attachment of proteins to liposomes, *Methods Enzymol.,* 149, 111, 1987.

44. **Zalipsky, S.,** Synthesis of end-group functionalized polyethylene glycol-lipid conjugate for preparation of polymer-grafted liposomes, *Bioconj. Chem.,* 4, 296, 1993.

45. **Zalipsky, S., Newman, M. S., Punatambekar, B., and Woodle, M. C.,** Model ligands linked to polymer chains on liposomal surfaces: application of a new functionalized polyethylene glycol lipid conjugate, *Polym. Mater. Sci. Eng.,* 67, 519, 1993.

46. **Holmberg, E., Maruyama, K., Litzinger, D. C., Wright, S., Davis, M., Kabalka, G. W., Kennel, S. J., and Huang, L.,** Highly efficient immunoliposomes prepared with a method which is compatible with various lipid compositions, *Biochem. Biophys. Res. Commun.,* 165, 1272, 1989.

47. **Mori, A. and Huang, L.,** Immunoliposome targeting in a mouse model: optimization and therapeutic application, in *Liposome Technology,* Vol. 3, 2nd ed., Gregoriadis, G., Ed., CRC Press, Boca Raton, FL, 1992, 153.

48. **Park, J. W., Hong, K., Carter, P., Kotts, C., Shalaby, R., Giltinan, D., Wirth, C., Asgari, H., Wood, W. I., Papahadjopoulos, D., and Benz, C.,** Development of anti-HER-2 immunoliposomes for breast cancer therapy, *Proc. Am. Soc. Clin. Oncol.,* 12, 118, 1993.

49. **Blume, G., Cevc, G., Crommelin, M. D. J. A., Bakker-Woudenberg, I. A. J. M., Kluft, C., and Storm, G.,** Specific targeting with poly(ethylene glycol)-modified liposomes: coupling of homing devices to the ends of the polymeric chains combines effective target binding with long circulation times, *Biochim. Biophys. Acta,* 1149, 180, 1993.

50. **Mori, A., Kennel, S. J., and Huang, L.,** Immunotargeting of liposomes containing lipophilic antitumor prodrugs, *Pharm. Res.,* 10, 507, 1993.

51. **Torchilin, V. P., Klibanov, A. L., Huang, L., O'Donnell, S., Nossiff, N. D., and Khaw, B. A.,** Targeted accumulation of polyethylene glycol-coated immunoliposomes in infarcted rabbit myocardium, *FASEB J.,* 6, 2716, 1992.

52. **Klibanov, A. L., Maruyama, K., Beckerleg, A. M., Torchilin, V. P., and Huang, L.,** Activity of amphipathic poly(ethylene glycol) 5000 to prolong the circulation time of liposomes depends on the liposome size and is unfavorable for immunoliposome binding to target, *Biochim. Biophys. Acta,* 1062, 142, 1991.

53. **Ahmad, I., Longenecker, M., Samuel, J., and Allen, T. M.,** Antibody-targeted delivery of doxorubicin entrapped in sterically stabilized liposomes can eradicate lung cancer in mice, *Cancer Res.,* 53, 1484, 1993.

54. **Ding, L., Samuel, J., MacLean, G. D., Noujaim, A. A., Diener, E., and Longenecker, B. M.,** Effective drug-antibody targeting using a novel monoclonal antibody against the proliferative compartment of mammalian squamous carcinoma, *Cancer Immunol. Immunother.,* 31, 105, 1990.

55. **Allen, T., et al.,** unpublished observations.

Chapter 21

DOXORUBICIN ENCAPSULATED IN POLYETHYLENE GLYCOL-COATED LIPOSOMES: INITIAL CLINICAL-PHARMACOKINETIC STUDIES IN SOLID TUMORS

Alberto Gabizon, Anthony Huang, Francis Martin, and Yechezkel Barenholz

TABLE OF CONTENTS

I. INTRODUCTION

Doxorubicin (DOX) is an anthracycline antibiotic widely used in the treatment of a variety of human cancers.[1] It is a very potent drug showing a strong association between dose and antitumor response. As with other cytotoxic drugs, DOX is nonselectively cytotoxic and its clinical use is limited by myelosuppression and stomatitis. Furthermore, the repeated use of DOX is drastically limited by a cumulative, nonreversible cardiac damage, which appears to be related to mitochondrial damage caused by the semiquinone radical of anthracyclines.[2] Clearly, any form of delivery that will shift the drug biodistribution into the tumor and away from the heart may result in a significant improvement of the therapeutic index of DOX. This goal has been pursued in a variety of ways from simple changes in the form of administration (continuous infusion as opposed to bolus injection[3]) through nonspecific delivery systems (drug-polymer conjugates) to sophisticated targeting devices (antibodies recognizing a membrane-associated antigen as vectors of the drug[5]).

Liposomes were also recognized more than a decade ago as promising delivery systems for DOX and analogs. The high affinity of DOX for a variety of negatively charged phospholipids, such as cardiolipin, was also an important catalyst of the initial studies in this field.[6] It was found in preclinical animal models that administration of DOX in liposome-associated form reduces the cardiac uptake of drug and the incidence and severity of histopathological observations correlated with cardiotoxicity.[7-8] However, these early studies with liposomal formulations of DOX did not show any drug targeting to tumor tissue, except for exceptional circumstances in which tumor cells diffusely infiltrated the liver.[9] In fact, a clinical-imaging study with radiolabeled liposomes suggested that in most cases liposome delivery could be detrimental for drug localization in tumors, given the dominant and rapid uptake of liposomal drug by the reticuloendothelial system (RES) of liver and spleen.[10] Another critical factor in liposome-mediated delivery is stability, i.e., the ability of liposomes to retain the drug payload upon circulation and dilution in the intravascular compartment.[11] Achieving a high degree of stability *in vivo* was another major obstacle with regard to early liposomal DOX preparations.[10]

In recent years, the development of new formulations of long circulating liposomes (Stealth® or sterically stabilized liposomes) with reduced uptake by the RES and enhanced accumulation in tumors has opened up a completely new dimension to the delivery of anthracyclines and possibly other cytotoxic drugs.[12,13] In studies in rodents and dogs with some of these new formulations, liposome-associated DOX has been shown to circulate with very long half-lives, in the range of 15 to 30 h.[14,15] When phospholipids with saturated acyl chains are used and DOX is encapsulated in the water phase, these preparations also show a high degree of stability in plasma.[16] An increased accumulation of drug in murine transplantable tumors and in ascitic tumor exudates has been reported using long circulating liposomes as doxorubicin carriers.[17,18] DOX encapsulated in long circulating liposomes also shows a superior therapeutic antitumor activity and decreased toxicity when compared to free DOX in a variety of mouse models.[8] Thus, long circulating liposomes appear to confer a double advantage as an anticancer drug delivery system: toxicity buffering as with other previous liposome formulations, and selective tumor accumulation leading to an enhanced antitumor activity.

In the following pages, we will review the results of our initial clinical studies with a formulation referred to hereafter as Doxil, which contains a polyethylene glycol (PEG) derivatized phospholipid that has been shown to confer optimal prolongation of vesicle circulation time in animal models.[19-22] Other chapters of this book deal in detail with the preclinical pharmacology of this and other similar formulations of long circulating liposomes. We will focus here on reviewing the data obtained in a recently published study comparing the pharmacokinetics of DOX and Doxil in cancer patients,[23] and discuss as well part of our current clinical experience with Doxil.

II. METHODOLOGY

Doxil* is a liposome preparation containing DOX provided by Liposome Technology Inc. (Menlo Park, CA), and whose main lipid components are hydrogenated soybean phosphatidylcholine, cholesterol, and polyethylene glycol (M_r 1900 to 2000) derivatized distearoyl phosphatidylethanolamine. DOX is encapsulated in the liposome internal aqueous space at a drug-to-phospholipid ratio of approximately 150 µg/µmol. The liposomes are suspended in 10% sucrose. The Gaussian mean vesicle size as measured by dynamic laser light scattering is in the range of 80 to 120 nm. The material used in the study presented here was stored in frozen form (-10 to $-20°C$) at a concentration of 2 mg DOX per ml. Currently, a highly stable, liquid storage form of Doxil has become available for clinical studies.

The characteristics of patients entered to this study are described in Table 1. As mentioned above, the main goal of this study was to compare the pharmacokinetics of DOX given in free form to that of DOX given in the liposomal formulation of Doxil. To avoid the problem of interpatient variability, a group of seven patients received both forms of treatment sequentially (free DOX in the first course of treatment, and Doxil in a second course of treatment with a 3-week interval). To further characterize the pharmacokinetics of Doxil, a second group of nine patients received Doxil upfront without prior treatment with free DOX. Two dose levels were chosen to be tested: the high dose, 50 mg/m², is close to the recommended dose for free DOX as single agent,[24] the low dose, 25 mg/m², allows for a twofold difference from the high dose, a useful factor to look at the dose dependence of pharmacokinetic parameters.

---

* Recently, the name of Doxil has been changed to DOX-SL.

TABLE 1
Patient Characteristics

Number of Patients

Male/female	6/10
Age (years)	
Median	59.5
Range	38–73
Zubrod performance status	2(1–3)
Prior chemotherapy Yes/No	12/4
Type of tumor	
Breast	6
Non-small cell lung	3
Ovarian	3
Mesothelioma (pleura)	1
Mesothelioma (peritoneum)	1
Sarcoma (soft tissue)	1
Pancreas	1
Total	16

The methodology involved in the pharmacokinetic studies, including sampling, drug extraction and drug measurement in plasma, urine, and effusion samples, and pharmacokinetic analysis has been described in detail.[23]

III. RESULTS AND DISCUSSION

A. CLINICAL OBSERVATIONS

Table 2 presents the number and distribution of courses by dose. Both forms of treatment, free DOX and Doxil, were given by bolus injection which generally lasted 5 to 10 min. In four instances of Doxil administration, there was an acute reaction during injection characterized by facial and neck flushing, general discomfort, and in one extreme case, respiratory difficulty. The patients were hemodynamically stable. This reaction resolved within minutes by discontinuing or reducing the rate of injection. The patients recovered completely. No clinical or radiographic evidence of pulmonary embolism was detected. Although acute reactions such as backpain have been reported after liposome infusion, the characteristics of the reaction

TABLE 2
Dosage and Courses of Treatment

Dosage	Number of Courses (number of patients)	
	Doxil	Free Doxorubicin
25 mg/m^2	9(8)[a]	3(3)
50 mg/m^2	45(15)[b]	4(4)
Total	54(15)	7(7)
Median no. of courses per patient (range)	3(1–9)	1

[a]   One course was discontinued due to acute reaction, resulting in a given dose of only 10 mg/m^2.

[b]   The scheduled dose (50 mg/m^2) was reduced by 10–20% in 4/15 patients, corresponding to 8/45 courses, due to Grade 3 toxicity (stomatitis, leukopenia) in three cases, and one case of hand-foot syndrome. All four patients requiring dose modification had been extensively pre-treated with chemotherapy.

observed here with Doxil appear to be unique. Since the suspension vehicle (sucrose 10%) is unlikely to be involved, the reaction must be due to a direct or indirect effect of the liposomes on vessel neuroreceptors. A physical stimulus resulting from the rapid injection of a particulate suspension, or a chemical stimulus derived from one of the surface liposome components are among the possible mediators of this reaction. At any rate, our further experience with more than 100 courses of treatment indicates that the occurrence of this reaction can be almost completely prevented by administration of Doxil in a 1-h drip infusion after dilution in 150 to 250 ml Dextrose 5%.[32]

Nausea, although frequent, was generally mild and delayed (i.e., 24 h post-treatment). Doxil-related vomiting occurred only in isolated instances and in a minority of patients. Oral antiemetics were effective in palliating nausea and vomiting. Intravenous antiemetic premedication is probably unnecessary in most cases. Delayed nausea is probably explained by the fact that the drug needs first to be released from the liposomes to become bioavailable and emetogenic.

Leukopenia was generally mild and always afebrile. Neutrophils and lymphocytes were usually affected to a similar degree. Nadir white blood cell (WBC) counts were observed on day 14 post-injection with complete recovery by day 21. No significant thrombocytopenia or anemia related to Doxil were observed at any dose level. There appears to be no cumulative damage to the blood counts with repeated courses of Doxil although the number of patients given prolonged treatment with Doxil is still too small to draw any conclusions. Figure 1A shows the WBC count along treatment with Doxil in a patient receiving a first course of free DOX, 50 mg/m^2, followed by nine courses of Doxil, 50 mg/m^2.

Undoubtedly, the most severe and frequent side effect of Doxil is stomatitis. Patients heavily pretreated with chemotherapy were specially susceptible. Stomatitis developed between days 7 to 14 post-treatment and resolved within 7 d or less. The oral mucosa, tongue, and in some cases the oropharynx were involved. Superimposed oral candidiasis was important in only one patient who was receiving dexamethasone treatment concomitantly. Stomatitis is the dose-limiting toxicity of DOX treatments given by continuous infusion.[3] The occurrence of stomatitis in Doxil-treated patients suggests that Doxil may mimic a continuous infusion. In this case, slow release of drug from liposomes in the intravascular compartment accounts for an important pathway of clearance of major pharmacological consequences. A less likely alternative is that stomatitis may result from increased deposition of liposome-associated drug in the oral mucosa.

A rather unique and unforeseen type of toxicity seen in Doxil-treated patients is the hand-foot syndrome, also referred to as palmar-plantar erythrodysesthesia.[25] This is a desquamating, painful dermatitis primarily affecting hands and feet. In this study, this complication was seen in two patients. Recent findings in other groups of patients receiving Doxil have confirmed that the above observation is a Doxil-related side effect.[32] The hand-foot syndrome was seen after two or more courses of Doxil at doses of at least 50 mg/m^2 and may be related to a cumulative toxic effect on the skin. This syndrome is a known side effect of various chemotherapy regimens.[25] Samuels et al.[26] reported a 53% incidence of hand-foot syndrome in a group of 17 patients receiving continuous infusion of DOX for a mean period of 118 d and with a mean dose rate of 3.7 mg/m^2/d.[26] Although this form of toxicity is not life-threatening, our current experience indicates that recovery is longer than for other forms of toxicity (myelosuppression, stomatitis) and may take several weeks. Here again, as in the case of stomatitis, it is still unclear whether toxicity results from intravascular slow drug release or from liposome-mediated deposition of drug in the skin. However, there are preclinical data indicating that Stealth liposomes accumulate in large amounts in the skin.[27] This would support the proposition that skin toxicity in humans is due to an increased deposition of liposome-associated drug in the skin.

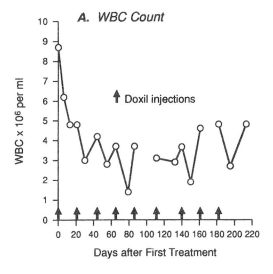

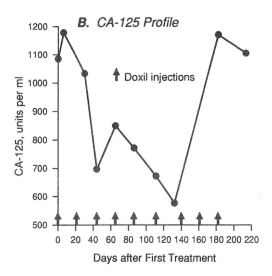

FIGURE 1. Female patient (#8), 47 years old, suffering from recurrent ovarian carcinoma. She was initially treated with six courses of carboplatin/etoposide chemotherapy with partial remission of disease lasting for 7 months. Upon relapse, she received three courses of cisplatin/cyclophosphamide chemotherapy with no response. Three weeks prior to Doxil chemotherapy, she received free DOX (50 mg/m^2) resulting in a WBC nadir of 1600/µl. (A) WBC count, (B) CA-125 (tumor marker) blood levels during Doxil treatment.

In this study and in other ongoing studies with Doxil,[28,32] there was no decrease of the cardiac left ventricle ejection fraction, nor has any other clinical indication of cardiotoxicity been detected. However, the number of patients who have received large cumulative amounts of Doxil (>450 mg/m^2) is still too small to provide any valid indications on the issue of cardiotoxicity. Likewise, there were no significant alterations of liver function, kidney function, and coagulation blood tests related to Doxil treatment.

This study was designed primarily to measure free and liposomal DOX pharmacokinetics with some patients receiving a course of free DOX prior to Doxil therapy, making the respective contribution of each form of treatment on response difficult to determine. Within the confines of this setting, two objective, non-measurable, antitumor responses of 6 to 7 months duration were documented. In a male patient with peritoneal mesothelioma, a decrease

TABLE 3
Pharmacokinetic Parameters (Median Values)

Treatment Parameter	Total Doxorubicin-Equivalents in Plasma		Total Doxorubicin-Equivalents in Plasma	
	Doxil 25 mg/m² (N = 8)	Doxil 50 mg/m² (N = 14)	Doxorubicin 25 mg/m² (N = 3)	Doxorubicin 50 mg/m² (N = 4)
C_0 (mg/l)	12.6	21.2	3.3	5.9
1st $t_{1/2}$ (h)	3.2	1.4	0.07	0.06
2nd $t_{1/2}$ (h)	45.2	45.9	8.7	10.4
AUC_0^∞ (mg·h/l)	609	902	1.0	3.5
CL (l/h)	0.08	0.09	45.3	25.3
V_{ss} (l)	4.1	5.9	254	365
MRT (h)	62.7	65.0	5.2	11.8

Abbreviations: AUC, area under the concentration × time curve from 0 to infinity; CL, clearance; V_{ss}, volume of distribution at steady state; MRT, mean residence time.

Adapted from Gabizon, A., et al., *Cancer Res.*, 54, 987, 1994.

of size of the peritoneal implants and of ascites was found in computed tomography (CT) scan. In a patient with ovarian cancer (Figure 1B), a sustained reduction of the blood levels of a tumor marker (CA-125) was detected. Both patients had been pretreated with a variety of chemotherapeutic regimens and relapsed. Additional antitumor responses to Doxil have been observed outside this study in six patients suffering from breast cancer (two patients), non-small cell (adenocarcinoma type) lung cancer, renal cell carcinoma, and pleural mesothelioma.[28,32]

B. PHARMACOKINETICS OF DOXIL AND COMPARISON WITH FREE DOX

Table 3 presents the pharmacokinetic parameters for DOX after administration of Doxil or free DOX. After administration of Doxil, between 19 to 57% of the drug was cleared from plasma with an initial half-life of 1 to 3 h. The remaining drug was cleared much more slowly with a second half-life of approximately 45 h. This second phase accounted for 43 to 81% of the total dose and for more than 95% of the total area under the curve (AUC). Clearance and volume of distribution values were low at both dose levels of Doxil and differed substantially from those of free DOX by more than 100-fold.

As seen in Table 3, median values for both AUC and C_0 increased proportionately with dose within the range tested (25 to 50 mg/m²). This suggests that no saturation of clearance occurs when the dose is raised from 25 to 50 mg/m². This is consistent with the preclinical observations showing that the clearance of Stealth liposomes is independent of dose.[29] Thus, the biphasic clearance of Doxil discussed above is probably not related to a saturation phenomenon. Other processes that accelerate drug clearance, such as drug leakage from circulating liposomes, may account for it. Obviously, saturation of clearance at higher doses of Doxil is still possible. This phenomenon has been documented in a recent preclinical report using DOX-loaded liposomes containing monosialoganglioside.[30]

As shown in Figure 2, the levels of total DOX and liposome-encapsulated DOX in plasma after Doxil administration are almost superimposable, stressing the fact that practically all of the circulating drug in plasma is in liposome-associated form. These observations also rule out the possibility of a sudden burst of drug release after injection, in contrast to what has been observed with other formulations of liposomal doxorubicin.[9] The lack of detection of any circulating free drug after injection of Doxil indicates that the leakage rate of drug from circulating liposomes is slower than the rate of clearance of free drug from plasma, thus

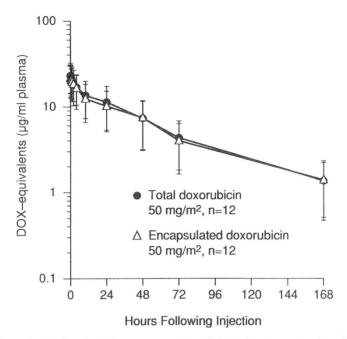

FIGURE 2.   Plasma levels of total and liposome-encapsulated DOX in Doxil-treated patients. Dose: 50 mg/m^2, N = 12. Liposome-associated drug accounts for practically all of the drug present in plasma. Liposome-associated drug was separated from nonliposome-associated drug using a Dowex resin.[14] These findings have been confirmed by fractionation of liposome-associated drug by chromatography of plasma through Biogel A15M columns.

preventing any accumulation of free drug in plasma. That comes as no surprise since *in vitro* studies in human plasma indicate that about 100 h are required for a 50% drug release from Stealth liposomes,[31] which is twofold slower than the half-life of Doxil in patients. If we assume leakage rates to be similar for the *in vivo* situation, one can conclude that a significant fraction of the drug is cleared from plasma by liposome-mediated transport into the peripheral tissue compartment.

As expected from a liposomal delivery system, the urinary excretion of DOX following Doxil administration was reduced and delayed as compared to that seen after treatment with free DOX (Figure 3). Encapsulation of DOX in liposomes would prevent its filtration by glomeruli because of the vesicle size and the stable retention of the drug inside the liposome. The reduced renal clearance of Doxil raises the possibility of an increase in dose intensity with liposomal treatment as compared to an equal given dose of free drug. Although no studies of biliary excretion have been done, preclinical data point at a reduced clearance of DOX from the liver when the drug is administered in Stealth liposomes.[14] Clearly, the pharmacokinetic changes preclude any comparison of dose intensity between free and liposomal drug on the basis of dose and schedule. The analysis of plasma AUC would also be wrong since most of what we are measuring is encapsulated drug which is yet to become bioavailable. Even if we were able to produce an accurate estimate of the AUC of free drug after Doxil injection, this measurement would ignore the tissue exposure resulting from direct liposome-mediated transport of drug to tissues. Thus, drug excretion curves are probably the best indicator of the change in dose intensity. With regard to metabolites, urine analysis provided us with another important clue. While no significant amounts of metabolites were found in plasma after Doxil treatment, the urine was, on the contrary, rich in metabolites representing the main metabolic pathways of DOX (Figure 4). This indicates that Doxil is being metabolized in a similar way to free DOX, although the rate of metabolite production is probably slower than the rate of excretion, thus preventing any significant accumulation in plasma.

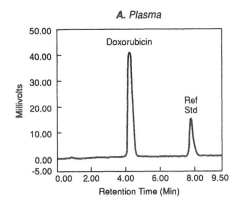
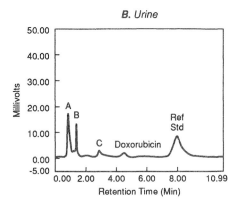

FIGURE 3. HPLC chromatogram of plasma and urine extracts obtained 24 h after injection of Doxil, 50 mg/m². DOX metabolites: **A** and **B** are the polar conjugates (glucuronide and sulfate) of the 4-demethyl,7-deoxyaglycone; **C** is doxorubicinol. The reference standard is daunorubicin. (Adapted from Gabizon, A. et al., *Cancer Res.*, 54, 987, 1994. With permission from Cancer Research Inc.)

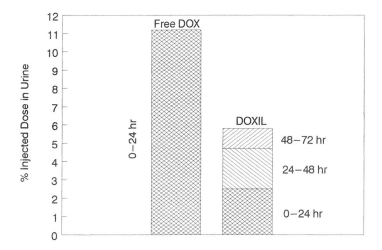

FIGURE 4. Urinary excretion of DOX and metabolites in free DOX and Doxil-treated patients. Dose: 50 mg/m². Median values shown, N = 4 for free DOX, N = 9 for Doxil. Drug measurements based on total fluorescence of DOX-equivalents.

The most relevant observation in preclinical studies of Stealth liposomes is the enhancement of drug concentration in the tumor interstitial fluid, which appears to be related to a liposome extravasation process through a hyperpermeable endothelium. We have examined the DOX levels in malignant effusions in patients receiving free DOX and Doxil. The sampling of these fluids was done as the best possible approximation to the drug concentration in the tumor interstitial fluid, using a relatively noninvasive method. Two important observations were made. The first one is that the accumulation of Doxil in malignant effusions behaves as a slow process peaking between 3 to 7 d after injection (Figure 5). This is in agreement with preclinical data on tumor localization of liposomes which also shows peak values several days after injection. The second observation is that the drug levels in malignant effusions were severalfold greater after Doxil treatment as compared to free DOX treatment (Figure 6). The differences between free DOX and Doxil are seen in both supernatants and cell pellets of effusions. These observations support the claim that long circulating liposomes are able to extravasate into human malignant effusions. Ultimately, the goal of drug delivery via Stealth liposomes is to achieve first-order targeting, which pharmacologically implies that the

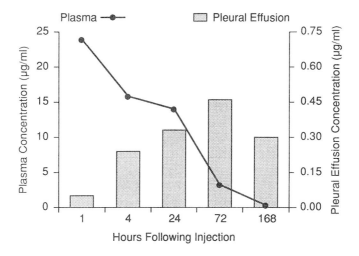

FIGURE 5. Delayed peak accumulation of drug in the malignant pleural effusion of a patient (#3, breast Ca) receiving Doxil, 25 mg/m^2. The bars with the right Y-axis represent the levels in pleural fluid. The line with the left Y-axis depict the parallel decline in plasma concentration.

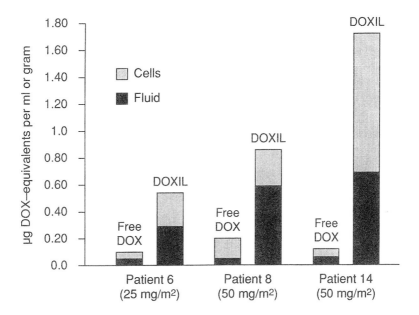

FIGURE 6. Enhancement of drug concentration in malignant effusions after Doxil treatment. Three patients with pleural effusions were treated successively with equal doses of free DOX and Doxil with a 3-week interval. Sampling was done between 4 to 24 h after injection of free DOX, and between 5 to 6 d after injection of Doxil. Patient 6 (breast Ca) received 25 mg/m^2. Patients 8 (ovarian Ca) and 14 (non-small cell lung Ca) received 50 mg/m^2. (Adapted from Gabizon, A. et al., *Cancer Res.,* 54, 987, 1994. With permission from Cancer Research Inc.)

liposomal drug accumulates in the interstitial fluid of tumor nodules followed by *in situ* release of drug.

IV. CONCLUDING REMARKS

Preclinical studies have shown that delivery of doxorubicin via Stealth liposomes results in prolonged stay of the drug in the intravascular compartment, enhanced drug accumulation in transplantable tumors, and superior therapeutic activity over free DOX. The results of this pilot study with Doxil, a Stealth liposome formulation of DOX, are consistent with preclinical findings and point at drastic changes in the pharmacokinetics of DOX, which appears to be controlled by the liposome carrier. The fact that the toxicity pattern of the drug is also changed with liposome delivery underscores the pharmacologic relevance of the changes in pharmacokinetic parameters. Phase I/II studies will have to address the issues of dose intensity and schedule optimization which may differ significantly from those of free DOX. The enhanced drug accumulation in malignant effusions is apparently related to liposome longevity in circulation, and, taken together with the preclinical data, suggests that liposomes may improve the delivery of DOX to human solid tumors. From the point of view of mechanism of drug delivery, a key issue in future pharmacokinetic studies is to elucidate the actual contribution of intravascular slow release vs. liposome transport to tissues to the clearance of liposomal drug from the circulation.

ACKNOWLEDGMENTS

Alberto A. Gabizon is the recipient of a research career development award from the Israel Cancer Research Fund. The work presented here was supported by Liposome Technology Inc. (Menlo Park, CA).

REFERENCES

1. **Young, R.C., Ozols, R.F., and Myers, C.E.,** The anthracycline antineoplastic drugs, *N. Engl. J. Med.,* 305, 139, 1981.
2. **Myers, C.,** Anthracyclines, in *Cancer Chemotherapy 8*, Pinedo, H.M. and Chabner, B.A., Eds., Elsevier, Shannon, 1986, 52.
3. **Hortobagyi, G.N., Frye, D., Buzdar, A.U., Ewer, M.S., Fraschini, G., Hug, V., Ames, F., Montague, E., Carrasco, C.H., Mackay, B. et al.,** Decreased cardiac toxicity of doxorubicin administered by continuous intravenous infusion in combination chemotherapy for metastatic breast carcinoma, *Cancer,* 63, 37, 1989.
4. **Duncan, R.,** Drug-polymer conjugates: potential for improved chemotherapy, *Anti-Cancer Drugs,* 3, 175, 1992.
5. **Trail, P.A., Willner, D., Lasch, S.J., Henderson, A.J., Hofstead, S., Casazza, A.M., Firestone, R.A., Hellstrom, I., and Hellstrom, K.E.,** Cure of xenografted human carcinomas by BR96-doxorubicin immunoconjugates, *Science,* 261, 212, 1993.
6. **Goormaghtigh, E. and Ruysschaert, J.M.,** Anthracycline glycoside-membrane interactions, *Biochim. Biophys. Acta,* 779, 271, 1984.
7. **Szoka, F.C.,** Liposomal drug delivery: current status and future prospects, in *Membrane Fusion,* Wilschut, J. and Hoekstra, D., Eds., Marcel Dekker, New York, 1991, 845.
8. **Gabizon, A.,** Liposomal anthracyclines, *Hematol. Oncol. Clin. North Am.,* 8, 431, 1994.
9. **Gabizon, A., Goren, D., Fuks, Z., Barenholz, Y., Dagan, A., and Meshorer, A.,** Enhancement of adriamycin delivery to liver metastatic cells with increased tumoricidal effect using liposomes as drug carriers, *Cancer Res.,* 43, 4730, 1983.
10. **Gabizon, A., Chisin, R., Amselem, S., Druckmann, S., Cohen, R., Goren, D., Fromer, I., Peretz, T., Sulkes, A., and Barenholz, Y.,** Pharmacokinetic and imaging studies in patients receiving a formulation of liposome-associated adriamycin, *Br. J. Cancer,* 64, 1125, 1991.
11. **Amselem, S., Cohen, R., and Barenholz, Y.,** In vitro tests to predict in vivo performance of liposomal dosage forms, *Chem. Phys. Lipids,* 64, 219, 1993.
12. **Papahadjopoulos, D., Allen, T.M., Gabizon, A., Mayhew, E., Matthay, K., Huang, S.K., Lee, K.D., Woodle, M.C., Lasic, D.D., Redemann, C., and Martin, F.J.,** Sterically stabilized liposomes: improvements in pharmacokinetics and antitumor therapeutic efficacy, *Proc. Natl. Acad. Sci. U.S.A.,* 88, 11460, 1991.
13. **Woodle, M.C. and Lasic, D.D.,** Sterically stabilized liposomes, *Biochim. Biophys. Acta,* 1113, 171, 1992.
14. **Gabizon, A., Shiota, R., and Papahadjopoulos, D.,** Pharmacokinetics and tissue distribution of doxorubicin encapsulated in stable liposomes with long circulation times, *J. Natl. Cancer Inst.,* 81, 1484, 1989.
15. **Mayhew, E.G., Lasic, D.D., Babbar, S., and Martin, F.J.,** Pharmacokinetics and antitumor activity of epirubicin encapsulated in long-circulating liposomes, *Int. J. Cancer,* 51:302, 1992.
16. **Gabizon, A., Barenholz, Y., and Bialer, M.,** Prolongation of the circulation time of doxorubicin encapsulated in liposomes containing a polyethyleneglycol-derivatized phospholipid: pharmacokinetic studies in rodents and dogs, *Pharm. Res.,* 10, 703, 1993.

17. **Gabizon, A.,** Selective tumor localization and improved therapeutic index of anthracyclines encapsulated in long-circulating liposomes, *Cancer Res.,* 52, 891, 1992.

18. **Huang, S.K., Mayhew, E., and Gilani, S. et al.,** Pharmacokinetics and therapeutics of sterically stabilized liposomes in mice bearing C-26 colon carcinoma, *Cancer Res.,* 52, 6774, 1992.

19. **Allen, T.M., Hansen, C., Martin, F., Redemann, C., and Yau-Young, A.,** Liposomes containing synthetic lipid derivatives of poly(ethyleneglycol) show prolonged circulation half-lives in vivo, *Biochim. Biophys. Acta,* 1066, 29, 1991.

20. **Klibanov, A.L., Maruyama, K., Torchilin, V.P., and Huang, L.,** Amphipathic polyethyleneglycols effectively prolong the circulation time of liposomes, *FEBS Lett.,* 268, 235, 1991.

21. **Senior, J., Delgado, C., Fisher, D., Tilock, C., and Gregoriadis, G.,** Influence of surface hydrophilicity of liposomes on their interaction with plasma protein and clearance from the circulation, *Biochim. Biophys. Acta,* 1062, 77, 1991.

22. **Blume, G. and Cevc, G.,** Liposomes for sustained drug release in vivo, *Biochim. Biophys. Acta,* 1029, 91, 1990.

23. **Gabizon, A., Catane, R., Uziely, B., Kaufman, B., Safra, T., Cohen, R., Martin, F., Huang, A., and Barenholz, Y.,** Prolonged circulation time and enhanced accumulation in malignant exudates of doxorubicin encapsulated in polyethylene-glycol coated liposomes, *Cancer Res.,* 54, 987, 1994.

24. **O'Bryan, R.M., Baker, L.H., Gottlieb, J.E., Rivkin, S.E., Balcerzak, S.P., Grumet, G.N., Salmon, S.E., Moon, T.E., and Hoogstraten, B.,** Dose response evaluation of adriamycin in human neoplasia, *Cancer,* 39, 1940, 1977.

25. **Lokich, J.J. and Moore, C.,** Chemotherapy-associated palmar-plantar erythrodysesthesia syndrome. *Ann. Intern. Med.,* 101, 798, 1984.

26. **Samuels, B.L., Vogelzang, N.J., Ruane, M., and Simon, M.A.,** Continuous venous infusion of doxorubicin in advanced sarcomas, *Cancer Treat. Rep.,* 71, 971, 1987.

27. **Huang, S.K., Lee, K.-D., Hong, K., Friend, D.S., and Papahadjopoulos, D.,** Microscopic localization of sterically-stabilized liposomes in colon carcinoma-bearing mice, *Cancer Res.,* 52, 5135, 1992.

28. **Gabizon, A., Isacson, R., Libson, E. et al.,** Clinical studies of liposome-encapsulated anthracyclines, *Acta Oncol.,* in press, 1994.

29. **Allen, T.M. and Hansen, C.,** Pharmacokinetics of Stealth versus conventional liposomes: effect of dose, *Biochim. Biophys. Acta,* 1068, 133, 1991.

30. **Parr, M.J., Bally, M.B., and Cullis, P.R.,** The presence of GM1 in liposomes with entrapped doxorubicin does not prevent RES blockade, *Biochim. Biophys. Acta,* 1168, 249, 1993.

31. **Gabizon, A., Pappo, O., Goren, D., Chemla, M., Tzemach, D., and Horowitz, A.T.,** Preclinical studies with doxorubicin encapsulated in polyethyleneglycol-coated liposomes, *J. Liposome Res.,* 3, 517, 1993.

32. **Gabizon, A.,** unpublished results.

Chapter 22

PHARMACOKINETICS AND TUMOR LOCALIZATION OF DOX-SL (STEALTH® LIPOSOMAL DOXORUBICIN) BY COMPARISON WITH ADRIAMYCIN IN PATIENTS WITH AIDS AND KAPOSI'S SARCOMA

Donald W. Northfelt, Lawrence Kaplan, Julie Russell, Paul A. Volberding, and Francis J. Martin

TABLE OF CONTENTS

I. INTRODUCTION

Reports of the development of Kaposi's sarcoma (KS) in previously healthy young homosexual men in 1981 marked the onset of the epidemic of acquired immunodeficiency syndrome (AIDS) in the U.S.[1-3] KS is the most common neoplasm in patients with human immunodeficiency virus (HIV) infection, having been diagnosed in 15 to 20% of all patients with AIDS in the U.S.[4] Although the percentage of AIDS patients developing KS is declining,[5] the prevalence continues to rise, and recent studies suggest that patients now being diagnosed with KS have more severe disease and less favorable prognoses than those diagnosed early in the epidemic.[6,7]

Curative therapy for AIDS-KS does not exist. Such therapy may not be possible until a means of reversing HIV-induced immunodeficiency is found. Until recently, AIDS-KS was

0-8493-8383-8/95/$0.00+$.50

rarely life-threatening; most patients with AIDS and KS ultimately died of opportunistic infections that developed as a result of profound immunodeficiency. More recently, however, patients are developing extensive visceral KS, and dying of systemic disease.[7] This may be a consequence of a decrease in deaths due to opportunistic infections.

Systemic chemotherapy has not increased survival in this population.[8] Therefore, palliation is usually the goal of treatment for AIDS-KS. Therapy is commonly used to eliminate or reduce cosmetically unacceptable lesions, to reduce painful or unsightly edema or lymphadenopathy, and to shrink symptomatic oral lesions. In addition, visceral involvement in KS, especially pulmonary KS, continues to be a particularly difficult clinical problem for which therapies have proven less effective than for cutaneous disease.[9,10] Because it is palliative in nature, treatment of KS in a patient with AIDS must be planned not only to produce tumor regression, but also to minimize detrimental effects on the patient's immune function and overall clinical status.

KS tumors respond well to a number of commonly employed chemotherapeutic agents, including the anthracycline antibiotic, doxorubicin. However, doses of doxorubicin sufficient to achieve a response often lead to myelosuppression, the single-dose limiting toxicity associated with the use of doxorubicin. Myelosuppression in AIDS patients compounds the underlying immune deficiency caused by HIV infection and may increase the incidence and severity of opportunistic infections. Although KS tumors may respond to treatment with single or multiagent chemotherapy, recurrence is common because the underlying immunodeficiency is progressive. Thus, multiple courses of therapy are required to control the disease. The use of a myelosuppressive drug in this setting is likely to compromise the patient further.

Gill et al.[11] reported the results of a comparative trial of ABV (Adriamycin, bleomycin, and vincristine) and the standard Adriamycin dose in the treatment of AIDS-related KS. ABV was significantly more efficacious but was associated with significant toxicities, particularly affecting peripheral nerves, lung, and skin. Low-dose Adriamycin alone was significantly less effective.[11]

Liposomes, microscopic phospholipid spherules, have been shown in animal models and early human trials to reduce certain toxic effects of encapsulated antitumor agents, including doxorubicin. It is believed that rapid uptake of doxorubicin-loaded liposomes by fixed macrophages (residing primarily in liver), followed by slow release of the drug into the bloodstream, ameliorates host toxicities associated with high peak plasma concentrations of this agent. Although this liver deposition phenomenon is useful in reducing side effects, the intrinsic activity of doxorubicin against systemic tumors is not favorably influenced by encapsulation in conventional liposomes. A new type of liposome formulated to contain surface-grafted segments of polyethylene glycol have been shown to circumvent uptake by hepatic macrophages and to circulate for prolonged periods in the bloodstream of rodents and dogs (so-called "Stealth®" liposomes). Radioactively labeled Stealth liposomes have been shown to selectively enter implanted animal tumors and a direct correlation between blood circulation half-life and tumor uptake has been established. In a preclinical model of KS,[13] Stealth liposomes containing colloidal gold particles have been shown to selectively accumulate in both early- and late-stage cutaneous lesions.[14]

DOX-SL™ is a long circulating liposomal formulation of doxorubicin. Following intravenous injection of DOX-SL into tumor-bearing mice, doxorubicin levels measured in tumors are substantially higher than those seen in animals receiving comparable doses of unencapsulated drug.[12] Based on these observations it is believed that DOX-SL may provide an opportunity for selectively delivering doxorubicin to tumor sites, including KS lesions. If this concept is validated in patients suffering from KS, DOX-SL therapy may provide disease palliation at relatively low doses by distributing a greater proportion of an injected dose of doxorubicin to KS lesions.

The ability to deliver effective palliative therapy with very low toxicity is especially important in a condition such as AIDS-related Kaposi's sarcoma, in which cure is currently not possible and myelotoxicity is particularly intolerable. In a pilot Phase I pharmacokinetic trial in cancer patients, DOX-SL was found to be well tolerated at doses of 25 and 50 mg/m^2 and tumor selective uptake was demonstrated.[15] The purpose of the current study is to confirm that DOX-SL has an acceptable toxicity profile in an AIDS-KS population and that localization of doxorubicin in KS lesions is achieved by Stealth liposomal delivery. The long-term therapeutic objective of DOX-SL therapy in this setting is to develop an effective single-agent maintenance regimen to slow the progression of KS based on the rationale that selective delivery of drug to KS lesions will lead to durable responses at relatively low, well-tolerated doses.

II. STUDY OBJECTIVES

- To compare plasma pharmacokinetics of DOX-SL and Adriamycin in AIDS-KS patients at three dose levels: 10, 20, and 40 mg/m^2
- To compare uptake of doxorubicin in KS lesions following administration of equal doses of DOX-SL and Adriamycin
- To assess the safety of a single dose of DOX-SL at 10, 20, and 40 mg/m^2 in AIDS-KS patients not restricted from taking anti-HIV medications and prophylactic therapy for opportunistic infections

III. INVESTIGATIONAL PLAN

A. STUDY DESIGN

The study was designed to allow for comparison of pharmacokinetic parameters and tumor localization of doxorubicin in patients randomized to receive equal doses of Adriamycin or DOX-SL. At the time of writing this interim report a total of 16 patients have completed the study: six patients at the 10- and 20-mg/m^2 level and four at 40 mg/m^2 (at study completion two additional patients will be enrolled at the 40 mg/m^2 level). Patients in each dose group were prospectively randomized to receive either DOX-SL or Adriamycin as a first dose. Blood samples were collected for up to 4 d following treatment and doxorubicin plasma concentrations measured. Three days after the first injection a portion of a representative cutaneous KS lesion was removed by excision from each patient and doxorubicin levels measured. Toxicity assessment was conducted prestudy and each week for 3 weeks following the first treatment. Three weeks following administration of the first dose, patients were crossed over to the alternative treatment and followed for an additional 4 weeks for safety evaluation only.

B. DRUGS, ADMINISTRATION AND DURATION OF TREATMENT

DOX-SL (Stealth® Liposomal Doxorubicin Hydrochloride Injection) was provided by Liposome Technology, Inc., Menlo Park, California. The composition of the lipid components is hydrogenated soybean phosphatidylcholine, cholesterol, methyl polyethylene glycol carbamate of distearoyl phosphatidylethanolamine, and alpha-tocopherol in the weight ratio 3/1/1/0.01. The concentration of doxorubicin hydrochloride in the product is 2 mg/ml with a drug-to-lipid ratio of 125 mg doxorubicin hydrochloride per gram total lipid. Greater than 95% of the drug is encapsulated in the liposomes. The mean particle diameter of the liposomes is 85 to 100 nm. There was no further dilution of DOX-SL prior to administration. DOX-SL was stored in frozen form (–10 to –50°C).

Adriamycin RDF® (doxorubicin HCl) was obtained from Adria Laboratories (Columbus, Ohio). The drug was reconstituted to 2 mg/ml in 0.9% sterile saline for injection prior to administration.

Both drugs were given by slow (15-min) bolus injection through a peripheral arm vein. Pharmacokinetic parameters and tumor localization were assessed following the first course and toxicity assessed following both courses (a "course" being one injection with a 3-week interval following drug administration).

C. PATIENT POPULATION

Patients with positive HIV serology and biopsy-proven KS were selected for entry into this study. Other criteria for selection included patients with at least one cutaneous KS lesion with no prior local or systemic therapy removable in part or totally by punch biopsy or excision and a Karnofsky performance status ≥60%. Lab values included hemoglobin >9.5 gm/dl, neutrophil count >1500 cells/mm^3, platelet count >100,000 cells/mm^3, bilirubin <1.25 × upper limit of normal, creatinine <1.25 × upper limit of normal, SGOT <3 × upper limit of normal, PT and PTT <1.4 × INR, serum calcium <10.5 mg/dl.

All patients understood and signed an informed consent statement which complies with Food and Drug Administration (FDA) regulations and had been reviewed and approved by the Institutional Review Board of the University of California/San Francisco General Hospital.

Patients were excluded who presented with pulmonary KS, active opportunistic infection, or significant documented cardiac or pericardial disease, severe pulmonary disease, or significant peripheral vascular disease. Ongoing therapy, including maintenance therapy for an opportunistic or other infection, and use of prophylaxis for prevention of *Pneumocystis carinii* pneumonia were allowed.

D. PLASMA PHARMACOKINETICS

Doxorubicin content in plasma samples was determined by measuring doxorubicin fluorescence in alcohol extracts of the plasma. A conventional spectrofluorometer was used to make the measurements and a calibration curve was established with doxorubicin reference material dissolved in the same solvent used for extraction of doxorubicin from plasma. Doxorubicin reference material was doxorubicin hydrochloride, Farmitalia Carlo Erba (Milan, Italy) lot 8019D631. This material had been qualified as a reference standard against USP doxorubicin hydrochloride reference standard and contained 955 mg doxorubicin/g dry weight.

Sample preparation — Human EDTA-plasma was prepared for analysis by mixing 0.15 ml plasma with 2.85 ml acid IPA (isopropanol/0.75 N HCl, 90/10 v/v). Protein precipitates were removed by centrifugation at 3000 rpm for 30 min.

Spectrofluorometric measurement — The clear supernatant was read on a Shimadzu RF-540 spectrofluorometer. The excitation wavelength was set to 470 nm and fluorescence emission was read at the doxorubicin emission maximum of 590 nm. The slit widths were 10 nm and the instrument was equipped with a red-sensitive photomultiplier. The instrument was zeroed with acid IPA. Doxorubicin concentration in the samples was determined from a calibration curve of doxorubicin reference standard prepared in acid IPA. Quality control (QC) check standards were prepared at three relevant concentration levels (0.4, 4.0, and 15.0 μ/ml) by spiking doxorubicin-free human EDTA-plasma with doxorubicin reference standard. QC check standards were included in every analysis run.

Validation — The assay was validated for linearity, precision, accuracy, specificity, and sensitivity. Linearity was determined from 0.4 to 20.0 μg/ml and with 3-point and 5-point calibration curves. In either case linearity was excellent with correlation coefficients of 0.9999. Method precision ranged from 4.1% relative standard deviation (RSD) for low doxorubicin concentrations (0.4 μg/ml) to 1.9% (RSD) for high doxorubicin concentrations (15 μg/ml). Accuracy was determined from analysis of QC check standards at low, medium, and high doxorubicin concentrations as described above by comparison of the known doxorubicin content (nominal value) to the assay result. The deviation from the nominal values was less than 10% in all cases. Specificity was determined from 10 plasma samples obtained from HIV-positive patients and demonstrated no interference from concomitant medications or

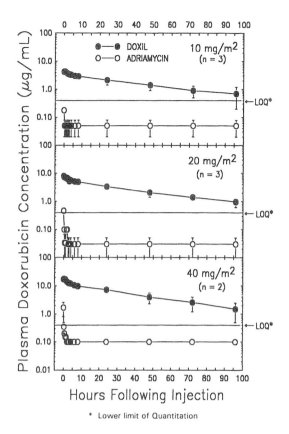

FIGURE 1. Plasma concentration of doxorubicin injected as a free drug (open circles) or encapsulated in the sterically stabilized (Stealth) liposomes (solid circles) at dose levels 10, 20, and 40 mg/m² (top to bottom, respectively). Number of patients is shown in brackets.

endogenous compounds. Sensitivity, expressed as the lower limit of quantitation (LOQ, the lowest amount that can be measured accurately), was 0.4 µg/ml. The limit of detection (LOD), defined as the lowest concentration that can be distinguished from the background signal, was 0.1 µg/ml. Plasma concentrations of Adriamycin-treated patients quickly dropped below the LOQ, in some cases below the LOD (Figure 1). Samples assaying below the LOD were reported as zero, samples above the LOD but below the LOQ were assigned an assay value and reported with the qualification that values do not represent accurate quantitative information. For purposes of graphic presentation of Adriamycin pharmacokinetics, arithmetic averages of reported values were plotted (Figure 1). Note that in this figure averaging of values below the LOD (reported as 0) and above LOD (0.1 or above) resulted in graphical data points below the LOD.

Derivation of pharmacokinetic parameters — Doxorubicin plasma concentration data were analyzed using a nonlinear least-squares data-fitting program (Strip, Micromatch, Inc., Salt Lake City, UT). Goodness of fit and model selection were based on the calculation of weighted residual sum of squares and the Akaike information criterion.[19] Correlation coefficients were ≥0.995.

E. MEASUREMENT OF DOXORUBICIN AND DOXORUBICINOL IN KS TISSUE BIOPSIES BY HIGH PERFORMANCE LIQUID CHROMATOGRAPHY (HPLC)

Doxorubicin and doxorubicinol (a metabolite of doxorubicin) were measured in KS lesions by extracting the disintegrated (enzyme-digested) tissue with methanol. Silver nitrate was added as an aqueous solution before extraction to improve the recovery of doxorubicin and doxorubicinol. The extract was purified and concentrated by solid-phase extraction (SPE). Doxorubicin and doxorubicinol were determined by reversed-phase HPLC with fluorescence detection and quantified from an internal standard (daunorubicin).

Sample preparation — KS tissue (50 to 200 mg), which had been stored at −50°C, was finely diced while still frozen and digested at 50°C for 3 h in a 20 times volume of a 5-mg/ml solution of protease and collagenase in a buffer of 0.2 *M* sodium phosphate at pH 6.2 with 2 m*M* calcium chloride. An internal standard (daunorubicin) was added as a methanol solution in a ratio of 40 ng/ml of enzyme broth. The digest was ground with a silanized glass Dounce homogenizer and then subdivided into 0.5-ml portions. Each portion received 100 µl of 1 *M* silver nitrate and then 4.5 ml of methanol. The resulting mixture was clarified by centrifuging at 3000 rpm for 15 min. The supernatant was diluted with 10 ml of water, centrifuged again at 3000 rpm for 5 min, and then applied to a C-18 SPE column (Analytichem C-18 Large Channel Reservoir by Varian Associates, Sunnyvale, CA) which had been prepped with a methanol and water wash. The column with retained sample was washed with 1 ml of water and the sample eluted with 1 ml of 80/20 v/v MeOH/(ammonium phosphate buffer 58 m*M*, pH 2.50 with 10 m*M* triethylamine). The eluant was blown dry with nitrogen at 40°C and resuspended in 250 µl of HPLC mobile phase. The resuspended sample was injected two times at 100 µl on a Hewlett/Packard 1090 HPLC system equipped with a Hewlett-Packard 1046A fluorescence detector.

Measurement by HPLC — The sample was separated on a Whatman Partisil ODS-3 5-µm column, 250 × 4.6 mm, heated to 40°C, eluted with a solution of MeOH/(ammonium phosphate 29.5 m*M* with 5 m*M* triethylamine, pH 2.50) v/v 45/58 at 1 ml/min. The analytes were detected by fluorescence at 245 nm excitation and 460 nm emission wavelength. Data acquisition and calculations were performed on a Hewlett-Packard work station.

IV. RESULTS

A. PATIENT DEMOGRAPHICS

Pharmacokinetic and tissue distribution results are based on 16 patients. Table 1 presents demographic information and AIDS risk categories for this group. Clinical observations and safety data are presented on the first 12 patients to complete the study. Dosages and courses of treatment are listed in Table 2.

B. PHARMACOKINETICS

Table 3 presents the pharmacokinetic parameters derived from plasma doxorubicin levels measured in the eight patients who had received DOX-SL as the first dose at the time of writing this preliminary report. Due to the rapid clearance of doxorubicin it was not possible to accurately calculate comparable pharmacokinetic parameters for patients receiving Adriamycin.

Table 4 summarizes literature values for pharmacokinetic parameters for Adriamycin at dose levels in the range of 19 to 56 mg/m^2.

Comparative plasma concentration vs time curves for DOX-SL and Adriamycin at the three dose levels (10, 20 and 40 mg/m2) are shown in Figure 1. Mean values ± standard deviation for three patients are presented in the 10 and 20 mg/m^2 groups and the average ± range for 2 patients in the 40 mg/m^2 group. At all three dose levels there is a striking difference between the plasma concentration vs time curves for doxorubicin compared to liposome-encapsulated form.

Doxorubicin levels in KS lesions measured 72 h postinjection in patients receiving both DOX-SL and Adriamycin are presented in Table 5. As reflected in the "selectivity index" (defined as the quotient of the doxorubicin level found in patients receiving DOX-SL divided by the doxorubicin level in patients receiving Adriamycin), substantially more doxorubicin was found in those patients given DOX-SL.

TABLE 1
Patient Demographics

	Number
Sex	
Male	16
Female	0
Risk Group	
Homosexual	12
Bisexual	1
Homosexual/Intravenous Drug User	2
Unknown	1
Total	16

Age		
Median	Range	
38	29–48	16

TABLE 2
Dosage and Courses of Treatment

	Number of Courses	
Dose Level	**DOX-SL**	**Adriamycin**
10 mg/m^2	6	6
20 mg/m^2	6	6
40 mg/m^2	4	4
Total	16	16

TABLE 3
Pharmacokinetic Parameters[a] for DOX-SL

Dose Level	C_o(mg·L^{-1})	Distribution Half-Life, $T_{1/2}\alpha_2$ (h)[b]	Clearance (L·h^{-1})	Volume of Distribution (L)	AUC$_{0\to\infty}$ (mg·h·L^{-1})
10 mg/m^2	3.0 ± 0.9	41.1 ± 6.8	0.14 ± 0.05	7.9 ± 2.4	151 ± 61
20 mg/m^2	6.9 ± 2.7	43.5 ± 13.8	0.17 ± 0.09	10.0 ± 4.4	277 ± 163
40 mg/m^2	17.6 ± 3.4	33.3 ± 8.4	0.15 ± 0.06	6.5 ± 1.05	580 ± 208

[a] C_o, plasma concentration at first time point (immediately following end of infusion); $T_{1/2}\alpha_2$, plasma half-life during the second (major) component of the distribution phase; Volume of distribution, volume of the central compartment; AUC, area under the concentration-time curve integrated from time 0 to infinity. Values are the means for n = 3 ± SD in the 10- and 20-mg/m^2 groups and average for n = 2 ± range for the 40-mg/m^2 group.

[b] As is evident from the kinetic curves presented in Figure 1, plasma distribution of doxorubicin following DOX-SL administration occurs in two distinct phases (designated $T_{1/2}\alpha_1$ and $T_{1/2}\alpha_2$). The second component accounts for the majority of the area under the time vs. plasma doxorubicin concentration curve. For this reason only the half-life which describes the rate of the second phase is shown here.

C. SAFETY: ADVERSE EVENTS

Hematologic abnormalities were the most frequent adverse events reported. Patient 11, who was entered under a protocol exemption for having a neutrophil count of 1100 cell/mm^3, experienced severe neutropenia (500 cells/mm^3) 2 to 3 weeks following the first study dose (DOX-SL at 20 mg/m^2). This same patient experienced severe neutropenia (400 cells/mm^3) following the second study "cross-over" dose (20 mg/m^2 Adriamycin). Patient 9, also entered under a protocol exemption for having a neutrophil count of 1100 cell/mm^3, experienced severe neutropenia (300 cells/mm^3) following administration of the second study dose (DOX-SL at 20 mg/m^2). Patient 10 experienced severe neutropenia (400 cells/mm^3) following his second study dose (DOX-SL at 20 mg/m^2). Patient 7, who was entered under a protocol exception for having a neutrophil count of 1000 cell/mm^3, experienced severe neutropenia 2 weeks following the first study dose (DOX-SL 20 mg/m^2). All hematologic adverse events reported in the DOX-SL group were afebrile and reversible.

Nonhematologic adverse events were infrequent and mild (Table 6). Patient 13 progressed to an opportunistic infection (cryptococcal meningitis) following administration of the first

TABLE 4
Pharmacokinetic Parameters[a] for Adriamycin

Dose Level (mg/m²)	C_o(mg·L⁻¹)	Distribution Half-Life, $T_{1/2}\alpha$ (min)	Clearance (L·h⁻¹)	Volume of Distribution (L)	AUC$_{0\to\infty}$ (mg·h·L⁻¹)	N Value[d]	Ref.
[b]19–22	1.36 – 3.02	ND	64.8 – 101.4	ND	0.37 – 1.20	5	16
[c]29	0.75 ± 0.35	10 ± 0.5	ND	ND	0.40 ± 0.13	9	17
[c]40–56	8.7 ± 2.7	2.46 ± 1.2	60.4 ± 23.4	1680 ± 840	ND	8	18

[a] C_o, plasma concentration at first time point (immediately following end of infusion); $T_{1/2}\alpha$, plasma half-life during the distribution phase; Volume of distribution, volume of the central compartment; AUC, area under the concentration-time curve integrated from time 0 to infinity.

[b] Values are ranges.

[c] Values are means + standard deviations.

[d] Number of patients in sample.

TABLE 5
Localization of Doxorubicin in KS Lesions

Dose Level (mg/m²)	Doxorubicin Concentration (μg/g tissue) Mean ± SD (n = 3)		Selectivity Index
	DOX-SL	Adriamycin	
10	2.06 ± 0.42	0.18 ± 0.07	11.4
20	1.61 ± 0.80	0.31 ± 0.16	5.2
40	7.11 ± 3.18*	0.72 ± 0.07*	9.9

* Average of two values ± range.

study dose (40 mg/m² Adriamycin) and was withdrawn from the study. The principal investigator withdrew patient 14 from the study following administration of the first study dose (40 mg/m² DOX-SL) because it was felt that administration of the second study dose was not in the best interests of the patient.

V. DISCUSSION AND CONCLUSIONS

A. SAFETY

A single dose of DOX-SL at 10 and 20 mg/m² is generally well tolerated in AIDS patients with KS. Severe adverse events were limited to hematological toxicities, most notably leukopenia and neutropenia. Episodes of neutropenia in the DOX-SL groups were afebrile and reversible. The frequency of adverse events increased in both the DOX-SL and Adriamycin groups as the dose was raised from 10 to 20 mg/m². There was little difference with respect to the total number of related or probably related adverse events between the Adriamycin and DOX-SL groups: 18 occurred in the DOX-SL groups compared with 14 in the Adriamycin groups.

B. PHARMACOKINETICS

The pharmacokinetic parameters of DOX-SL are significantly different from those of Adriamycin in AIDS patients with KS. As shown in Table 3, the half-life values of the major DOX-SL plasma distribution phase ($T_{1/2}\alpha_2$) ranges from 33 to 43 h. DOX-SL clearance is 0.14 to 0.17 l·h⁻¹ and AUC values range from 150 to 372 mg·h·l⁻¹. In contrast, as shown in Table 4,

TABLE 6
Nonhematologic Adverse Events
Related or Probably Related to
Study Drugs*

	DOX-SL Dose (mg/m^2)		Adriamycin Dose (mg/m^2)	
Adverse Event	10	20	10	20
Alopecia, mild	0	1	2	1
Fatigue, mild	0	1	0	0

Note: Listing includes patients in 10 and 20 mg/m^2
dose groups only.

Adriamycin at comparable dose levels has been reported to distribute to tissues with a plasma half-life of less than 10 min, clearance is considerably faster (in the range of 60 l·h^{-1}) and AUC values substantially lower (ranging from 0.4 to 1.2 mg·h·l^{-1}). The apparent volume of distribution for DOX-SL (<10 l) is only slightly greater than the estimated plasma volume, suggesting that encapsulated drug is restricted from entering the extravascular compartment during the plasma distribution phase. In contrast Adriamycin has been reported to distribute rapidly to all body tissues with an apparent volume of distribution in excess of 1500 l.

Following DOX-SL administration a linear increase in plasma C_o and AUC is observed with increasing dose and distribution kinetics independent of dose (Figure 1). DOX-SL appears to distribute to tissues in two distinct phases ($T_{1/2}\alpha_1$ and $T_{1/2}\alpha_2$): the first, which occurs within the first 5 h, is only slightly faster than a second prolonged phase, which takes place over a period of several days. When these data are compared with those of Adriamycin, it is apparent that, when encapsulated in the DOX-SL liposomes, doxorubicin distribution is controlled by the liposome carrier: Adriamycin distributes within minutes to a large volume (essentially all body tissues), whereas encapsulated drug distributes quite slowly to a much smaller volume (approximately restricted to the vascular compartment).

C. TUMOR LOCALIZATION

Seventy-two hours after injection, doxorubicin levels achieved in KS lesions in patients receiving DOX-SL are 5 to 11 times greater than those in patients given comparable doses of Adriamycin.

The mechanism by which liposome-encapsulated drug selectively enters KS lesions is not fully understood. Stealth liposomes of the same size and lipid composition as DOX-SL, but containing entrapped colloidal gold designed to serve as a marker to follow liposome distribution by light and electron microscopy, have been shown to enter solid colon tumors[12] implanted in mice and KS-like lesions in HIV-transgenic mice.[14] In the latter mouse model, transcytosis of liposomes from the lumen of blood vessels into the extravascular compartment of KS lesions and intracellular uptake of liposomes by spindle cells within lesions were observed[14]. Extravasation of liposomes may also occur by passage of the particles through endothelial cell gaps which have been reported to be present in certain solid tumors[20-22] and are known to be present in KS-like lesions.[13] These processes may contribute to the selective uptake of DOX-SL seen here.

D. CONCLUSIONS

The present observations are consistent with preclinical findings[12] and confirm that the "Stealth" coating allows liposomes containing doxorubicin to circulate for prolonged periods of time in plasma of AIDS patients with KS. The drug-loaded liposomes appear to selectively enter KS lesions, possibly by a process of extravasation and transcytosis. Once lodged in the tumor, the drug is presumably available to be released locally as the liposomes degrade and become permeable *in situ*.[23]

Based on evidence of selective accumulation of doxorubicin in KS lesions following DOX-SL administration and similar tolerance of DOX-SL compared to Adriamycin seen in this study, 10 to 20 mg/m^2 would appear to be a rational dose range to use in Phase I/II studies of DOX-SL in the treatment of KS in AIDS patients.

REFERENCES

1. **Friedman-Kein, A., Laubenstein, L., Marmor, M. et al.,** Kaposi's sarcoma and *Pneumocystis* pneumonia among homosexual men — New York City and California, *MMWR*, 39, 305, 1981.
2. **Hymes, K.B., Cheung, T.L., Green, J.B. et al.,** Kaposi's sarcoma in homosexual men: a report of eight cases, *Lancet*, 2, 598, 1981.
3. **Masur, H., Michilis, M.A., Green, J.B. et al.,** An outbreak of community acquired *Pneumocystis carinii* pneumonia: initial manifestations of cellular immune dysfunction, *N. Engl. J. Med.*, 305, 1431, 1981.
4. **Beral, V., Peterman, T.A., Berkelman, R.L. et al.,** Kaposi's sarcoma among persons with AIDS: a sexually transmitted infection?, *Lancet*, 1, 123, 1990.
5. **Rutherford, G.W., Schwarcz, S.K., Lemp, G.F. et al.,** (1989) The epidemiology of AIDS-related Kaposi's sarcoma in San Francisco, *J. Infect. Dis.*, 159, 71, 1989.
6. **Cusick, P.S., Moss, A.R., Bacchetti, P. et al.,** Serologic variables at diagnosis with Kaposi's sarcoma (KS): predictors of shortened survival. Abstracts: Fifth International Conference on AIDS. Montreal, 1989 Poster W.C.P. 38, 598, 1989.
7. **Peters, B.S., Beck, E.J., Coleman, D.G. et al.,** Changing disease pattern in patients with AIDS in a referral center in the United Kingdom: the changing face of AIDS, *Brit. Med. J.*, 302, 203, 1991.
8. **Volberding, P.A., Kusick, P., and Feigal, I.W.,** Effect of chemotherapy for HIV-associated Kaposi's sarcoma on long term survival, *Proc. ASCO*, 8, 3, 1989.
9. **Gill, P.S., Akil, B., Colletti, P. et al.,** Pulmonary Kaposi's sarcoma: clinical findings and results of therapy, *Am. J. Med.*, 87, 57, 1989.
10. **Kaplan, L.D., Hopewell, P.C., Jaffe, H. et al.,** Kaposi's sarcoma involving the lung in patients with acquired immunodeficiency syndrome, *J. AIDS*, 1, 23, 1988.
11. **Gill, P.S., Ranick, M., McCutchen, J.A. et al.,** Systemic treatment of AIDS related Kaposi's sarcoma: results of a randomized trial, *Am. J. Med.*, 90, 427, 1991.
12. **Papahadjopoulos, D., Allen, T.M., Gabizon, A., Mayhew, E., Matthay, K., Huang, S.K., Lee K.-D., Woodle, M.C., Lasic, D.D., Redemann, C., and Martin, F.J.,** Sterically stabilized liposomes: improvements in pharmacokinetics and antitumor therapeutic efficacy, *Proc. Natl. Acad. Sci. U.S.A.*, 88, 11460, 1991.
13. **Vogel, J., Henricks, S.H., Reynolds, R.K., Luciw, P.A., and Jay G.,** The HIV *tat* gene induces dermal lesions resembling Kaposi's sarcoma in transgenic mice, *Nature*, 335, 606, 1988.
14. **Huang, S.K., Martin, F.J., Jay, G., Vogel, J., Papahadjopoulos, D., and Friend, D.S.,** Extravasation and transcytosis of liposomes in Kaposi's sarcoma-like dermal lesions of transgenic mice bearing the HIV *tat* gene, *Am. J. Pathol.*, in press.
15. **Martin, F.J. and Gabizon, A.,** Human pharmacokinetics of Stealth liposomes containing doxorubicin, *J. Cell. Biochem. Suppl.*, 16E, 98, 1992.
16. **Eksborg, S., Stendahl, U., and Lönroth, U.,** Comparative pharmacokinetic study of Adriamycin and 4'Epi-adriamycin after their simultaneous intravenous administration, *Eur. J. Clin. Pharmacol.*, 30, 629, 1986.
17. **Eksborg, S., Strandler, H-S., Edsymr, F., Näslund, I., and Tahvanainen, P.,** Pharmacokinetic study of IV infusions of Adriamycin, *Eur. J. Clin. Pharmacol.*, 28, 205, 1985.
18. **Mross, K., Maessen, P., van der Vijgh, W.J.F., Gall, H., Bovem, E., and Pinedo, H.M.,** Pharmacokinetics and metabolism of Epirubicin and Doxorubicin in humans, *J. Clin. Oncol.*, 6, 517, 1988.
19. **Landaw, E.M. and DiStefane, J.J.,** Multiexponential, multicompartment, and noncompartmental modeling. II. Data analysis and statistical considerations, *Am. J. Physiol.*, 15, R665, 1984.
20. **Shubik P.,** Vascularization of tumors: a review, *J. Cancer Res. Clin. Oncol.*, 103, 211, 1982.
21. **Dvorak, H., Nagy, J.A., Dvorak, J.T., and Dvorak, A.M.,** Identification and characterization of the blood vessels of solid tumors that are leaky to circulating macromolecules, *Am. J. Pathol.*, 133, 95, 1988.
22. **Seymour, L.W.,** Passive tumor targeting of soluble macromolecules and drug conjugates, *Crit. Rev. Ther. Drug Carrier Syst.*, 9(2), 135, 1992.
23. **Lasic, DD.,** *Liposome: from Physics to Application*, Elsevier, New York, 1993, 540.

Chapter 23

EFFICACY OF DOX-SL (STEALTH® LIPOSOMAL DOXORUBICIN) IN THE TREATMENT OF ADVANCED AIDS-RELATED KAPOSI'S SARCOMA

Johannes R. Bogner and Frank-D. Goebel

TABLE OF CONTENTS

I. INTRODUCTION

Kaposi's sarcoma (KS) is the most frequent opportunistic neoplasm encountered in patients with acquired immunodeficiency syndrome (AIDS). About 20% of HIV-infected homosexual men present with KS as an index diagnosis of AIDS.[1,2] Complications of AIDS-related KS depend primarily on the stage of the disease and its pattern of clinical manifestations (cutaneous vs. visceral). Gastrointestinal and pulmonary KS are potentially life threatening. Complications such as bleeding, ileus, and respiratory failure contribute to morbidity and mortality caused by KS. Edema of face or limbs occurs along with more severe categories of KS. Several different antitumor chemotherapeutic agents have been used as systemic intervention for KS when chemotherapy was indicated.[3] Vinca alkaloids, doxorubicin, and bleomycin have been administered in several trials, either as a single agent or in different combinations.[3-8] Neither single-agent therapy nor any combination treatment has been satisfactory enough to be regarded as "standard" in systemic advanced KS. Although combinations such as doxorubicin, bleomycin, and vincristine (ABV) provide comparably high response rates, this regimen results in considerable toxicity and complication rates (about 60% of patients developed opportunistic infection following ABV therapy).[7,9]

Conventional doxorubicin used as a monotherapy at a dose of 15 mg/m^2 weekly resulted in a 10% partial response with no complete remissions being reported.[10] Pharmacologic data have been reported for Stealth liposomal encapsulated doxorubicin (DOX-SL™) showing long plasma half-life, an increased accumulation in tumor tissue compared with the same dose of free doxorubicin, and a decreased uptake by tissues such as liver, spleen, and bone marrow.[11-13] In animal models it has been reported that the prolonged circulation time of DOX-SL correlates with superior therapeutic effectiveness.[11] We have used DOX-SL to treat patients with advanced KS in an open dose-escalating trial. Our objectives were to evaluate efficacy and toxicity of DOX-SL and to evaluate whether results of a first Phase I/II trial would justify randomized trials comparing it to more traditional combination regimens.

A reliable method to quantify flattening of palpable cutaneous KS lesions during anti-KS therapy is not available. In an attempt to improve staging, we evaluated the use of an ultrasound technique to measure tumor volume. Cutaneous KS lesions show a characteristic sonographic pattern of low echogenicity in contrast to the hyperechoic surrounding subcutis.[14,15] Thus, ultrasound can be used to measure KS lesions in three dimensions and offers a means to derive accurate determination of tumor volume changes during therapy. Accordingly, another objective was to evaluate prospectively measurement of KS lesion thickness and volume by ultrasound measurement in order to generate more objective data on the course of lesions during an experimental therapy.

II. METHODS

A. STUDY POPULATION

Only AIDS patients with biopsy-proven advanced KS were eligible for the trial. Eligibility criteria (LTI trial 30-03, n = 42; patients with sonographic evaluation in LTI 30-11 and 30-12, n = 8) included severe KS presenting either as visceral KS or progressive disseminated cutaneous disease with edema of face or limbs or with oral lesions. Further inclusion criteria were age >18 years, Karnofsky status >50%, white blood cell count >2000 cells/µl, hemoglobin >10 g/dl, and platelets of >50.000 cells/µl. Only patients with a positive HIV antibody status (ELISA and Western blot) were eligible.

Patients were excluded who presented with acute opportunistic infections or non-Hodgkin's lymphoma, systemic chemotherapy or radiation of KS within 4 weeks prior to entry into the trial, major psychiatric illness, and cardiac failure. All patients reported here were recruited from the outpatient clinic of the Medizinische Poliklinik, Klinikum Innenstadt, University of Munich (Germany).

The patients gave written informed consent. The concomitant use of nucleosides with anti-HIV activity was not prohibited. Usual medication for prophylaxis of *Pneumocystis carinii* pneumonia and secondary prophylaxis of oral candidiasis was allowed.

B. EVALUATION AND CLASSIFICATION OF PARTICIPANTS

The baseline evaluation consisted of a medical history and a physical examination. All visible KS lesions were evaluated for surface area (expressed in mm^2) and nodularity. Additional sonographic measurements of tumor volume in a subset of eight patients in trials LTI 30-11 and 30-12 were performed according to our method reported previously.[16] The following laboratory evaluations were performed: complete white blood cell count (including a differential), platelets, hemoglobin, creatinine, blood urea nitrogen, bilirubin, alkaline phosphatase, transaminases, sodium, potassium, calcium, serum albumin, and electrophoresis, erythrocyte sedimentation rate, and immunocytology (T-cells, B-cells, CD4, CD8). Each patient received an electrocardiogram (12 lead), an echocardiogram, a chest roentgenogram, and abdominal sonography.

In patients with symptoms or signs of gastrointestinal KS, additional endoscopy was performed. Bronchoscopy was done in patients with pulmonary involvement as suspected by chest roentgenogram and/or pulmonary computed tomography.

In trial 30-03, patients were evaluated weekly. The weekly reevaluations consisted of a history and a blood count and, if clinically indicated, further diagnostic procedures were implemented. Biweekly, KS was assessed as stated for the evaluation at baseline. Patients reaching a cumulative dose of 300 mg/m^2 of DOX-SL were reevaluated by electrocardiogram and echocardiogram.

In trials 30-11/30-12 (subset of patients in whom sonographic measurements were performed) cycle length was 3 weeks. Evaluations of KS were obtained every 3 weeks at the end of each cycle.

Classification of HIV disease was carried out according to the definitions of the Centers for Disease Control. KS stages were determined as proposed by the AIDS-Clinical Trials Group (ACTG) using the *T*umor-*I*mmune system-*S*ystemic illness system, TIS.[17] At baseline, five target lesions which were judged to be representative in terms of size, distribution, and nodularity were selected, documented, and measured (30-03). Clinical response criteria were used as recommended by the ACTG:[17] complete response (CR) was defined as the absence of any detectable residual disease, including tumor-associated edema. In remaining macular lesions a biopsy documenting histological absence of malignant cells was required. The response was only rated complete if it lasted for at least 4 weeks.

A partial response (PR) was defined as minimum decrease of 50% in the sum of the areas of all previously existing lesions (surrogate: target lesions) lasting for at least 4 weeks or at least a 75% decrease in nodularity of all previously existing lesions. The rating of PR was only allowed if no new skin or oral lesions appeared and if KS-related edema did not worsen. Stable disease (SD) was any response not meeting the criteria for CR, PR, or progressive disease.

Progressive disease was defined as any occurrence of new lesions or the increase of more than 25% in the size of previously existing lesions or the increase of edema or effusions.

Patients were considered eligible for evaluation of antitumor response after at least two cycles of therapy (cycle = 2 week period from dosing to physical examination and laboratory reexamination at day 14).

C. TREATMENT SCHEDULE

1. Trial 30-03

Liposome-encapsulated doxorubicin was administered every 2 weeks intravenously at doses of 10 mg/m^2 (n = 10), 20 mg/m^2 (n = 29), and 40 mg/m^2 (n = 3). In case of grade 3 or 4 toxicity, therapy was interrupted until abnormal values or signs returned to at least grade 2 toxicity. In case of grade 3 or 4 neutropenia the administration of G-CSF was allowed. Patients in the 40-mg/m^2 stratum were scheduled to receive 20 mg/m^2 if a severe treatment-related toxicity occurred. Enrollment began in November 1991. Cutoff for analysis was June 30, 1993. After completion of the first six cycles all patients were eligible for treatment continuation to prevent relapse.

DOX-SL was supplied by Liposome Technology, Inc., Menlo Park, California.

2. Trials 30-11 and 30-12

All subjects were treated with 20 mg/m^2 every 3 weeks by intravenous line. As part of the two trials 30-11 and 30-12 the parameter "tumor volume" was introduced as assessed by sonographic measurement of 27 target lesions in eight patients. In each patient target lesions were defined and followed two- and three-dimensionally during the trial. Ultrasound was performed using a 7.5-MHz linear transducer (Kretz Combison 310A; Kretztechnik Wiesbaden, Germany).

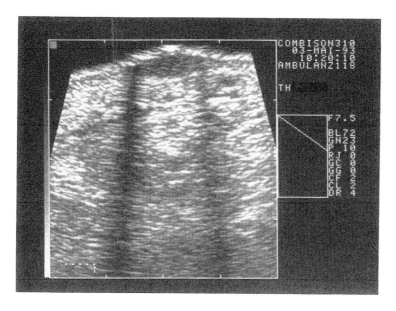

FIGURE 1. Representative example of ultrasound imaging of a cutaneous KS lesion: the hypoechogenic (dark) lesion (KS) is located within more echogenic subcutis (brighter).

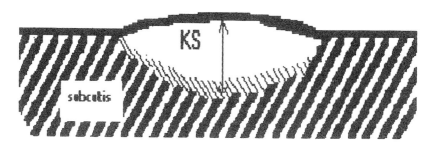

FIGURE 2. Diagram of ultrasonic measurement of cutaneous KS (7.5-MHz linear scanner): the surrounding subcutis is highly echogenic whereas the lesion appears hypoechogenic. Tumor thickness can be determined at the site of the maximum perpendicular diameter.

Sonomorphological characteristics of cutaneous KS lesions are low echogenicity as compared to the hyperechoic surrounding subcutaneous tissue and a relative homogenicity (Figure 1). Tumor thickness was measured at the site of the highest perpendicular diameter. Longitudinal and cross measurements were taken at the site of the greatest diameter of each lesion as determined by superficially visible lesion size (Figure 2).

Volumes were computed according to the formula for approximation of ellipsoids (length × width × thickness divided by 2).

In order to avoid systematic errors of possible interindividual differences, follow-up examinations were performed by the same examiner who had documented the baseline volume. Statistical analysis was performed by analysis of variance and Student's T-test.

III. RESULTS

A. BASELINE STAGING OF HIV INFECTION AND KS
1. Trial 30-03

All patients were males. The average age was 39.1 ± 8.5 years. Forty-one patients were Caucasian, one was Hispanic. At baseline, 25 patients (59%) were classified CDC IV C1+ IVD due to a history of an AIDS-defining opportunistic infection. Ten patients (24%) had

TABLE 1
TIS-Staging of KS in 42 Patients Receiving Liposomal
Doxorubicin (Trial LTI 30-03)

TIS Stage	Number of Patients	Percent
$T_1I_1S_1$	26	61.9
$T_0I_1S_1$	12	28.6
$T_1I_0S_1$	2	4.7
$T_1I_1S_0$	2	4.7

Note: T_0 = good risk (confined to skin and/or lymph nodes and/or minimal oral KS; T_1 = poor risk (tumor-associated edema or ulceration; KS in other viscera); I_0 = CD4-lymphocytes >200/µl; I_1 = CD4-lymphocytes <200/µl; S_0 = no history of opportunistic infection or thrush, no B-symptoms; S_1 = history of opportunistic infection or thrush; Karnofsky status <70%, other HIV-related illness.

AIDS-related complex and fulfilled the criteria for classification in CDC IV C2+IVD. Five patients (12%) were grouped CDC IV A+IVD and two patients (5%) had CDC IV D ("only" KS).

Classification of KS according to TIS stages[17] (Table 1) was as follows: poor risk evaluation (T_1) was present in 30 patients (71%) and a CD4 count below 200 cells/µl at baseline was found in 40 patients (95%). The median baseline CD4 count was 24 cells/µl.

Oral and gastrointestinal KS were present in 29 (69%) and 10 (24%) patients, respectively. Four patients had proven pulmonary KS.

2. Trials 30-11/30-12

Eight patients had completed ultrasonic measurements. Their mean age was 39.6 ± 8.5 years. KS staging was T_1 (poor risk) in all patients, I_1 (CD4 count below 200/µl) in all patients, and S_1 (opportunistic infections before KS) in six patients.

Oral or/and gastrointestinal KS was present in six and two patients, respectively. One patient had proven pulmonary KS.

B. EFFICACY

1. Tumor Response and Clinical Efficacy (30-03)

Forty patients were eligible for evaluation of efficacy, two patients (4.8%) did not receive more than one cycle of therapy (severe infection due to *Mycobacterium avium* complex, not apparent at baseline) and three cycles (due to lymphoma), respectively. In an intention to treat analysis, after six cycles (12 weeks), three patients (7.1%) had a complete response which was histologically confirmed (Table 2). A partial response was documented in 35 patients (83.3%). Stable disease was observed in three patients (7.1%). No patient showed progression of KS while on continuous therapy.

In all four patients with proven pulmonary involvement clinical improvement as rated by pulmonary X-ray and pulmonary function parameters could be observed (Figures 3A and 3B). However, complete resolution of pulmonary KS was not seen.

Clinical improvement in patients with gastrointestinal disease included resolution of abdominal pain, subileus, and bleeding. One patient suffered from intestinal hemorrhages, necessitating transfusion every other day. Within 1 week after the first infusion of liposomal doxorubicin the bleeding stopped and hemoglobin levels stabilized. Four of 29 patients with oropharyngeal KS had a complete resolution of lesions. A representative example is shown in Plates 1A and 1B.*

* All plates follow page 124.

TABLE 2

**Response of KS in 42 Patients Treated with
Lipsomal Doxorubicin Intravenously Every
2 Weeks (Trial LTI 30-03)**

	Dose Level (mg/m²)			
	10	20	40	Total
Response	n	n	n	n (%)
CR, complete response	0	3	n.a.[a]	3 (7.1)
PR, partial response	7	28	n.a.[a]	35 (83.3)
SD, stable disease	3	0	n.a.[a]	3 (7.1)
Progression	0	0	n.a.[a]	0 (0)

Note: n = 40 were evaluable after 6 cycles.

[a]   Not applicable, because all patients on 40 mg/m² had to
be switched to 20 mg/m² prior to evaluation.

2. Measurement of Tumor Volume (30-11/30-12)

Tumor volume as determined by ultrasonography was 556 ± 122 mm³ (range 22 to 2204 mm³) at baseline. After the sixth cycle the average volume was reduced to 42 ± 26 mm³ (range 0 to 624 mm³; $p < .01$, paired t-test; average value $\pm$ SEM, standard error of the means), see Figure 4.

Representative ultrasonic images prior to and after therapy are given in Plates 2A and 2B, respectively. Clinical rating of lesions suggested complete flattening, as shown in a representative example in Figures 5A and 5B. Response as rated by the criteria cited above was CR n = 1 and PR n = 7. No patient showed progression or stable disease while on therapy.

C. TOXICITY

Adverse events, toxicity, and occurrence of AIDS-defining events were monitored during the study period and follow-up period (trial 30-03), to attain additional information about long-term feasibility of therapy with DOX-SL. The median observation period (30-03) was 27 weeks (range 3 to 72 weeks). The median cumulative dose was 190 mg/m² (range 40 to 680). At this low dose level no clinical evidence of cardiac toxicity was observed.

During the whole treatment period (induction therapy and follow-up; median 25 weeks) four patients developed stomatitis (9%), 21 patients (50%) reported alopecia at least to some extent. Four out of those 21 showed complete alopecia (Table 3). The most frequent hematologic toxicity was neutropenia (Table 4). In a total of 17 patients (40%) grade 4 toxicity was observed at some point during the study. Another 12 patients showed grade 3 toxicity. The median baseline neutrophil count, however, was low (1600 cells/μl, range 1200 to 2400). Concomitant administration of G-CSF resulted in postponement of only 15% (55 of 368) of scheduled treatment administration.

D. DOSE-RELATED TOXICITY AND EFFICACY

In the high-dose group (40 mg/m²) limiting toxicity was observed in 3/3 patients after 6 weeks of treatment. One patient had complete alopecia and two developed grade 4 neutropenia. All patients in this group were switched to 20 mg/m², which was tolerated in subsequent courses.

Three patients initially assigned to the 10-mg/m² dose group were switched to 20 mg/m² because after interruption of chemotherapy for as few as 5 weeks new KS lesions became detectable. In two patients therapy had been interrupted due to intercurrent infections. In the

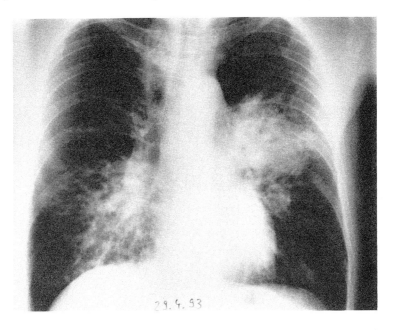

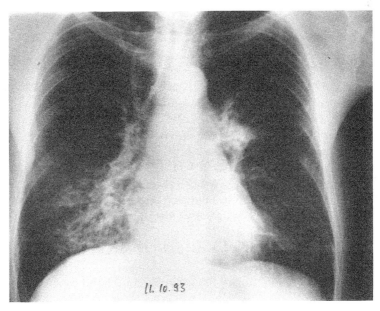

FIGURE 3.   Chest X-ray of a patient with biopsy-proven pulmonary KS; (A) prior to therapy, (B) after therapy.

third patient treatment had been postponed because he had temporarily chosen to stop therapy after an initial satisfactory result. However, 6 weeks later DOX-SL again was effectively administered at the dose level of 20 mg/m^2.

Nausea, stomatitis, and constipation were the most common adverse events, none of which was severe enough to terminate chemotherapy in the 10-and 20-mg/m^2 groups.

E. HIV-RELATED EVENTS AND INFECTIONS

AIDS-defining events observed while patients were on therapy (induction therapy and follow-up) were opportunistic infections (n = 23) and non-Hodgkin's lymphoma (n = 1). Six

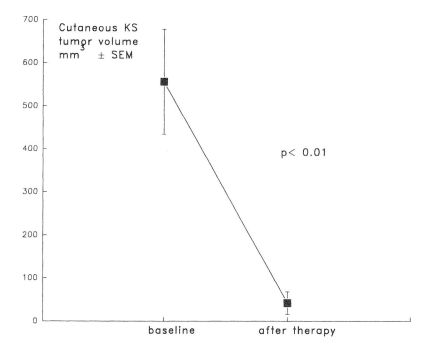

FIGURE 4.   KS volume as determined by cutaneous ultrasound prior to and after therapy. SEM, standard error of the means.

patients had ulcerating oral or genital herpes. Four had a relapse or first episode of candida esophagitis; in seven patients, MAI infection was detected and seven developed CMV manifestations (n = 4 CMV retinitis, n = 3 other CMV localizations). One patient developed cerebral toxoplasmosis. In two patients more than one of those infections were diagnosed.

Not AIDS-defining events were oral thrush (9 episodes), herpes simplex (8 episodes), pyomyositis (4 patients), and otitis media (n = 1). Three individuals experienced deep vein thrombosis. Nineteen of 40 individuals enrolled died during the observation period of 25 weeks (median) (n = 18) or were lost to follow-up (n = 1). Causes of death included MAI infection, wasting syndrome, NHL, and cerebral toxoplasmosis.

IV. DISCUSSION

This open-label dose-escalating trial showed a high response rate after single-agent DOX-SL therapy. The overall response rate (CR + PR) was 90%. Complete response is rarely observed in response to chemotherapy of advanced KS. However, after administration of single-agent DOX-SL at a dose of 20 mg/m^2 histologically proven CR has been achieved in three patients.

A very high percentage of patients had a partial response and in a substantial number of patients long-term treatment (median duration: 25 weeks) was feasible. Results of an ACTG protocol on single-agent conventional doxorubicin are available for 26 patients with comparable tumor burden and staging of HIV disease.[10] Doxorubicin was applied at a weekly dose of 15 mg/m^2. In the group of patients with advanced disease no CR was achieved. There was also no partial response as rated by the same criteria as applied in our study. Comparison of our data with results of other trials depends on comparability of response criteria. Dose levels used in conventional vs. liposomal doxorubicin therapy were higher in the conventional therapy: a monthly dose of 40 mg/m^2 of liposomal doxorubicin (20 mg/m^2 biweekly) as used

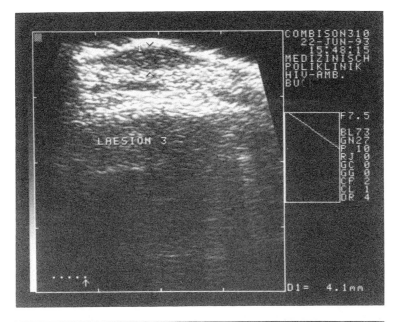

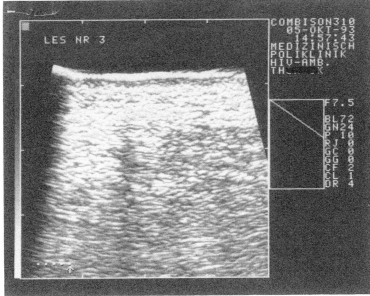

FIGURE 5.   Representative cutaneous lesion before (A) and after (B) therapy with "Stealth" liposomal doxorubicin.

in our trial 30-03 is less than the dose in the trial reported by Fischl and colleagues (60 mg/m² per month).[10]

Overall response in other single-agent studies ranged from 26% using vinblastine to 48% using bleomycin.[5,18,19]

The rationale for combining chemotherapeutic agents for treatment of AIDS-KS had been to enhance efficacy while keeping toxicity under control. Using a combination of doxorubicin, bleomycin, and vinblastine (ABV), Gelmann and colleagues reported an overall response (CR + PR) of 84%.[9] In another study, two dosages of doxorubicin, vincristine, and bleomycin were tested.[20] The overall response rate was 79%. Combination of bleomycin with vincristine (BV) in patients with neutropenia resulted in a 72% response rate.[8]

TABLE 3

**Adverse Events During Biweekly
Liposomal Doxorubicin Therapy
(During Median Observation
Period of 25 Weeks)**

	Dose Level (mg/m^2)			
	10	20	40	Total
Response	n	n	n	n (%)
Nausea	0	1	3	4 (9.5)
Stomatitis	0	5	2	7 (16.6)
Alopecia (complete)	0	3	1	4 (23.8)
Constipation	0	3	2	5 (11.9)
Cardiotoxicity	0	0	0	0

TABLE 4

**Hematologic Toxicity in Patients Receiving
Biweekly Liposomal Doxorubicin (During
Median Observation Period of 25 Weeks)**

	Dose Level (mg/m^2)			
	10	20	40	Total
Response	n	n	n	n (%)
Neutropenia				
Grade 4 <500/µl	2	13	2	17 (40.4)
Grade 3 500–999/µl	3	9	0	12 (28.5)
Grade 2 1000–1499/µl	3	6	0	9 (12.5)[a]
Grade 1 1500–2000/µl	2	2	0	4 (9.5)[a]

[a]   Patients with baseline count >2000/µl.

The frequency of opportunistic infections which occurred in association with these three combination regimens ranges between 61 and 89% of all treated patients.[4-8,10,18]

DOX-SL, as shown in our results, was within the same range of efficacy of combination therapies. The rate of opportunistic infections was 54%. Whether these data indicate that single agent DOX-SL at a dose level of 20 mg/m^2 is superior to combination chemotherapy of AIDS-KS in patients with advanced disease still has to be determined. At present, this hypothesis is being tested in two ongoing randomized multicenter trials comparing efficacy and toxicity of DOX-SL vs. the combination of Adriamycin, bleomycin, and vincristine (ABV) and bleomycin and vincristine (BV), respectively.

In a patient population with advanced AIDS-KS *and* a severe immunodeficiency (baseline CD4 count median 24/µl) a certain number of opportunistic infections is to be expected, even if no chemotherapeutic intervention for KS is performed. A concomitant antiretroviral therapy in all patients could further reduce the number of opportunistic infections. In patients with neutropenia using combination therapy of liposomal doxorubicin and antiretroviral nucleosides might result in an improvement of immunological function and survival.

Early evaluation of DOX-SL pharmacokinetics suggested a high proportion of the encapsulated doxorubicin is delivered to KS tumors.[12] The long circulation period of these new liposome formulations incorporating a synthetic polyethylene glycol- derivatized phospholipid has a pronounced effect on liposome tissue distribution. A considerable increase in the pharmacological efficacy can be achieved.[12] Liposomes carrying polyethylene glycol on the surface and the high vascularization of KS lesion in combination with an abnormal permeability of capillaries result in a proportionally higher drug deposition in tumor tissue as compared to other tissues.[11,21] Therefore, lower toxicity was anticipated as compared to conventional doxorubicin.

The most common toxicity was neutropenia. The concomitant use of G-CSF was not prohibited by the protocol. With G-CSF support in some patients only 14% of all scheduled cycles had to be postponed. The use of G-CSF in this open trial may therefore indirectly contribute to the high overall response rate. Most episodes of neutropenia occurred late in the course of therapy or during the follow-up treatment. Only a minor influence on response rates can therefore be expected.

During DOX-SL treatment measurements of the surface area of KS lesions (in two visible dimensions) may suggest only minor reduction in size while sonographic evaluation can indicate extensive remission. Hemosiderin (iron) deposits in the skin may remain visible for extended periods of time despite successful destruction of tumor cells. Therefore, ultrasonic assessment appears to be more appropriate: tumor volume rather than pigmentation area can be evaluated. The response criteria for KS evaluation developed by the Oncology Committee of the ACTG include that the character of KS lesions is noted and evaluated.[17] Rather than using some grading of "nodularity", which requires personal judgement by the investigator, we sought to find a method that relies on an objective measurement. Ultrasound has been proposed as a harmless diagnostic aid to clinical examination of proliferative vascular lesions of the skin, like lymphangiomas and hemangiomas.[14] Our findings suggest that ultrasound is a useful method for follow-up of growth and remission of cutaneous KS. Residual pigmentation due to iron deposition in skin is not affected by chemotherapy and its presence confounds attempts to measure KS lesion size — often leading to an underestimate of response to therapy. Response of KS lesions followed by ultrasound evaluation appears to report the effect of chemotherapymore accurately. As we have shown, reduction of tumor volume as measured by ultrasound was corroborated by histological evidence of lack of tumor cells.[16]

As a rule, siderophages are not abundant in macules or patches of KS.[22] The lesions seen before treatment showed clinical features of nodularity and histological features characteristic for the plaque stage of KS, combining the structure of patch and nodular lesions. On lesions that were clearly nodular prior to therapy (clinically and as determined by ultrasound) punch biopsy was performed after treatment.[16] The histologic features of early patch stages of KS were detectable. Nevertheless, these lesions showed an increased number of siderophages in the dermis as well as slight fibrosis. Thus, the response detected by ultrasound (shrinking of volume to zero) was confirmed histologically after therapy. We therefore suggest that clinical investigators consider ultrasound to assess remission of cutaneous KS in clinical trials designed to evaluate new treatment regimens.

Overall, liposomal doxorubicin was well tolerated and the only dose-limiting toxicity was neutropenia. DOX-SL at dose levels of 10 and 20 mg/m^2 is safe and effective for the treatment of advanced AIDS-KS. Controlled trials comparing DOX-SL to conventional combination chemotherapy (ABV and BV) are presently underway.

ACKNOWLEDGMENT

We are indebted to M. Liebschwager, E. Kolbe, Ch. Rauch, A. Truebenbach, H. Liess, C.K. Schewe, U. Liegl, S. Spaethling, M. Held, and P. Sandor who also cared for the patients.

REFERENCES

1. **Des Jarlais, D.C., Stoneburner, R., Thomas, P., and Friedman, S.R.,** Declines in proportion of Kaposi's sarcoma among cases of AIDS in multiple risk groups in New York City, *Lancet,* II, 1024, 1987.

2. **Drew, W.L., Mills, J., Hauer, L.B. et al.,** Declining prevalence of Kaposi's sarcoma in homosexual AIDS patients paralleled by fall in cytomegalovirus transmission, *Lancet,* I, 66, 1988.

3. **Kahn, J.O., Northfelt, D.W., and Miles, S.A.,** AIDS-associated Kaposi's sarcoma, in *AIDS Clinical Review 1992,* Volberding, P. and Jacobson, M.A., Eds., New York, 1992, 263.

4. **Kaplan, L., Abrams, D., and Volberding, P.,** Treatment of Kaposi's sarcoma in acquired immunodeficiency syndrome with an alternating vincristine-vinblastine regimen, *Cancer Treat. Rep.,* 70, 1121, 1986.

5. **Lassoued, K., Clauvel, J.P., Katalama, C. et al.,** Treatment of the acquired immune deficiency syndrome related Kaposi sarcoma with bleomycin as a single agent, *Cancer,* 66, 1869, 1990.

6. **Gill, P.S., Rarick, M., McCutchan, J.A. et al.,** Systemic treatment of AIDS-related Kaposi's sarcoma: results of a randomized trial, *Am. J. Med.,* 90, 427, 1991.

7. **Laubenstein, L.J., Krigel, R.L., Odajnyk, C.M. et al.,** Treatment of epidemic Kaposi's sarcoma with etoposide or a combination of doxorubicin, bleomycin and vinblastine, *J. Clin. Oncol.,* 2, 115, 1984.

8. **Gill, P.S., Rarick, M., and Bernstein-Singer, M.,** Treatment of advanced stage Kaposi's sarcoma using a combination of bleomycin and vincristine, *Am. J. Oncol.,* 13, 315, 1990.

9. **Gelmann, E.P., Longo, D., Lane, H.C., et al.,** Combination chemotherapy of disseminated Kaposi's sarcoma patients with the acquired immune deficiency syndrome, *Am. J. Med.,* 82, 456, 1987.

10. **Fischl, M.A., Krown, S.E., Boyle, K.P. et al.,** Weekly doxorubicin in the treatment of patients with AIDS-related Kaposi's sarcoma, J. AIDS, 6, 259, 1993.

11. **Vaage, J., Mayhew, E., Lasic, D., and Martin, F.,** Therapy of primary and metastatic mouse mammary carcinoma with doxorubicin encapsulated in long circulating liposomes, *Int. J. Cancer,* 51, 942, 1992.

12. **Papahadjopulos, D., Lasic, D.D., Redemann, C. et al.,** Sterically stabilized liposomes: improvements in pharmacokinetics and antitumor therapeutic efficacy, *Proc. Natl. Acad. Sci. U.S.A.,* 88, 11460, 1991.

13. **Bally, M.B., Nayar, R., Masin, D., Cullis, P.R., and Mayer, L.D.,** Studies on the myelosuppressive activity of doxorubicin entrapped in liposomes, *Cancer Chemother. Pharmacol.,* 27, 13, 1990.

14. **Betti, R., Nessi, R., Blanc, M. et al.,** Ultrasonography of proliferative vascular lesions of the skin, *J. Dermatol.,* 17, 247, 1990.

15. **Bogner, J.R., Held, M., and Goebel, F.D.,** Cutaneous ultrasound for evaluation of Kaposi sarcoma, *J. AIDS,* 6, 530, 1993.

16. **Bogner, J.R., Zietz, C., Held, M. et al.,** Ultrasound as a tool to evaluate remission of cutaneous Kaposi sarcoma, *AIDS,* 7, 1081, 1993.

17. **Krown, S.E., Metroka, C., and Wernz, J.C.,** Kaposi's sarcoma in the acquired immune deficiency syndrome: a proposal for uniform evaluation, response, and staging criteria, *J. Clin. Oncol.,* 7, 1201, 1989.

18. **Volberding, P.A., Abrams, D.I., Conant, C.M. et al.,** Treatment of epidemic Kaposi's sarcoma in the acquired immunodeficiency syndrome, *Ann. Intern. Med.,* 103, 335, 1985.

19. **Mintzer, D.M., Real, F.X., Jovino, L. et al.,** Treatment of Kaposi's sarcoma and thrombocytopenia with vincristine in patients with the acquired immunodeficiency syndrome, *Ann. Intern. Med.,* 102, 200, 1985.

20. **Gill, P.S., Rarick, M., Espina, B. et al.,** Advanced acquired immunodeficiency syndrome-related Kaposi's sarcoma, *Cancer,* 65, 1074, 1989.

21. **Rahman, A., Treat, J., Roh, J.K. et al.,** A phase I clinical trial and pharmacokinetic evaluation of lipsome-encapsulated doxorubicin, *J. Clin. Oncol.,* 8, 1093, 1990.

22. **Gottlieb, G.J. and Ackerman, A.B., Eds.,** *Kaposi's Sarcoma: Text and Atlas,* Philadelphia, 1988.

CONCLUSION AND AFTERWORD

Danilo D. Lasic

After reviewing and proofreading the chapters, I'd like to add a few more thoughts, some novel developments, and a conclusion to the whole book.

Recent physical measurements completely confirm the picture described in this volume. In contrast to some reports in the literature which claim that liposomes can be made from bilayers with higher than 20 mol% of longer PEG-chains-containing lipids, neutron reflectivity measurements by Kuhl, Smith, Steinberg, and Israelachvili have shown a picture consistent with the one shown in Chapters 4 through 8. The thickness of the top layer in these experiments correlates well with the thicknesses determined by several other methods (as reviewed in Reference 1) and the small angle neutron scattering results (Auvray, Auroy, and Lasic, in preparation) in which very similar mushroom and brush extensions were obtained and the bilayer destabilization at higher mol% of PEG-lipids[2] was reconfirmed. The former results can be therefore understood as artifacts of insufficient hydration or incomplete system characterization. PEG-lipids, however, can form monolayers[3] (see also Chapter 8) and foam[4] as well as, in mixtures with other lipids, emulsions and microemulsions with peculiar properties.[5,6] In aqueous solutions PEG-lipids form micelles.[7] Small angle neutron scattering and cryoelectron micro-scopy have shown that micelles are rather small, from 11 to 20 nm for PEG molecular weight range from 750 to 5000 Da. (Frederik, Lasic, and Auvray, in preparation). Further theoretical and experimental analyses of these systems will be presented in forthcoming papers by Evans[8] and Sackmann.[9] The monolayer part of the latter work also showed a phase segregation and pancake-mushroom polymer transition in the low coverage regime. The intermixing of lipids occurred only when the depletion layer was formed at higher fractions of PEG-lipids.[9]

Recent synchrotron scattering studies in multilayers carried out by Safinya's group show that at very low coverages, with polymer (PEG-DMPE) densities much less than the monolayer coverage density, the polymer coated membranes exhibit a new long-range repulsive interlayer interaction unrelated to any polymer brush or mushroom effect, but arising from undulation induced antidepletion forces. These interactions which are present in very dilute multilayers (interlayer spacings ranging from 2 to 6 radii of gyration of PEG polymer) arise only when the membrane-tethered PEG-lipids are freely diffusing on the liquid membrane surfaces.[10,11] This group is also studying PEG-lipid containing microemulsions and liquid gels with unusual characteristics.

At present we still believe that optimal biological stability of PEG-coated liposomes is around mushroom brush transition (see Figure 8 in Reference 1). However, this may not be true for other, more rigid and less anisotropic polymers, such as found in biological systems. I hope more researchers will study adsorption of (well-defined) proteins on such surfaces. As expected from mechanical considerations of the bilayers,[1,7] the anchoring of PEG moiety is very important too.[12,13]

As expected (p. xii),[14,15] new polymers were discovered which render similar invisibility to liposomes: poly (2-X-2-oxazolidine), X = methyl or ethyl with degree of polymerization around 50, were shown to behave very similarly to PEG.[16] Similar effects were also shown by some vinyl polymers, such as poly (acrylamide) and poly (vinyl pyrrolidone).[17] Further developments in synthetic chemistry are improvements in antibody conjugation techniques. Recent work in Allen's lab shows good activity of stealth immunoliposomes loaded with doxorubicin against hematological malignancies when using an internalizing antibody.[18] Increased immunospecificity of antibody-conjugated stealth liposomes was observed also by Barenholz and colleagues.[19,20]

While several laboratories are further exploiting stealth liposomes with various drugs, such as improved delivery of anticancer agents by hyperthermia (where Papahadjopoulos, Dewhirst, and colleagues have found that hyperthermia-induced extravasation is important for further enhancing the therapeutic efficacy of the encapsulated drugs),[21,22] increased efficacy of Amphotericin B in long circulating liposomes against fungal infections and visceral leishmaniasis, and gentamicin or ceftazimide in bacterial infections,[23] the major interest is now in clinical studies of doxorubicin in humans. DoxSL, a renamed Doxil, was administered to about 1000 patients with Kaposi sarcoma which, in general, showed favorable tolerance and response. Moreover, around 100 patients with solid tumors, including non-small cell lung cancer, are in Phase I and II clinical studies. The anti-tumor activity is very encouraging with many responses while the altered biodistribution may result in a rather unique skin toxicity in some patients, which will have to be further investigated. Further developments may include stealth liposomes as artificial oxygen carriers[24,25] and, perhaps, in DNA plasmid and antisense oligonucleotide delivery for gene therapy.

In conclusion, it is nice to know that this multi- and interdisciplinary research has resulted, as this book can testify, not only in novel opportunities, new model systems, and prolongation of liposome circulation times but also in the prolongation of human life.

REFERENCES

1. **Lasic, D.D.,** Sterically Stabilized Vesicles, *Angew. Chem. Int. Ed. Eng.,* 33, 1685, 1994.
2. **Lasic, D.D., Woodle, M.C., Martin, F., and Valentincic, T.,** Phase diagram studies of "stealth-lipid"-lecithin mixtures, *Periodicum Biologorum,* 93, 287, 1991.
3. **Buerner, H., Winterhalter, M., and Benz, R.,** Surface Potential of Lipid Monolayers with Grafted polyethylene Glycols, *J. Colloid Interface Sci.,* 168, 183, 1994.
4. **Lal, J. and Auvray, L.,** unpublished, 1993. Briefly, PEG-lipids can form black foam film with thickness from 3 to 7 nm for PEG range from 750 to 5000 Da. Films were not very stable but Frankel's law was obeyed. The formation itself demonstrates that ordering of polymer chains has to take place. PEG-lipids in mixtures with lecithin (and cholesterol), however, form extremely stable foams (see Reference 1).
5. **Détappe, V. and Lasic, D.D.,** unpublished, 1992. Microemulsions could be made either by swelling PEG-lipid micelles or by titrating normal microemulsions with these lipids. Stable emulsions could be formulated by sonication or homogenization of oil, water, lecithin, and PEG-lipids. By varying concentrations homogeneous samples with particle sizes from 70 to 200 nm could be prepared while at larger sizes (>500 nm) quite heterogeneous suspensions were obtained. In some cases, co-surfactants were also used.
6. **Auvray, L. et al.,** in preparation.
7. **Lasic, D.D.,** *Liposomes: from physics to applications,* Elsevier, Amsterdam, 1993, 298.
8. **Klingerberg, D., Szoka, F., and Evans, E.,** Stability of polymer grafted membranes in concentrated solutions of nonadsorbing free polymer, *Macromolecules,* submitted.
9. **Baekmerk, T.R., Elender, G., Lasic, D.D., and Sackmann, E.,** Conformational transitions in monolayers of phospholipid-polyetheleneoxide lipo-polymers and interaction forces with solid surfaces, in preparation.
10. **Warriner, H.E., Idziak, S.H.J., Kraiser, K.E., and Safinya, C.R.,** Polymer coated biomembranes compromised of phospholipids and polymer-lipids, *Bull. Am. Phys. Soc.,* 39, 198, 1994.
11. **Warriner, H.E., Idziak, S.H.J., Kraiser, K.E., and Safinya, C.R.,** A new interaction in polymer coated fluid multilamellar membranes, in preparation.
12. **Silvius, J.R. and Zuckermann, M.J.,** Interbilayer transfer of phospholipid anchored macromolecules via monomer diffusion, *Biochemistry,* 32, 3153, 1993.
13. **Parr, M.J., Ansell, S.M., Choi, L., and Cullis, P.R.,** Factors influencing the retention and chemical stability of PEG-lipid conjugates incorporated into large unilamellar vesicles, *Biochim. Biophys. Acta,* 1195, 21, 1994.
14. Reference 7, page 293.
15. **deGennes, P.G. and Brochard, F.,** private communication, 1992.
16. **Woodle, M.C., Engbers, C.M., and Zalipsky, S.,** New Amphipatic Polymer-Lipid Conjugates forming Long-Circulating Reticuloenothelial System-Evading liposomes, *Bioconjugate Chem.,* 5, 493, 1994.
17. **Torchilin, V.P., Shtilman, M.I., Trubetskoy, V.S., Whitman, K., and Milstein, A.M.,** Amphiphilic vinyl polymers effectively prolong liposome circulation times *in vivo, Biochim. Biophys. Acta,* 1195, 181, 1994.
18. **Lopez-deMenezes, D., Polarsky, L., and Allen, T.M.,** Selective cytotoxicity of immunoliposomes to B-lymphocytes, Abstract, AACR, Toronto, 1995.

19. **Bolotin, E.M., Cohen, R., Bar, L., Emanuel, N., Ninio, S., Lasic, D.D., and Barenholz, Y.,** Ammoniumsulfate gradients for efficient and stable remote loading of amphiphatic weak bases into liposomes and ligandoliposomes, *J. Liposome Res.,* 4, 455, 1994.

20. **Emanuel, N., Kedar, E., Bolotin, E.M., Smorodisky, N.I., and Barenholz, Y.,** Therapeutic feasibility of ligando-liposomes containing doxorubicin against tumor associated antigens, submitted, *Cancer Res.*

21. **Huang, S.K., Stauffer, P.R., Hong, K., Guo, G.W.W., Phillips, T.L., Huang, A., and Papahadjopoulos, D.,** Liposome and hyperthermia in mice increase tumor uptake and therapeutic efficacy of doxorubicin in sterically stabilized liposomes, *Cancer Res.,* 54, 2186, 1994.

22. **Gaber, M.H., Ning, W.Z., Hong, K., Huang, S.K., Dewhirst, M., and Papahadjopoulos, D.,** Thermosensitive liposomes: extravasation and release of contents in tumor microvascular networks, *Can. Res.,* submitted.

23. **Bakker-Woudenberg, I.A.J.M. and van Etten, E.W.M.,** Liposomes with prolonged circulation as carriers of antimicrobial agents in bacterial and fungal infections, in *Book of Abstracts,* Conference on Liposomes in Biomedical Research, Berlin, October 1994.

24. **Zheng, S., Beissinger, R., Sherwood, R., McCormick, D.L., Lasic, D.D., and Martin, F.,** Liposome encapsulated hemoglobin, *J. Liposome Res.,* 3, 578, 1993.

25. **Bangham, A.D.,** Liposomes: realizing their promise, *Hospital Practice,* December 1992, p. 51.

INDEX